American Environmentalism

Philosophy, History, and Public Policy

American Environmentalism

Philosophy, History, and Public Policy

J. Michael Martinez

CRC Press
Taylor & Francis Group
Boca Raton London New York

CRC Press is an imprint of the
Taylor & Francis Group, an **informa** business

CRC Press
Taylor & Francis Group
6000 Broken Sound Parkway NW, Suite 300
Boca Raton, FL 33487-2742

Printed on acid-free paper
Version Date: 20130212

International Standard Book Number-13: 978-1-4665-5970-7 (Hardback)

Library of Congress Cataloging-in-Publication Data

Martinez, J. Michael (James Michael)
American environmentalism : philosophy, history, and public policy / James Michael Martinez.
pages cm.
Includes bibliographical references and index.
ISBN 978-1-4665-5970-7
1. Environmentalism--United States--History. 2. Environmental policy--United States--History. 3. Environmental protection--United States--History. 4. United States--Environmental conditions. I. Title.

GE197.M3 2013
363.70973--dc23 2013003669

Visit the Taylor & Francis Web site at
http://www.taylorandfrancis.com

and the CRC Press Web site at
http://www.crcpress.com

This book is dedicated to Christopher Kainan Carter,
my grandson and fellow nature lover

Shall I not have intelligence with the earth? Am I not partly leaves and vegetable mould myself?

—Henry David Thoreau, *Walden; or, Life in the Woods*

It's hard for the modern generation to understand Thoreau, who lived beside a pond but didn't own water skis or a snorkel.

—Loudon Wainwright, musician, actor, humorist

Contents

SECTION III POSTMODERN AMERICAN ENVIRONMENTAL POLITICS AND POLICY

List of Figures

Introduction and Acknowledgments

Nature knows no indecencies; man invents them.

—**Mark Twain,** ***Mark Twain's Notebook***[1]

One blustery December afternoon in 1976, shortly before my fourteenth birthday, my mother decided the price of Christmas trees was entirely too high. She was a traditionalist who enjoyed the splendor of the holiday season, but she was also a single mother raising a child on a modest income. To combat what she viewed as usurious pricing, mom resolved to procure a fir by cutting out the middleman. We would snag our own Christmas tree free of charge. She would transport my best friend, John Harris, and me to a patch of abandoned acreage and release us with instructions to cut a suitable specimen. That is how two pimply-faced adolescents sporting shag haircuts, Fonzie-style leather jackets, and Converse high-top sneakers found themselves stumbling around lost in the woods near Effingham, South Carolina.

After purchasing matching axes from our local K-Mart retail store, mom found a deserted-looking dirt road visibly lacking "No Trespassing" signs. She parked her dented Chevy Nova on the shoulder. We were deep in the pine forests of South Carolina at least 20 miles from our house. I was unfamiliar with the isolated terrain.

Lighting a cigarette, she pointed. "Go on, boys," she said. "This is as good a place as any. Go on into the woods now. You can burn off some of that youthful energy by cutting us a tree."

John and I did as we had been instructed— with one minor deviation. Had we simply walked in a straight line away from the Nova and cut a tree within sight of the automobile, all would have been well. Yet we were young and dumb. As we trudged away from the road, one of us pointed and exclaimed, "Now, *that's* a Christmas tree!"

"No, man—look over there. That one's even taller!"

Invariably, after we raced to the new tree, we spied an even more impressive candidate lurking farther away in the distance. By the time we had settled on a mutually agreeable fir, we had lost sight of both my mother and her automobile. The forest was thick and we were disoriented; we could spy no distinctive landmarks from our journey. Every tree, ironically, resembled every other tree. We could have cut any one we stumbled upon without venturing into the thicket.

Alas, we had traveled such a long distance that calling my mother's name only met with silence.

Mom had finished smoking her cigarette. Realizing we had been absent for an inordinate length of time, she sat in the car blowing the horn to attract our attention. When that desperate action failed to produce results, she drove up and down the dirt road with the windows rolled down, calling our names. In a long-ago era before cell phones and GPS tracking systems had become ubiquitous, my mother was worried and apoplectic. What should she do? The sun had not set, but the sky clearly was fading to a burnt orange glow, suggesting that dusk was on the horizon.

Fearing my mother's wrath if we were lost *and* failed to procure a Christmas tree, John and I chopped down the intended target. It was a minor miracle that we accomplished the task without inadvertently severing a limb from either the tree or ourselves. Dragging the newly acquired prize in the direction of where we thought the road was located— which turned out to be in the opposite direction— we suddenly remembered we had not eaten anything in at least three hours. (Adolescent boys must consume prodigious quantities of foodstuffs at frequent and regular intervals if they are to maintain their faux insouciance, a crucial component of their omnipresent narcissism.) Ravenous, frantic to find the car, and growing progressively frightened of the dark woods enveloping us, John and I screamed ourselves hoarse for someone— anyone—to rescue us from our self-created plight.

We might have stumbled around all night in the scary forest, at least until National Guard units parachuted in to find us in the light of day, were it not for a serendipitous event. As it turned out, the land was not abandoned. The owner, a local octogenarian misanthrope renowned for his less-than-friendly manner and his uncanny resemblance to numerous movie villains and serial killers, happened upon my mother parked on the dirt road. After he listened to her tale of woe, he agreed to reconnoiter the area in search of two young interlopers. An hour later, John and I spotted the headlights of his pickup truck slicing through the dark night as he swerved along a small logging road adjacent to his land.

In a fairy tale, the story would have featured a happy ending. Our savior would have been in touch with his inner child after having undergone a recent Ebenezer Scrooge-like transformation. He would have reunited us with my mom, bid us adieu in the spirit of Christmas, and sent us on our way toward a bright, successful future. In the real world, however, he drove us— along with our tree— to meet up with mom, who sat in her Nova crying hysterically. Furious that we had cut down a tree on his land without permission, the would-be Ed Gein agreed not to turn us over to the local police if mom would surrender all the cash on her person. It was the money from her Christmas bonus she had intended to use for our presents.

Mom drove us away from the dirt road, sadder and wiser, with the Christmas tree strapped to the top of the car— somehow, we had remembered to procure the requisite twine— and $183 missing from her wallet. It was a princely sum at the time— far more money than she would have spent on a tree had we simply purchased one, as we usually did, from the seasonal Christmas-tree lot situated six blocks from our house.

As a life-long nature lover, I have enjoyed countless hours hiking in the woods, spending the night on camping trips with the Boy Scouts, rafting down wild western rivers, transporting family and friends to national parks for recreation and conviviality, and reflecting on my life and its meaning while nestled in an isolated forest enclave. Yet that evening lost in the woods showed me a different side of nature. It was hardly dramatic, "movie-of-the-week" stuff. Thankfully, no one drank urine to survive, hacked off his arm, devoured another human being, or starved to death, but the experience demonstrated how quickly nature could switch from friend to foe. The forest did not change in the few hours we were lost, but our interpretation of the event changed, which influenced our relationship with nature. To my amazement, the character of nature altered in direct proportion to my growing sense of panic. It was Mother Nature's fault we got lost; she tricked us. Our own imbecility and poor sense of direction were not factors.

Nature has fascinated humanity for millennia. Sometimes—as in my case during that interminable, moonless night in December 1976—she has been anthropomorphized and cast as a villainess in a morality play featuring intrepid (or stupid) human beings who must navigate through dark, treacherous forests where who-knows-what-evil lurks before they emerge into the bright light of orderly, structured, manufactured communities that banish the chaos and disorder of the wilderness. At other times, nature has served as a scenic backdrop for our most profound reflections on life, love, spirituality, God, and existential meaning, a secluded sanctuary from the bustling crowds, smog, and stressful vicissitudes of postmodern urban life. Radical naturalists contend that nature is intrinsically valuable and should not depend on human characterizations of evil and good to determine its worth. Whatever else that can be said, nature serves as a *tabula rasa*—a loving mother, a vengeful bitch-goddess, a nurturing deity, a broad canvas for human action, an inanimate object, a dumping ground.

For adventuresome Americans of an individualist creed, the natural environment represents an exciting mystery to be explored in the quest for a better life. When young Christopher McCandless, the protagonist of Jon Krakauer's bestselling 1996 book *Into the Wild*, and later the subject of a Hollywood film, marched into the Alaskan wilderness in 1992, he expressed the longing that many Americans feel for a life lived away from the hustle and bustle of modern urban clusters. Echoing Henry David Thoreau's famous challenge to "live deep and suck out all the marrow of life" and thereby avoid sinking into quiet desperation, McCandless wrote to a friend that "so many people live within unhappy circumstances and yet will not take the initiative to change their situation because they are conditioned to a life of security, conformity, and conservatism, all of which may appear to give one peace of mind, but in reality nothing is more damaging to the adventurous spirit within a man than a secure future." The antidote to the sickness of the human spirit for McCandless, as for Thoreau, was to seek out nature. "The very basic core of a man's living spirit is his passion for adventure. The joy of life comes from our encounters with new experiences, and hence there is no greater joy than to have an endlessly changing horizon, for each day to have a new and different sun."[2]

For persons less inclined to indulge their aesthetic sensibilities than an idealistic young man (who eventually starved to death when his zeal for natural experiences exceeded his survival abilities), the natural environment must be protected because it holds utilitarian value; that is, a clean, healthy environment allows the citizenry to rely on the fruits of the earth to satisfy human desires. A beautiful sunset is pleasant to behold, but a mouthful of clean air is essential to life. To the pragmatic American, environmental conservation, preservation, and protection are not necessarily intrinsically valuable activities, but they are instrumentally necessary endeavors if human beings are to continue their ascension along the evolutionary scale.[3]

Notice the emphasis on a uniquely "American" worldview. American environmentalism—its philosophy, history, and public policy repercussions—initially seems to be an artificial and provincial limitation on a quintessentially global concern. If pollution respects no boundaries and if natural resource preservation and the problems of environmental degradation require innovative, multifaceted, systemic solutions spanning across numerous countries and continents, focusing on the United States at the exclusion of 95 percent of the world's population appears needlessly myopic. The point is well taken. Yet the United States, as is the case with many issues, has developed a distinct perspective on environmentalism that bears examination apart from global considerations. The eminent historian Walter Russell Mead once observed, "We Americans have always believed that we inhabit a special universe, or at least a special part of the ordinary one." Nowhere is this insight better illustrated than in the development of the American brand of environmentalism.[4]

Notice also that the term "environmentalism" in this book refers to the ideas and concepts as well as the principal participants in an ongoing debate about whether the natural environment should be conserved, preserved, and protected and, if so, the means by which the goal can be accomplished. Because the topic is enormous and encompasses so much history as well as a vast array of conceptual constructs, the narrative necessarily paints with a broad brush across a wide canvas. Walter Rosenbaum, a well-known commentator on environmental politics, once observed, "Environmentalism has never been a church of one creed." To understand the multiple issues and schools of thought, it is necessary to explore the topic from several perspectives. Each chapter of *American Environmentalism* could be expanded into a standalone book. Indeed, a curious reader can peruse the endnotes and list of references for a multitude of sources on the respective topics. The purpose here is to introduce the general reader to the major issues while leaving exhaustive coverage of the subject matter to the works cited as source material.[5]

Protecting the natural environment and promoting sustainability have become important objectives for U.S. policymakers and citizens at the dawn of a new millennium, but achieving such goals presents myriad challenges for even the most committed environmentalist. A major difficulty arises owing to the range of diverse perspectives among participants in the ongoing debate about what constitutes protection of the natural environment and, for that matter, how the term "sustainability" should be defined. Two respected commentators on environmental issues, Charles H. Eccleston and Frederic March, noted in their book *Global Environmental Policy: Concepts, Principles, and Practice* that "sustainability" is "often misused as a feel-good buzzword" devoid of meaning. To adopt a sustainable lifestyle, preferably without much research or sacrifice, consumers make choices they believe to be important. "The public may be motivated to choose paper sacks over plastic ones and reusable diapers over disposable ones," the authors wrote. "They may even scale down from large sport utility vehicles to smaller ones." Dilettantes, thrilled to be "part of the solution and not part of the problem," smugly congratulate themselves on their "sustainable" decisions. Yet Eccleston and March rightly observed that consumer choices about packaging and household products, even if they are not based on erroneous information gleaned from a five-minute Internet search or from chatting with quasi-informed pseudo-friends via social media, "are highly unlikely to even remotely offset the scale of environmental destruction in even one region."[6]

If consumer choices do not mitigate environmental destruction on a macro level, the question arises as to how conscientious citizens can preserve and protect natural resources. Good intentions should count for something. The argument here is that understanding the range of environmental issues is the first step toward developing efficacious public policy. Accordingly, this book seeks to clarify the salient variables by exploring the philosophical, historical, and public policy issues in American environmentalism. The central problem is whether competing interests can be reconciled while developing consistent, coherent, effective public policy to regulate uses and protection of the natural environment without destroying the national economy. This is a complex question, with no easy answers. Accordingly, the book reviews a range of possible solutions. Although other published works delve into the philosophy of sustainability, the history of American environmentalism, or the politics of the environmental movement, *American Environmentalism: Philosophy, History, and Public Policy* explores each issue in turn.

The book is divided into three parts. Section I, "Philosophical Foundations of American Environmentalism," delves into key normative concepts that undergird American perspectives on nature by providing an overview of philosophical ideas and concepts found in the western intellectual tradition, the presuppositions inherent in neoclassical economics, and the various anthropocentric (human-centered) and biocentric (earth-centered) positions on sustainability. Section II, "Historical Foundations of American Environmentalism," traces the evolution of attitudes about nature from the time of the Ancient Greeks through Europeans in the Middle Ages and the Renaissance, the Enlightenment and the American

Founders, the nineteenth and twentieth centuries, and up to the present. Section III, "Postmodern American Environmental Politics and Policy," examines the political landscape as non-governmental organizations (NGOs), industry leaders, and government officials struggle to balance industrial development with environmental concerns. Because some issues overlap among the three parts of the book, I use my judgment in determining where a discussion is most appropriate. For example, a discussion of Rousseau's views on nature could appear in Section I or Section II since the Genevan philosopher is an important figure in both the development of western philosophy as well as the history of the Enlightenment era. I discuss Rousseau in Section II because Section I involves a basic review of general philosophical topics. With that review completed, an extended discussion of how environmental philosophy developed historically is appropriate in Section II.

The book is descriptive, not prescriptive; that is to say, it does not deconstruct philosophical schools of thought, competing historiographies, or policy models, nor does it offer new recommendations for systemic change. The goal is to understand the landscape of environmentalism, not conquer it. To the extent that a policy model is offered, it can be characterized as "incremental pluralism." Decisions about conserving, preserving, and protecting the natural environment are made by government actors (elected officials and unelected public administrators), but NGOs influence the slow, iterative process through lobbying and media campaigns. The final chapter offers general advice for moving forward, but certainty is conspicuously absent. As Voltaire once remarked, "doubt is not a pleasant condition, but certainty is absurd."[7]

The underlying thesis is that life in a twenty-first-century industrialized state requires an accommodation between competing interests, including anthropocentric and biocentric schools of thought. Is this a laudable goal or a fool's errand? Is a dichotomous choice—pro-environment or pro-industry—a *sine qua non* of environmentalism or an oversimplified litmus test of ideological purity propagated by parties pursuing a partisan agenda? Can such a question be posed without betraying one's policy orientation? I fear that too often parties positioned on both sides of the ideological divide argue over how many angels can dance on the head of a pin. Would it not be preferable to focus attention on genuine, large-scale environmental problems involving systemic issues? Why not devote time, energy, and resources to understanding and resolving genuine, pressing environmental problems in lieu of following the well-worn path of bromides and histrionic doomsday claims about relatively minor, ancillary issues? Environmental concerns are only understood on a broad, multiscalar level, and efficacious public policy is only developed after the issues are identified and properly framed.

Let us be clear at the outset: Outrageous claims, silly misrepresentations, bogus arguments, absurd contentions, and overblown prophesies of impending calamities are bandied about by many parties on all sides of the debate—industry spokespeople, elected representatives, unelected regulators, concerned citizens, and environmental NGOs alike. No one comes to the debate with clean hands or "white knight" status, despite media claims to the contrary. Politics and science have become so intertwined in American environmentalism that separating the two becomes an almost-impossible feat of Sisyphean complexity. In lieu of descending into this morass, I wonder if I can circumvent the silliness to explore the crucial issues through a more focused, disciplined approach. Rather than engage in acrimonious debate over minutiae, as so often occurs in the context of "green" claims, I seek to recast the issues. This effort may be quixotic, but how else to cut the Gordian knot?

After recasting the debate, it is clear that Americans want to protect the environment, but they want to maintain their standard of living at the same time. As well-known environmental ethicist Mark Sagoff wrote in a famous 1981 essay, "At the Shrine of Our Lady of Fatima, or Why Political Questions Are Not All Economic":

> I used to buy mixers in returnable bottles—but who can bother to return them? I buy only disposables now, but, to soothe my conscience, I urge my state senator to outlaw one-way containers. I love my car; I hate the bus. Yet I vote for candidates who promise to tax gasoline to pay for public transportation. I send my dues to the Sierra Club to protect areas in Alaska I shall never visit.... And of course I applaud the Endangered Species Act, although I have no earthly use for the Colorado Squawfish or the Indiana bat. I support almost any political cause that I think will defeat my consumer interests. This is because I have contempt for—although I act upon—those interests. I have an "Ecology Now" sticker on a car that leaks oil everywhere it's parked.[8]

Sagoff identified a critical insight into American environmentalism. To cite the well-worn colloquialism, we want to have our cake and eat it, too. Human beings in general—and Americans in particular—cherish contradictory values. Deep inside many a human heart lies a desire to be a "good" person and live up to a noble creed. Yet nobility can be a suffocating shroud. The late comedian Lewis Grizzard once confessed that he supported animal rights, but he also wanted to enjoy a good steak now and again. A citizen wants to protect the natural environment, but he or she must still scratch out a living and raise a family in a society constructed on a foundation of anthropocentric Utilitarianism driven by neoclassical economic precepts.

Let us therefore acknowledge our contradictory impulses: We seek to protect natural resources and ensure long-term ecosystem health, but we do so with the realization that individuals must still live an "American life," which involves consuming resources and seeking employment in a consumer economy. Wholesale, systemic changes may be a desirable ethos—although even this conclusion is debatable—but the neoclassical economic regime is unlikely to be displaced in the near future. A distance exists between the ideal and the practical.

Speaking of acknowledgments, with each book I write, I am reminded of my debt to the supporting *dramatis personae* involved. I especially appreciate the advice and assistance I received from Lara Zoble, associate editor at CRC Press/Taylor & Francis Group, as well as Laurie Schlags, project coordinator at CRC Press, and Jay Margolis, project editor. In fact, many talented professionals at the press were instrumental in producing this book.

I owe a debt of gratitude to the numerous archivists and researchers who aided my research efforts, especially Kimberly David at the Library of Congress and Ellen Sandberg at the Granger Collection, New York. The interlibrary loan staff at the Horace W. Sturgis Library, Kennesaw State University, proved to be extremely helpful as well. I appreciate the marketing assistance I received from Liz Kula of webdesignsbyliz.com. She created my website and has kept it up to date as I have struggled to build my career. Liz has been an invaluable asset.

My colleagues at Dart Container Corporation have been supportive, especially Frank Liesman, vice president of Legal & Government Affairs, who authorized me to devote departmental time and money toward completing the manuscript in a timely manner. Robert C. Dart, the company's chief executive officer, kindly provided the working conditions necessary for researching and writing the work. In alphabetical order, I extend my heartfelt thanks to other Dart Government Affairs & Environment Department (GAED) members and engineering professionals for insights and inspiration: Todd Atkins, Jonathan Choi, Pam Dolbee, Ray Ehrlich, Fred Forrester, Kim Frankovich, Kevin Kennedy, Scott Klipa, Humberto Koike, Harry Kopp, Jim Lammers, Ruben Nance, Gerardo Pedra, Greg Perkowski, Susana Sanchez, Cheryl Schmidt, AnnMarie Treglia, and Michael Westerfield. Jeff Nelson from Dart Container Sales Company (DCSC) was a valuable source of information about his father, the late Wisconsin Senator Gaylord Nelson, chief architect of Earth Day 1970. Chuck Redmon, manager of National Account Sales at DCSC, proved to be a reliable and valuable sounding board for ideas. I would be remiss if I also did not acknowledge Anne D. Mularoni, who served as the Dart GAED manager during the first 14 years of my tenure in the department. Under Anne's patient tutelage, I learned much about American environmentalism. I will be forever grateful for her wise counsel and sound judgment.

Industry colleagues—especially Jennifer Killinger, Suzanne Morgan, Jane Mott-Palmer, and Sarah Visser—provided tremendous help as well. I also appreciated the willingness of several high-profile environmental experts to provide endorsements on the back cover.

As always, my friends and family have been incredibly supportive. My long-time friend from law school, Keith W. Smith, proved to be an invaluable asset as he steered me to the literature on Christian perspectives on nature. I also thank Loren and Polly Mead; Martha and Dick Pickett; Bob and Peggy Youngblood; the late Charles DuBose and his wife, Glenda; Wallace and Leila Jordan; Dr. William D. Richardson of the University of South Dakota; Dr. Ronald McNinch of the University of Guam; Sheila Traub; and the indomitable Shirley Hardrick.

Most of the epigraphs appearing in this book were in the public domain, with one notable exception. As a long-time popular music fan, I wistfully recall days and nights listening to the rock-music group The Doors in my college dormitory. A poster of Jim Morrison, the group's lead singer, hung on the wall over my bed during my senior year. I appreciate permission to use partial lyrics from the following source in Chapter 7: "When the Music's Over"; words and music by The Doors; Copyright ©1967 Doors Music Co.; copyright renewed; all rights reserved; used by permission; reprinted by permission of Hal Leonard Corporation. The song appeared on the group's sophomore album *Strange Days*, released in September 1967.

I have attempted to present all sides in the environmental debate. My purpose has been to illuminate the salient issues as opposed to producing a polemic or a diatribe for or against a particular position. Needless to say (but I will say it, anyway), any errors, omissions, or inadvertently strident declarations are my responsibility alone. In fact, this book does not represent or reflect the views and opinions of Dart Container Corporation, Kennesaw State University, the University of South Dakota, the University of Georgia, or any other institution with which I am now or ever have been affiliated. Recognizing that not everyone will be satisfied with the book, I can only offer my assurances that the intent is to provide a reasonable accounting of the topic even if the work does not definitively resolve the issues raised.

Notes

1. Quoted in Caroline Thomas Harnsberger, *Mark Twain at Your Fingertips: A Book of Quotations* (Mineola, N.Y.: Courier Dover Publications, 2009), 320.
2. Quoted in Jon Krakauer, *Into the Wild* (New York: Villard, 1996), 56–57.
3. For an excellent discussion of the salient issues, see, for example, Roger Scruton, *How to Think Seriously about the Planet: The Case for an Environmental Conservatism* (New York and Oxford: Oxford University Press, 2012), 183–208.
4. Walter Russell Mead, *Mortal Splendor: The American Empire in Transition* (Boston: Houghton Mifflin, 1987), 3.
5. Walter A. Rosenbaum, *Environmental Politics and Policy*, 4th ed. (Washington, D.C.: Congressional Quarterly Press, 1998), 28.
6. Charles H. Eccleston and Frederic March, *Global Environmental Policy: Concepts, Principles, and Practice* (Boca Raton, Fla.: CRC Press, 2011), xxxvii.
7. Voltaire is quoted in Robert Bryce, *Power Hungry: The Myths of "Green" Energy and the Real Fuels of the Future* (New York: Public Affairs Books, 2010), 285.
8. Mark Sagoff, "At the Shrine of Our Lady of Fatima, or Why Political Questions Are Not All Economic," in *The Environmental Ethics & Policy Book*, 2nd ed., Donald VanDeVeer and Christine Pierce, editors (Belmont, Calif.: Wadsworth, 1998), 303.

About the Author

A member of the bar in Georgia and South Carolina, J. Michael Martinez began his career in the private practice of law. He later earned a PhD in political science and a second PhD in public administration. As a master's student at the Georgia Institute of Technology during the 1990s, he focused his studies on environmental public policy, eventually publishing peer-reviewed journal articles on hazardous and radioactive waste management as well as environmental ethics. His scholarly articles on a variety of subjects have appeared in numerous academic publications, including *Administration & Society, The American Review of Public Administration, The Georgia Historical Quarterly, The Journal of Environment & Development, The Journal of Medicine and Philosophy, The Journal of Policy History, Nonprofit Management & Leadership, Politics & Policy,* and *The South Dakota Law Review*, among others.

Dr. Martinez has co-edited and contributed chapters to three academic texts: *Ethics and Character: The Pursuit of Democratic Virtues* (with William D. Richardson and Kerry R. Stewart, 1998), *Confederate Symbols in the Contemporary South* (with William D. Richardson and Ron McNinch-Su, 2000), and *The Leviathan's Choice: Capital Punishment in the Twenty-First Century* (with William D. Richardson and D. Brandon Hornsby, 2002). He also has written and published three books on Southern history, focusing on the Civil War and Reconstruction: *Life and Death in Civil War Prisons* (2004), a selection of the History Book Club; *Carpetbaggers, Cavalry, and the Ku Klux Klan: Exposing the Invisible Empire During Reconstruction* (2007); and *Coming for to Carry Me Home: Race in America from Abolitionism to Jim Crow* (2012). His other books include *Administrative Ethics in the Twenty-First Century* (with William D. Richardson, 2008), *Public Administration Ethics for the*

Twenty-First Century (2009), and *Terrorist Attacks on American Soil: From the Civil War Era to the Present* (2012).

Since 1992, Dr. Martinez has worked in the Government Affairs & Environment Department at Dart Container Corporation, a leading manufacturer of disposable foodservice products. He also teaches political science, criminal justice, and public administration courses as a part-time faculty member at Kennesaw State University, the University of South Dakota, and the University of Georgia, respectively. As an adjunct faculty member during the 1990s, he taught political science at Georgia State University and environmental law at Atlanta's John Marshall Law School. From 1988 until 1992, he was on the staff of the Southern States Energy Board, a nonprofit interstate compact organization that researches energy and environmental laws and regulations.

Visit him on the Internet at www.jmichaelmartinez.com.

I

PHILOSOPHICAL FOUNDATIONS OF AMERICAN ENVIRONMENTALISM

Chapter 1

Philosophical Ideas and Concepts

> The earth, he said, has a skin; and this skin has diseases. One of these diseases, for example, is called "humanity."
>
> **—Friedrich Nietzsche, *Thus Spoke Zarathustra*[1]**

Nature, according to some ontological philosophers, exists as a reality apart from the human animal. Long before humankind emerged, the natural world evolved over numerous millennia as evolutionary adaptations siphoned off weaker plants and animals, leaving more resilient genera in their wake. Frequent upheaval became the standard for natural processes and a terrestrial balance that man eventually altered. Epoch upon epoch came and went before a human sound was uttered or a creature walked upright and sought to exercise an unfettered will.

It was not a foregone conclusion that a sentient, thinking, adapting, rational being would fend off predators and debilitating disease to emerge as a dominant planetary force. Yet despite long odds, *Homo sapiens* crawled up from the muck of the primordial ooze, gradually adapting to the environment and overcoming previously insurmountable obstacles. The new animal was unlike any predecessor. It transcended what had come before: The size and capacity of the brain, an ability to use opposable thumbs, and its inquisitive and acquisitive characteristics distinguished the creature from its competitors, ensuring its place atop the food chain. The animal was equipped with an ingenuity and viciousness that would alter natural history in myriad ways still not fully understood and perhaps not altogether salutary. Biocentrists contend that if human beings had never existed, it would not have been necessary to invent them.

Nature required most species to live with clearly delineated limitations, yet human beings were dedicated to transcending boundaries. As humans constructed a civilization predicated on overcoming natural impediments, they viewed themselves as superior to the environment. They dedicated themselves to conquering nature and harnessing her resources. To be fully human was to use one's superior brain and dexterity to outpace all would-be rivals in the race for acquiring food and encouraging procreation. Man was not a part of nature; he was above nature. He was not a creature dependent on context. He was a creature who created context, who manipulated the world around him to suit his purposes, and who tapped undeveloped resources and fashioned an improved environment. Nature was not something to be enjoyed and celebrated, an intrinsically valuable feature of the external world. It was a thing to be subjugated as an instrument of human happiness.[2]

Concepts of Nature in the Western Intellectual Tradition

Lynn White, Jr., famously observed that the western intellectual tradition is based on an implicit understanding of nature as instrumentally valuable. God created the earth for human beings to exploit. Men must establish dominion over nature, taming it and improving it as necessary to meet human needs. This emphasis on nature's utility and subservience to man has reverberated throughout western culture.[3] In the *Phaedrus*, Plato recalled that his teacher Socrates refused to travel beyond the boundaries of the city-state, Athens, because the surrounding countryside was not worth exploring. Only the affairs of men were important. Nature served as a backdrop, and a poor one, for human endeavors.[4]

The Christian perspective emphasized the instrumental character of nature. In Genesis, as God created the earth, He formed man in His own likeness. Man walked among the vegetation and creatures that God had created, and He "let them [men] have dominion over the fish of the sea, and over the birds of the air, and over the cattle, and over all the earth, and over every creeping thing that creeps upon the earth." God commanded in Genesis 1:28, "Be fruitful and multiply, and fill the earth and subdue it; and have dominion over the fish of the sea and over the birds of the air and over every living thing that moves upon the earth." In Genesis 2:19–20, God solidified man's exalted position as the highest earthly creature by allowing human beings to classify the things of the earth as men saw fit.

> So out of the ground the Lord God formed every beast of the field and every bird of the air, and brought them to the man to see what he would call them; and whatever the man called every living creature, that was

> its name. The man gave names to all cattle, and to the birds of the air, and to every beast of the field....

Eastern philosophy developed in a different cultural milieu than the western intellectual tradition and never evinced an antagonistic environmental ethic. The human form might appear as a different creature in the next life; he dared not alienate other animals or subvert the natural context in this life. The forest that a man cut today might be the forest the reincarnated human would need to sustain him tomorrow. Yet the calculation was more than a consequentialist ethic. Even before shuffling off the mortal coil, human beings are not so different from other forms of life: They take in food, depend on the earth for crops, defecate, bleed, rise from the soil, and return to ash at the end of a brief, tumultuous existence. To view *Homo sapiens* as a unique consciousness superior to the surrounding world, impervious to its tribulations, is to succumb to a dangerous hubris that will destroy the natural environment and, with it, the human animal.

For all its virtues as an environmental ethos, eastern philosophy was not hegemonic or static. The multitudes, whether they live in Great Britain or Outer Mongolia, must still be fed, their crops planted and harvested, their houses constructed from the earth's resources. If the environment was not an impediment to be conquered through ever-more efficient industrial processes, neither were all members of eastern society satisfied to eke out a subsistence living to avoid diluting pristine natural resources. Nature was instrumentally valuable even if eastern thinking was not as willing as occidentals were to scorch the earth in service of a theoretically higher purpose. Setting aside the propensity of some biocentrists to tout the myth of eastern environmental harmony, all civilizations developed a plan for harnessing the resources of the earth. The difference was in degree, not kind.[5]

Philosophers of the Far East frequently revered nature, in word if not always in deed, but western thinkers unabashedly reserved their accolades for human mastery over the earth. Consider Shakespeare's famous oration from *Hamlet*:

> ...[T]he earth, seems to me a sterile promontory, this most
> excellent canopy, the air, look you, this brave
> o'erhanging firmament, this majestical roof fretted
> with golden fire, why, it appears no other thing to
> me than a foul and pestilent congregation of vapours.
> What a piece of work is a man! how noble in reason!
> how infinite in faculty! in form and moving how
> express and admirable! in action how like an angel!
> in apprehension how like a god! the beauty of the
> world! the paragon of animals![6]

The words can be interpreted in several ways. Superficially, Shakespeare uses the scene to express the melancholy Dane's discontent with his life. Hamlet knows he should appreciate the beauty of the world, but he is depressed owing to his indecision about avenging his father's death. The words also can be read as a commentary on the barrenness of nature unless it is seen through human eyes. Hamlet characterizes the utility of nature according to his moods. A contented Hamlet is pleased to frolic under "this most excellent canopy" and enjoy "this brave o'erhanging firmament, this majestical roof fretted with golden fire." A dour, depressed Hamlet denounces the earth as "a sterile promontory" and a breath of fresh air "appears no other thing to me than a foul and pestilent congregation of vapours." Nature does not change, but its beauty changes as it is filtered through a noble, reasonable, angelic consciousness. Man is the only "piece of work" that genuinely matters in the cosmos.

Consciously or unconsciously, the lines from *Hamlet* reflect Psalm 8:3–8 from the Christian Bible:

> When I consider your heavens,
> the work of your fingers,
> the moon and the stars,
> which you have set in place,
> what is mankind that you are mindful of them,
> human beings that you care for them?
> You have made them a little lower than the angels
> and crowned them with glory and honor.
> You made them rulers over the works of your hands;
> you put everything under their feet:
> all flocks and herds,
> and the animals of the wild,
> the birds in the sky,
> and the fish in the sea,
> all that swim the paths of the seas.

In the traditional narrative, Christianity's ascendancy over Paganism was a triumph of light and virtue over darkness and depravity. Paganism viewed human beings as spiritually one with nature. To fell a tree or dam a river was to risk enraging the higher power, whether that was construed as an amorphous Nature, with a capital "N," or multiple cosmic deities. The goal was to tread lightly or risk the wrath of Nature striking back at her transgressors, as she sometimes did with violent weather and natural disasters.

When Christianity supplanted Paganism in many parts of the world, the reverence of nature born of superstition and primitivism became a sign of the heathen. A heathen would be converted with the Word or the sword, but one way or another, Paganism would be vanquished. The unholy man could not be allowed to revel in his illusions, for the natural world was dark and chaotic, as anyone who has been lost in a strange forest at night without a map, compass, or light source learns to his chagrin. Christianity offered its adherents a compelling alternative—light in lieu of darkness, order instead of chaos, the Word in place of the Void. Christian belief became a means of overcoming ignorance and, by extension, taming the wilds of Nature.[7]

Not everyone has subscribed to this interpretation. With the advent of the modern environmental movement beginning in the 1970s, some Christian apologists argued that the biblical tradition, especially the humanistic passages found in the New Testament, does not promote a hostile environmental ethic. They took umbrage at Lynn White, Jr.'s thesis that Christianity rationalizes all manner of malfeasance in the name of unconstrained human dominion over the earth. Rather than blame Christianity for an ecological crisis, they suggested that White and his adherents look to the origins of modern science and technology as well as western man's emphasis on short-term economic gain as culprits. Neoclassical economics and the destructive effects of the Industrial Revolution are the roots of the modern conundrum. God wants human beings to exercise wise environmental stewardship, which does not justify the wholesale destruction of natural resources. In short, the traditional narrative represents a fundamental misreading of biblical passages. In their book *Pollution and the Death of Man*, Francis A. Schaeffer and Udo W. Middelmann succinctly summarized the Christian ecological perspective this way:

> It [i.e., the earth] belongs to God, and we are to exercise our dominion over these things not as though entitled to exploit them, but as things borrowed or held in trust. We are to use them realizing that they are not ours intrinsically. Man's dominion is under God's dominion.[8]

Rights

After conquering the chaos of the natural world, human beings devised plans for living within a collectivity. Primitive humans depended on might to make right. The concept of divine right later propelled some men above others inside an authoritarian regime. Eventually, a concept of rights developed to limit government controls.

Defending "inalienable" rights has become de rigueur in the postmodern era, especially among partisans who loudly complain about their infringement by offenders variously identified as the government, political parties, the wickedly biased media, factions hostile to liberty, or, as a catch-all, the "powers that be." When pressed to explain the concepts of rights, the shrillest defenders often fall silent or mumble about "the things I am entitled to." It is little wonder the topic invites such passionate discourse as well as unbridled confusion. The concept of "rights" generally is discussed not by defining the term, but by discussing various types of rights.[9]

A claim right, for example, allows the holder of that right to expect another person or entity to perform an act or refrain from performing an act that affects the claimant. A liberty right, by contrast, provides the holder with an opportunity to take action but does not affect other parties directly. In another sense, persons may exercise positive rights or negative rights. Positive rights provide the holder with permission to undertake actions while negative rights allow the holder to be left alone without interference.

In a manner of speaking, a right is a kind of promissory note that allows the bearer to act or be protected from interference as he or she moves through space and time. To possess and act on a right is to send a message to would-be interlopers that the self-actualized person who holds the right intends to call for payment under the promissory note. For positivists, a right presented to an authoritative governmental body requires, at a minimum, the body to undertake procedural steps ensuring that the rights-holder will be protected from arbitrary action that potentially harms his or her interests.[10]

A discussion of rights invariably leads to a search for the source of rights. The American Founders added a bill of rights to the U.S. Constitution to ensure that questions about whether citizens could be protected from specific government actions would be resolved with minimal ambiguity, although the scope and interpretation of those rights remain part of a perennially divisive political debate. Other sources of rights tend to be nebulous. The debate over rights is especially contentious when nature or God is cited as the ultimate source, for perceptions of nature and God differ radically among the populace. More to the point in the context of environmental issues, appeals to natural rights beg the question of what makes something part of the "natural order of things" and therefore presumably desirable.[11]

Natural Rights and Natural Law

Natural rights spring from natural law, a concept with no precise meaning. Natural law is difficult to discuss because it has been explained in myriad ways across a broad expanse of western history and culture. To understand the twenty-first century interpretation, it is necessary to recall a time when Christianity spread from the twilight of the Roman Empire into the medieval period. In the late Middle Ages, nation-states emerged in Western Europe. In some instances, these developments were complementary, but upon occasion an authoritarian government faced a defiant clergy that instructed the flock to render unto Caesar the things that were Caesar's and reserve other tribute to the church. A schism between the law of the regime and "higher" law—whether "natural," "God-given," or both—gradually widened during the Enlightenment epoch.

Strong regimes were constructed to fortify against the dark of night where marauding barbarians launched vicious attacks against crumbling empires. It was not an irrational fear in a world where the lack of a strong, well-equipped government might lead to chaos and disorder. When Hobbes wrote of the "state of nature" during the seventeenth century, he envisioned it to be a chaotic, violent, brutish world populated by savages. Rational beings would do whatever they could to remove themselves from that horrific condition. Even a flawed, potentially abusive nation-state was preferable to existence among the unwashed hordes. Nineteenth-century thinkers, paraphrasing Tennyson, would bemoan the horrors of "nature red in tooth and claw."[12]

The quest for a strong government that would protect the citizenry from the viciousness of nature eventually lent credence to the divine right of kings, a self-serving doctrine that monarchs employed to justify all manner of authoritarian mischief. A king ruled his subjects based on a hereditary principle that was not subject to vote or dissent save in rarefied settings stage-managed by the crown. To a contrarian who would assail the principle, a clever ruler set a two-tiered rhetorical trap: To argue against the king was to risk eternal damnation, for the king was God's earthly emissary. If this rationale were not enough to protect the *status quo ante*, a second concept prompted obedience without appeal to an afterlife: To undermine the king was to undermine the nation and risk the destruction of all manmade institutions, thereby plunging the kingdom into chaos. Even a clear-eyed materialist such as Hobbes recognized that a flawed sovereign was preferable to no sovereign at all. Only an irrational being would risk returning to a state of nature—no matter how heavy-handed the monarch proved to be.

With the expediency of divine right masquerading as principle, a seemingly impenetrable authoritarian edifice towered above western political thought until no less an architect of democracy than John Locke tapped into a long-standing tradition to tear down one set of walls while simultaneously constructing another. The concept of natural law had existed for centuries when Locke came of age. It did not quite rise to the level of a coherent political philosophy or a well-defined doctrine, but the notion persisted.[13]

The Ancient Greeks spoke of natural law as immutable principles that form the ur-stuff, or context, of life. Every idea or institution of man rests on a foundation of a "higher" natural law, as Aristotle explained. For the Stoics, the universe was indifferent; the "natural order of things" existed and had to be followed. To act in accordance with the orderly nature of the universe was to live a virtuous life. Thus, in the context of a human life, people are born, live, experience pain and hardships, and die. Virtue requires that human beings recognize these unchanging characteristics of life and accept them without complaint. Subsequent thinkers sought to understand what propelled natural law. For early Christians, God's hand directed nature. The universe was not indifferent, and "evil" existed insofar as man was unable to comprehend and act upon the divine plan.[14]

Whatever its source, natural law relies on several fundamental presuppositions. First, a series of objective norms exists and antedates the development of human laws and institutions. The norms can be discovered and tested as any empirical claim can be verified using scientific measures. Human conventions and institutions ground regimes, but they do not posit a claim superior to natural law. When a conflict occurs between "higher" natural law and the laws of man, the former must take precedence. This articulation lends credence to Locke's claim that any regime that violates natural law is illegitimate and can be altered; it serves as a rationale for the U.S. Declaration of Independence and its assertion of self-evident truths that all men are created equal and possess inviolable rights to life, liberty, and the pursuit of happiness.

Second, natural law is based on reason, even if the ultimate source of reason is not always clear to human beings. For non-Christian natural-rights theorists, the immutable laws of cause-and-effect, if understood correctly, necessitate a particular conclusion based on a dispassionate assessment of the evidence. As an example from biology, the body must be nurtured with food or it will suffer and ultimately perish. This conclusion is supported by accumulated data about what constitutes useful foodstuffs. The data indicate which items are beneficial for the growth, development, and maintenance of the body and which items trigger deleterious effects that harm or poison the body. The rules about proper nutrition have been developed through trial-and-error over time. Eventually, the careful observer articulates conclusions about the relationship between the foods put into the body and the consequences that result. Reason permits human beings to understand the "laws of nature" and act accordingly.

Yet reason is not always self-evident. As an expression of divine will in Christian theology, natural law is not flawed or intertwined with biases, contradictions, and compromises, as human laws are. It is a perfect expression of God's infinite power to construct a just world. If natural law appears imperfect or inscrutable, it is because human beings do not possess the capacity for understanding divine will. Reason exists in nature, but man is not always privy to the secrets of the universe. To paraphrase Cassius, the flaws are not in the stars, but in ourselves.

A third insight is a normative claim that most natural law supporters accept explicitly or implicitly. Because a thing exists and its existence can be deemed reasonable—even if the reason is not always clear to infallible humans—natural law can be classified as "good." The concept of nature is inherently valuable. This value judgment makes a virtue of necessity. It assumes that because existential limitations exist and the universe is an orderly place, the limitation therefore must be "beneficial." If the limitation were not beneficial, it would not exist. To allow a limitation to exist as an unplanned flaw in the schemata is to call into question the harmony of the natural universe. Even illness or death can be seen as positive. An individual human being might quake in fear at the thought of perishing, but a species prevails. As evolution advances and consciousness improves, each succeeding generation is an improvement over its predecessors. The dead body returns to the earth to provide nourishment for the soil that will spawn later generations. For those of a religious bent, the body decays while the spirit ascends onto a higher plane that is superior to life on earth. In either case, natural law is philosophically desirable.[15]

Natural law has been assailed by numerous sources, especially postmodernists, existentialists, and iconoclasts such as Friedrich Nietzsche. Their attacks go to the heart of the natural law conundrum. Consider the logical inconsistencies of each presupposition. If natural law is predicated on objective principles, which factors qualify them as "objective"? A phenomenon that seems "objectively true" today sometimes is found to be demonstrably false later. In one epoch, diseases such as malaria and yellow fever were presumed to be caused by noxious vapors and soiled linens. Later, as microbiologists probed causes of infection, they learned that bacteria and germs were the culprits. Modern medicines and treatments could repel microorganisms, but vaccine development required humans to overcome a seemingly "natural" condition. During the seventeenth and eighteenth centuries, Newtonian science was hailed as a persuasive explanation for the heretofore mysterious operations of the natural world—surely the "truth" had been isolated for all time—only to be radically modified by Einstein's theory of relativity during the twentieth century.[16]

Objectivity is not as "objective" as the classical philosophers once believed. Scientific progress relies on a series of halting, iterative changes. Characterizing natural law as "objective" does not resolve the dilemma of whether nature ought to form the basis for a value judgment. Similarly, proclaiming an act "natural" begs the question—natural for whom? Is it "natural" to die of diseases that would be prevented or treated if the proper vaccine were administered? If so, administering the vaccine can be deemed unnatural because it goes against the "natural order of things." An animal that otherwise would have died manages to persevere owing to the intervention of a new drug. For most human beings, such a triumph over nature is deemed an incremental step toward progress. Labeling an event or activity "natural" does not resolve the difficulty in understanding the penumbra of natural law.

Even if a definition of "natural" can be agreed upon, the conclusion that "natural is good" does not follow as a logical imperative. Value judgments are unnatural in the sense that they are decisions reached by human beings. They do not represent an ontological reality. If natural law predates human beings, human assessments of natural law are necessarily subject to the flaws inherent in any human enterprise. Nature is not good, bad, right, or wrong. Events occur; cause triggers effect. The imperfect human animal, not nature, places valuations on phenomena.[17]

Utilitarianism

Utilitarianism arguably is the most influential alternative to natural law as a philosophical construct for man's understanding of nature. The doctrine appeals to a variety of decision-makers, even those not predisposed to engage in philosophical reflection, because it appears to be intuitively obvious. Just as some natural law supporters defend their position as a compelling explanation for the operation of the world as it is, not as we might wish it to be, a Utilitarian fashions a theoretical defense from a conclusion about the nature of man.

Utilitarians argue that rational decision-makers act based on a principle of utility, which dictates that human beings seek the greatest happiness, or the greatest pleasure, from choices among competing alternatives. Except for the rare ascetic, mystic, holy man, or mentally challenged individual, human beings weigh the costs and benefits of each choice before deciding on a selection that appears to provide the greatest benefit for the least cost. That is, a typical individual seeks to maximize pleasure and minimize pain to the greatest extent possible. Because the selection often is made based on incomplete or inaccurate information, the calculation may be flawed, but the intent is to select a choice that satisfies a human need to advance one's self-interest while avoiding undue hardship or pain.

The English philosopher Jeremy Bentham (Figure 1.1) famously championed Utilitarianism, which he based on David Hume's theory of moral sentiments that human beings possess an instinctive understanding of which acts are useful. John Stuart Mill, Bentham's godson and intellectual heir, modified Bentham's justification of Utilitarianism by asserting that the wisdom of individual choices is dictated by the character of the people who make decisions. Rational beings frequently look past their own short-term happiness to pursue long-term goals that presumably lead to future happiness. A person of good character chooses a course of action that is not necessarily pleasurable to the masses. He chooses "higher" pleasures based on his experiences with pleasure and pain. He saves money as a rainy-day contingency instead of indulging in current frivolity. He attends school and struggles to master difficult intellectual topics because he believes this strategy will lead to greater pleasure in the future. The person of good character sometimes chooses to satisfy the desires of a larger group even if the choice harms his own choices or damages the interests of a minority.[18]

Because Utilitarianism seeks the greatest amount of happiness for the largest number possible, it has served as an underlying principle for democratic regimes. Policymakers recognize they cannot satisfy the needs of all constituents. In a large republic filled with multiple interests and numerous, often competing, constituencies, no optimal solutions exist. If a large enough number of citizens can be minimally satisfied, an elected official can cobble together a workable coalition that allows effective public policy to be developed and implemented.

Utilitarianism has not suffered from a dearth of critics. Defining "happiness" and measuring its presence or absence are the major points of contention. Critics argue that ethics based on maximizing a person's individual preferences are base and undermine society by encouraging appetite-driven human beings to consider their selfish proclivities above all else. Mill's rejoinder is to cite the need for civic virtue: A proper Utilitarian considers the happiness of other persons—the greatest amount of happiness in society as an aggregate collection of individual preferences. Good Utilitarianism is not solely concerned with the individual as the appropriate unit of analysis. The philosophy need not become a convenient rationale for unbridled selfishness. The philosophy requires persons to calculate happiness based on an impartial judgment divorced from self-interest. A genuine Utilitarian calculation

Figure 1.1 The English philosopher Jeremy Bentham was a well-known proponent of Utilitarianism. (Courtesy of the Library of Congress)

occasionally requires an individual to sacrifice his own happiness for the happiness of a larger group.[19]

Detractors contend that this numbers-based theory highlights the deficiencies of democratic government. A majority seeks pleasure instead of pain—this insight is hardly astonishing—but pleasure-seeking does not necessarily form a justifiable basis for philosophical wisdom or, for that matter, good government. Avoidance of pain in pursuit of pleasure is a base calculation that can lead to a base society. Concepts such as "just" or "right" do not depend on the desires of a majority. The tyranny of a majority can lead to oppressive public policy and abhorrent personal conduct all in the name of the higher good or the greatest good for the greatest number. Utilitarianism presupposes that choices are almost completely a matter of calculating costs and benefits and selecting a policy that maximizes benefits and minimizes costs.

Comparisons between competing policy choices can become exceedingly complex. The classical Utilitarian suggests that choices can be reduced to discrete units that can be examined side by side. Such an analysis requires no small leap of imagination to square the circle. How does one weigh and evaluate depths of preferences, for example? A large group may favor a policy slightly while a much smaller group may strongly prefer the alternative. If the choice is considered a matter of discrete units, the larger group will always triumph at the expense of the minority. If a weight is applied, by what criteria will some discrete units be evaluated as "more valuable" compared with competing units? The exercise soon degenerates into a subjective debate that was supposed to be resolved by a more or less objective calculation in the first place.[20]

Consider a common public policy issue—industrial development versus environmental preservation. Developers argue that an increasing population and the rise of the middle class throughout the world require aggressive policies allowing them to clear away forests and consume natural resources at an advanced rate. In the absence of such consumption, the infrastructure for supporting the citizenry will be absent and extreme shortages will result. Industrialization is a vast, ravenous, never-satiated machine that devours an ever-increasing quantity of natural resources. Environmentalists contend that aggressive resource consumption without careful natural resource protection will exacerbate a variety of ecological problems, including soil erosion, water shortages, and global warming. If resources are not carefully preserved and protected, human civilization will collapse. This debate can be viewed as the classic tradeoff among competing options inherent in any highly industrialized society.

Utilitarianism, the philosophical champion of rational cost-benefit calculations, can be used to justify either position. Developers can calculate the resources necessary to sustain a growing population in the short term and leave it to future generations to handle subsequent environmental problems. Who knows what technological advances and scientific tools will be available to citizens in coming centuries? A strict weighing of costs and benefits using current technology is difficult when the potentially negative consequences are not likely to occur, or the probability of their occurrence is so far removed that the consequences cannot be tabulated accurately. Environmentalists dispute this calculation and argue that based on their own cost-benefit analysis, prudent resource management is required immediately so that resources are not exhausted. The problems of future generations must be dealt with now because without wise environmental protection in the present, there will be no resources available for those persons who come after the current generation passes from the scene.[21]

Aside from the difficulty in valuation and measurement, Utilitarianism assumes that individuals are capable of reaching rational decisions. Yet even rational beings make decisions based on irrational factors. Pride, greed, envy, lust, fear, and revenge are a few of the numerous irrational, passionate emotions that can lead to misguided, occasionally horrific decisions. In some situations, as free people everywhere have repeatedly learned to their detriment, it is possible that persons would willingly sacrifice their liberty on the altar of the greatest good. In a totalitarian state, individuals might be willing to endure abuses and surrender their civil liberties in hopes that they will be protected from a state of nature.[22]

Complicating this debate is the belief that natural law and Utilitarianism can exist as complementary concepts. According to social compact theorists, natural law forms the basis for rights enjoyed by human beings. Because the committed Hobbesian recognizes the need to escape a short, brutish life in a state of nature, he willingly surrenders a measure of natural rights to create a regime that will protect him from the disastrous effects of natural anarchy. The champion of Lockean principles contends that the citizen can insist on changes within the regime and even dissolve the regime under extreme circumstances. The thoroughgoing Utilitarian takes up the argument by explaining how the regime will satisfy the populace without risking a potential return to a natural state. In this way, a democratic regime is constructed on a rickety foundation of natural law and Utilitarianism.

Utilitarianism has been the predominant philosophical justification for the liberal democratic state and the default philosophy for generations of thinkers, but its critics have undermined its stability. By the early twentieth century, as socialist movements gained momentum in parts of Europe and Asia, the old theories appeared antiquated and in need of shoring up. In his 1903 work *Principia Ethica*, G. E. Moore attempted to do exactly that. His objective was nothing less than to revise classical Utilitarianism and improve on its numerous deficiencies.

According to Moore, Utilitarians, including Jeremy Bentham and John Stuart Mill, argued for the greatest good for the greatest number, but they did not explain what they meant by "good" except to assume that "goodness" referred to "pleasure" without resolving the difficulties associated with precisely defining the terms. Pleasure and good may be identical, but not necessarily so. Moore suggested that a value judgment such as "good" is an ambiguous concept that cannot be defined outside of a specific context. Different circumstances, backgrounds, and levels of understanding or access to information mean that individuals are always guessing at what choices and actions are desirable. Because the good cannot be known first-hand, "good" acts must be judged by examining consequences and determining whether they achieve a desired outcome. The question is whether something is intrinsically good, that is, good for its own sake, or instrumentally good because it accomplishes another goal. Since the former question can never be answered definitively, the latter must be addressed. Utilitarianism is an imperfect heuristic tool for answering the second query—nothing more, nothing less.[23]

Kantianism as an Alternative to Utilitarianism

Moore's defense of Utilitarianism sought to rehabilitate the philosophy by forestalling an examination of the "good" apart from considering outcomes. The emphasis on outcomes as determining the worth of an action drew objections in some quarters because a consequentialist ethic changes with the circumstances. Renowned eighteenth-century skeptic David Hume was especially adamant in the adaptability of Utilitarianism. In his quest to avoid metaphysical pretensions, Hume contended that theoretical constructs lacking practical utility consist of "sophistry and illusion" and must be "committed to the flames." Yet a Utilitarian calculation alone is a necessary but insufficient foundation for ethics. Feeling, or taste, also is required because rational knowledge alone does not provide guidance on assigning moral praise or blame. Utility and feeling together influence the manner in which individuals arrive at ethical judgments. If human beings were not naturally benevolent—if they did not exhibit sympathy—they would not make ethical judgments because such judgments would be of no value. Hume's perspective can be thought of as a philosophy of "modified self-interest and a confined benevolence." His philosophy also contained the same deficiencies that all Utilitarian-based ethical theory must confront. If the utility of the philosophy determines its value, the standards are different for different cultures and people. It is difficult to establish authoritative guidelines for behavior because the standards constantly change over time.[24]

The relativism implicit in Utilitarianism triggered a reaction among philosophers who believed that philosophical and ethical standards must be absolute. Perhaps the most influential reaction came from Immanuel Kant (Figure 1.2), a German philosopher who claimed that Hume's "hypothetical imperative" awakened him from his "dogmatic slumbers." Kant was disturbed by the emphasis on an ever-shifting, situational standard. Utilitarianism permitted individuals to rationalize any terrible behavior according to the situation. In championing an endlessly flexible philosophical construct, Utilitarians embraced only a philosophy of veiled expediency.

Kant countered that if man is to be free to act as a rational, moral agent, he must obey absolute rules of conduct. In a world where a hypothetical imperative exists, man is little more than a lower-order animal because he can undertake actions without reflection or analysis. Kant theorized that a universal moral law exists, which he labeled the "categorical imperative." The imperative is a standard

Figure 1.2 The great deontological philosopher Immanuel Kant argued against Utilitarianism. (Courtesy of the Library of Congress)

of behavior for free, rational, moral agents under all circumstances without respect to space and time. "Act only on that maxim whereby thou canst at the same time will that it should become a universal law," he wrote. The categorical imperative is the Golden Rule where a person should do unto others as he would have them do unto him. If a person chooses to undertake an action, he must ask himself a question—"would I want to live in a society where everyone acts in this manner?"

If the answer is "no," the person must refrain from acting. If the answer is "yes," the person should act.

The categorical imperative has another meaning as well, according to Kant. Rational beings must never be treated as a means to an end. This insight undermines the Utilitarian desire to seek the greatest good for the greatest number by reducing preferences to discrete units that can be compared and traded. If rational beings are ends in themselves, their welfare cannot be bargained away. A choice among alternatives becomes a question of duty. Duty does not change owing to changing circumstances. The determination of a duty is not subject to bargaining and negotiation among competing and variable value judgments.[25]

In the context of environmentalism, the Hume–Kant debate between utility and duty can be analogized as the debate between persons who regard nature as instrumentally valuable and those who contend it is intrinsically valuable. This point will be developed in Chapter 3, but suffice to say that the debate concerns whether natural resources should be treated as any other commodity and therefore subject to traditional economic valuations or whether natural resources are unique and should be protected as part of humankind's duty to nature without regard to tradeoffs among competing claims. The dispute between the instrumental school of thought (anthropocentrism) and the intrinsic school of thought (biocentrism) lies along this fault line.

Friedrich Nietzsche disparaged "old Kant" for the old man's relentless emphasis on duty. If a person's duty can be identified (which is not always the case) and the person chooses to act accordingly, why is the act deemed "ethical"? Adhering to a rigid understanding of one's obligations sometimes can stifle creativity and limit human achievement. Emerson's observation that "a foolish consistency is the hobgoblin of little minds, adored by little statesmen and philosophers and divines" suggests that performing duty for duty's sake is not praiseworthy, but shortsighted and narrow-minded.[26]

In his famous essay "On a Supposed Right to Lie Because of Philanthropic Concerns," Benjamin Constant famously developed a devastating rejoinder to Kant. If performing one's duty requires an unwavering devotion to principle, what happens when two principles conflict? Constant's example involved two principles generally accepted by most adherents of western tradition: the duty to tell the truth and protect human life as precious. Yet if a desperate individual fleeing from a murderer seeks assistance, and a helpful bystander points to a suitable hiding place, how should the bystander respond when, minutes later, the murderer appears and inquires as to the whereabouts of his intended victim? If the Good Samaritan tells the truth, he must disclose the hiding place, which probably will lead to the death of the desperate individual. The admission, while it does not legally implicate the bystander in the murder, nonetheless is a direct and proximate cause of the crime. It appears to violate the Samaritan's ethical duty to protect human life as precious. If the bystander does not disclose the hiding place, he has violated his duty to tell the truth. A conscientious bystander might explain that the lie was necessary to save a life, but the lie violates Kant's principle about telling the truth, which is supposed to be an inviolable duty. For Kant, duty is absolute or it is not. The minute a person rationalizes deviations from the principle, no matter how well meaning or praiseworthy the rationalization, he slips into the muddy waters of the hypothetical imperative along with all the unwashed Utilitarians.[27]

Ethical Egoism and Social Darwinism

The debate assumes another guise as well. Egoism, a fundamental precept behind Utilitarianism, seeks to understand human motivations and beliefs by examining the self as the unit of analysis. Thus, psychological egoism is an empirical claim about how the world operates. The human animal is inherently self-interested and therefore acts in ways that advance self-interest even in instances where the common good is undermined. As an alternative, altruism suggests that self-interested beings occasionally act to promote an other-directed goal, which explains why philanthropists donate money anonymously or soldiers sacrifice their lives to save their peers. Yet a dedicated adherent of ethical egoism argues that such seemingly selfless acts ultimately advance a self-interested goal because the person undertaking the action receives gratification or believes he earns a higher reward in the next life. All actions, no matter how seemingly noble, are undertaken with the purpose of benefiting the self-interested ego.

A related concept, ethical egoism, is a normative theory. Whereas psychological egoism claims to describe how people in fact behave, ethical egoism suggests that self-interest ought to be the motive. Detractors denounce the theory as a poor justification for selfishness, but ethical egoism can be defended by suggesting that self-interest is a broader concept than mere selfishness. A parent who nurtures a child and establishes a stable home is acting in the interest of his family, but he also is improving civic and community life. Presumably, the citizen who pays his taxes, teaches his children values of self reliance and independence, and actively participates in political and civic activities contributes to the good of the whole. The entrepreneur who accepts risks in the interests of reaping financial rewards creates jobs and investment opportunities for other persons. The ethical egoist contends that pursuing a policy of what Alexis de Tocqueville called "self-interest, properly understood" leads to benefits extending across the broad canvas of the social fabric.[28]

By contrast, Social Darwinism suggests that even as self-interest can be beneficial, it can produce casualties. Adapted from Charles Darwin's biological theories, Social Darwinism can be thought of as both an empirical and a normative concept. Empirically, some well-meaning souls do not succeed despite their best efforts. Using analogies of the family and business endeavors, it is clear that the well-meaning parent who sends his children to good schools, sets an example, and engages in civic virtue may discover that the children misbehave, drop out of school, or participate in criminal behavior. Despite his best efforts, the family unit disintegrates. The entrepreneur may discover that for all his business acumen, superior marketing schemes, and professional contacts, the venture fails and he must declare bankruptcy. The Social Darwinist shrugs and mutters, "It is the survival of the fittest." A contentious debate rages to this day about what constitutes the "fittest," but the Social Darwinist points to the natural world as an empirical example. It is not always clear what makes an animal thrive over its competitors, but history is filled with examples of species that have survived through countless millennia. As a normative concept, Social Darwinism suggests that the fittest ought to survive. Weakness—whether among animal species or in the social realm—must be eradicated so that desirable traits endure.

Social Darwinism suffers from the same logical difficulty that applies to natural law or consequentialism, namely the problem of judging the desirability of a concept by examining outcomes. The theory assumes that because a thing survives while its competitors do not, the survivor must therefore be superior to the organisms that failed to adapt to an ever-changing environment. By mixing an empirical observation with a normative concept—"the survivor can be judged as 'good' because it self-evidently exists"—Social Darwinism makes a virtue of necessity. A survivor is "good" and a victim of natural selection is "bad." Such value judgments are based on the presupposition that the selection is "natural" and therefore, by definition, virtuous.

Yet the human animal constantly intervenes in "natural" processes and stacks the deck toward a favored outcome. If a human being does not wish to sleep outside in a natural condition, he seeks employment to earn money to purchase a home. The home is equipped with modern conveniences, including central heating and air conditioning, so that the undesirable aspects of the natural environment can be moderated to maximize comfort. If insects and rodents seek refuge inside the home, the incensed property owner contacts a pest control company to eradicate the offending population. The human animal discards waste that must be pumped into the environment, whether it is in the form of excrement, leftover foodstuffs, chlorofluorocarbons, carbon dioxide, or a corpse at life's end.

The Darwinist argues that the creation of this artificial environment is the advantage a human animal enjoys owing to its superior adaptability. Existing at the top of the food chain allows the superior species to modify his world in accordance with his wishes. Yet this theory does not consider whether the burdens created by a superior animal can and should be borne by other animals. Moreover, if individual members of the species imperil the species collectively through environmental degradation, for example, can and should the burdens be more evenly distributed? This question goes to the heart of the natural-resource allocation problem.[29]

Environmental Justice

Defenders of environmental justice argue that a separate and distinct environmental ethic must be developed. The philosophical doctrines discussed in this chapter do not provide satisfactory guidance for developing an effective theory of environmentalism. In addition, most influential philosophical doctrines within the western intellectual tradition ignore the disproportionate distribution of burdens. Rights theorists desire a mechanism for assigning and protecting rights, but they usually presume that rights extend only to persons and that unequal distributions are not especially problematic as long as arbiters of disputes are in place and are well-functioning. Yet the concept of environmental justice suggests that lower animals and other living things possess rights. In addition, the manner in which rights are distributed is important. The core question concerns how those rights are balanced against human rights. The balancing act remains contentious.[30]

Utilitarianism argues that the goal should be the greatest aggregate happiness for all persons, but it does not specify what should happen when one segment of society bears a disproportionately large share of the burden while another segment enjoys a larger share of the benefits. If Utilitarianism does not employ a mechanism for judging among and between depths of preferences, it also does not provide guidance on resolving equity questions. Maximizing pleasure in the aggregate may seem to be an intuitively defensible objective until one realizes that one segment enjoys a large share of the pleasure and another segment endures a large share of the pain. If redistributing the burdens and benefits provides unhappiness and perhaps reduces aggregate happiness, Utilitarianism fails to provide an effective solution when happiness and equity clash.

Kantianism heralds the importance of duty, but it does not provide a clear mechanism for determining who owes a duty to whom. Individuals undoubtedly owe duties to other individuals, but do they owe duties to non-humans as well? This exposition initially sounds odd, but it raises a valid point. If duty is eternal, a burden that must be borne without compromise, what happens when duties extend beyond human beings to lower animals and to nature? Eventually, duties will bump up against duties. In egregious cases, fulfilling my duty to one entity precludes fulfilling my duty to another entity. Building on Benjamin Constant's insight, Kantianism does not provide a mechanism for resolving disputes that arise owing to overlapping, contradictory duties.

For their part, ethical egoists and Social Darwinists champion self-interest, but at what scale should it be measured? I may choose to advance my self-interest today, but in doing so I may harm my self-interest in the future by consuming natural resources without replenishing them. The interested self is not necessarily a selfish soul, but it is not a long journey from "self-interest, properly understood" to "myself above all others." This short journey can impose devastating consequences on other human beings, animals, the regime, and the planet.

The environmental justice movement finds fault with each traditional school of thought and seeks to recast the philosophical debate. Rather than assume the human being is the unit of analysis and consider rights, duties, and costs from that perspective, "environmental burdens and benefits" must be afforded weight. Although some environmentalists prefer a radical shift so the human animal is no longer the primary consideration but one among many competing units of analysis, such a seismic shift is not required. Even the conservative anthropocentric philosopher who believes that nature is important because it is instrumentally useful for humanity can argue for a broader perspective.

The standard analytical framework for capitalism is to construct a decision matrix based on costs, benefits, and return on investment (ROI). If each variable can be translated into a standard unit so it can be compared with other standard variables, decision-makers can assess options. Presumably, the option that promises the greatest ROI—that is, it maximizes benefits while minimizing costs—is the preferred option. In some cases, a decision-maker may take a calculated risk that promises a higher-than-normal ROI, but it may impose higher-than-normal costs if the plan does not succeed. Owing to incomplete or faulty information as well as the vagaries of the outside world, the cost-benefit calculation is usually a gamble of sorts. The neoclassical economists' position is discussed in Chapter 2.

Environmental justice advocates contend that the traditional cost-benefit analysis fails to consider environmental degradation adequately because the calculation of negative externalities does not factor in the misallocation of resources. Economists sometimes assign a value to environmental resources, including pollution, but the valuation assumes that natural-resource destruction can be calculated merely as a diminution in value, treating nature as another fungible commodity. Setting aside the question of aesthetics, a cost-benefit analysis assumes that natural resources can be evaluated without considering the scale of degradation—for example, a polluted river over time leads to the loss of species which, over the course of *x* number of years, can lead to unforeseeable changes in habitat—but the calculation ensures that short-term financial gains will trump questions about long-term systemic destruction.[31]

Mainstream economists explicitly or implicitly subscribe to the Hobbesian and Lockean concepts of a social contract. The core value of any contract is a free exchange of promises. One party offers something of value to another party in exchange for consideration. If each party believes the exchange to be beneficial, the parties experience a meeting of the minds. Subsequent transactions become a matter of adjustments to the original contract. This arrangement underscores economic and political choices made by members of the collectivity. Yet it presupposes that all parties freely and fairly participate in the exchange.

In his famous treatise *A Theory of Justice*, John Rawls questioned the nature of exchanges because the traditional analysis does not adequately consider the unequal distribution of goods and services. If some parties are unequal and cannot freely and fairly participate in the marketplace, the foundational assumption that all parties consent to the exchange is suspect. The difficulty, according to Rawls, is that parties know their relative positions in the marketplace. If they were forced to develop a system of consensual exchange without advance knowledge of their positions, rational actors would confront distribution questions immediately. If I am required to determine the rules of exchange within a hypothetical regime without knowing whether I will be rich or poor, I will design a system such that any exchange among parties will not allow a disadvantaged party to be harmed beyond a certain point. This principle ensures that the regime does not inadvertently become a system of extreme haves and have-nots.[32]

Environmental justice advocates insist that Rawls's theory can be used to modify traditional economic and political analyses about nature. Rather than assign rights to human parties and allow free trade among parties without ensuring an equal distribution of resources—and without providing a safety net for parties unable to compete—rules must be established to equalize the playing field and recognize differences not fully captured by mainstream economic analysis. Rawls assumed that parties acting in an original position possess incomplete information about their self-interest. Therefore, they design a distribution system that protects the least-advantaged members because a decision-maker might discover he is among the least advantaged. By analogy, we operate under a veil of ignorance when we engage in environmental decision-making. We must protect the natural environment so we do not permit our lack of information about long-term environmental effects to harm our sometimes-hidden interests.

In the context of international relations, the ramifications are monumental. Well-developed, wealthy nations that enjoy a surplus of resources and a decided advantage in producing and consuming new technologies look askance at their brethren who are only now engaged in modern industrialization. Burgeoning consumption rates in developing nations can lead to horrific environmental degradation associated with dirty operations, crippling injuries to workers, and displacement of traditional farming and hunting societies in favor of a capitalist model—developments that cause leaders of highly industrialized nations to blanch. The cry goes out: "How dare China, India, or Brazil destroy forest resources, overplant the soil, and pollute rivers? Have they learned nothing from history about the calamitous effects of natural-resource destruction?"

These nations have learned much from their supposedly more advanced and industrialized neighbors, but perhaps it is not the lesson that wealthy nations desire. A hundred years ago, the United States constructed giant factories that belched out black smoke and polluted rivers, producing hazardous and in some cases radioactive wastes that remain problematic. The result was the establishment of an economic powerhouse. If environmental degradation is the price to pay so that a developing nation can rise up from poverty, feed, clothe, house, and educate its citizens as well as enjoy a seat at the table of power and influence among the countries of the world, then the benefits are worth the costs. This type of thinking represents a significant threat to the development of a new environmental ethic.[33]

Conclusion: Where Do We Go from Here?

Negative environmental externalities were not fully understood one hundred years ago. As modern research has demonstrated the costs of environmental degradation, citizens have decried the natural-resource despoliation that threatens the health, safety, and welfare of human beings. Yet raising a hue and cry does not a satisfactory resolution make. Aside from radical biocentrists or extreme Luddites, few Americans are willing to forgo the benefits of modern technology, including advanced medicines and medical procedures, rapid and relatively inexpensive modes of transportation, efficient means of communication, and a plethora of entertainment options. The cliché often bandied about is to deride vehement environmentalists by professing a strong aversion to returning to mud huts. Yet practicing effective environmental stewardship, as this book discusses in coming pages, need not require an either–or choice: Either continue to consume resources without altering current behavior or eschew all modern conveniences and revert to life in a pre-industrialized state. Exercising dominion over the earth does not provide license for wholesale, unfettered destruction. A median position can be developed.[34]

In searching for a median position, the philosophical ideas and concepts underlying American environmentalism are difficult to assess because they are based on a hodge-podge of competing, sometimes contradictory positions. As every novice philosophy student soon realizes, no clear path leads through the morass of conflicting theories toward enlightenment, or at least to a clear, consistent statement of immutable, unimpeachable values. If a stand must be taken, a position must be constructed on a foundation that may not consist of solid bedrock, but one that can be assiduously defended, nonetheless. To construct this original position, core questions must be addressed to develop a philosophy of environmentalism.

The first consideration is whether non-humans possess rights that must be respected by humans. Whether those rights are derivative and instrumental, as in an anthropocentric framework, or inalienable and intrinsic, as in a biocentric conception, they must exist and be acknowledged by substantially everyone if rights and responsibilities are to be determined. Without a rights-assignment, the earth becomes a prop in a stage play of humanity. *Homo sapiens* struts across the proscenium using and discarding items with no regard save for whether the thing can be replenished as necessary. Absent non-human rights, nature is nothing more or less than another fungible good to be consumed. Thus, if non-human entities possess rights, they can and must be accorded philosophical relevance, a seat at the table of consideration. The nature and extent of those rights are subjects of no small contention, but their existence is a requirement for a robust environmental ethic.

Assuming *arguendo* that non-human entities do not possess rights, in a sense they begin at an enormous disadvantage. Rights provide a starting point for all the negotiations that follow. An entity that possesses rights can petition—perhaps through a fiduciary—to protect and assert those rights. Violations can be penalized.

Absent a system of rights, a standard of valuation must be established, preferably by consensus. This requirement becomes a sticking point among competing schools of thought. For proponents of industrial development, a value can be assigned to the natural environment according to traditional neoclassical principles. The valuation can rise or fall depending on market principles and external forces that alter human preferences. For environmentalists who contend that "nature is different," neoclassical economists miss the point.

To understand this crucial distinction and thereby answer the question of "Where do we go from here?" Chapter 2 summarizes the neoclassical-economic position. Chapter 3 explores the environmentalists' rejoinder, focusing especially on the range of positions from traditional cost-benefit analysis through deep ecology. Chapter 10 discusses factors to assess in constructing a path forward.

Notes

1. Friedrich Nietzsche, *Thus Spoke Zarathustra*, Graham Parkes, translator (New York and Oxford: Oxford University Press, 2005), 113.
2. Numerous sources document the rise of the human animal. See, for example, Gordon H. Orians, "Nature & Human Nature," *Daedalus* 137, 2 (Spring 2008): 39–48; Stuart L. Pimm, *The Balance of Nature* (Chicago: University of Chicago Press, 1991); Arthur J. Robson, "Evolution and Human Nature," *The Journal of Economic Perspectives* 16, 2 (Spring 2002): 89–104; Kenneth Schmitz, "Towards the Reciprocity of Man and Nature: Receptivity, Normativity, and Procreativity," *Nova et Vetera* (English Edition) 10, 1 (Winter 2012): 81–94.
3. Lynn White, Jr., "The Historical Roots of Our Ecological Crisis," *Science* 155, 3767 (10 March 1967): 1203–1207.
4. Plato, *Phaedrus* (Millis, Mass.: Agora Publications, 2009), 57, 92.
5. See, for example, J. Baird Callicott and Roger T. Ames, editors, *Nature in Asian Traditions of Thought: Essays in Environmental Philosophy* (Albany: State University of New York Press, 1989); Ronald L. Massanari, "A Problematic in Environmental Ethics: Western and Eastern Styles," *Buddhist-Christian Studies* 18, 1 (1998): 37–61; Orians, "Nature & Human Nature," 40.
6. William Shakespeare, *The Tragedy of Hamlet: King of Denmark* (New York: Airmont Books, 1965), Act II, Scene ii, 64–65. For a discussion of ecology in Shakespeare's works, see, for example, Sharon O'Dair, "'To Fright the Animals and To Kill Them Up': Shakespeare and Ecology," *Shakespeare Studies* 39 (October 2011): 74–83.
7. Benjamin Kline, Ph.D., *First Along the River: A Brief History of the U.S. Environmental Movement*, 3rd ed. (Lanham, Md.: Rowman & Littlefield, 2007), 4–5.
8. Francis A. Schaeffer and Udo W. Middelmann, *Pollution and the Death of Man* (Wheaton, Ill.: Crossway, 1970), 69. See also, for example, Calvin DeWitt, *The Environment and the Christian: What Does the New Testament Say about the Environment?* (Grand Rapids, Mich.: Baker Academic, 1991); J. Patrick Dobel, "Stewards of the Earth's Resources: A Christian Response to Ecology," *Christian Century* 94, 2 (October 12, 1977): 906–909; Raymond E. Grizzle, Paul E. Rothrock, and Christopher B. Barrett, "Evangelicals

and Environmentalism: Past, Present, and Future," *Trinity Journal* 19, 1 (Spring 1998): 4–14; Peter J. Hill, "Environmental Theology: A Judeo-Christian Defense," *Journal of Markets and Morality* 3, 2 (Fall 2000): 158–172.

9. Andrew Heywood, *Political Ideas and Concepts: An Introduction* (New York: St. Martin's Press, 1994), 138.
10. Heywood, *Political Ideas and Concepts*, 138–141; Kenneth R. Hoover, *Ideology and Political Life*, 2nd ed. (Belmont, Calif.: Wadsworth, 1994), 86–87.
11. Edward S. Corwin, "The 'Higher Law' Background of American Constitutional Law," *Harvard Law Review* 42, 2 (December 1928): 152–153.
12. Hoover, *Ideology and Political Life*, 22–30; Donald VanDeVeer and Christine Pierce, "An Introduction to Ethical Theory," in *The Environmental Ethics & Policy Book*, 2nd ed., Donald VanDeVeer and Christine Pierce, editors (Belmont, Calif.: Wadsworth, 1998), 29–32.
13. Heywood, *Political Ideas and Concepts*, 149–151; Hoover, *Ideology and Political Life*, 31–34.
14. W. T. Jones, *The Classical Mind*, 2nd. ed. (New York: Harcourt Brace Jovanovich, 1970), 201–202, 220–222; 326–333.
15. Heywood, *Political Ideas and Concepts*, 110–113; Jones, *The Classical Mind*, 330–331; VanDeVeer and Christine Pierce, "An Introduction to Ethical Theory," 29–32.
16. Friedrich Nietzsche, *Beyond Good and Evil*, Walter Kaufmann, translator (New York: Vintage, 1966), 15–17; VanDeVeer and Christine Pierce, "An Introduction to Ethical Theory," 31–32.
17. Orians, "Nature & Human Nature," 39–48.
18. The literature on Utilitarianism is voluminous. See, for example, Jeremy Bentham, *An Introduction to the Principles of Morals and Legislation* (New York: Hafner, 1948); David E. Cooper, *World Philosophies: An Historical Introduction* (Cambridge, Mass.: Blackwell, 1996), 348–349.
19. John Stuart Mill, *On Liberty* (Arlington Heights, Ill.: AHM Publishing, 1947), 9.
20. Mortimer J. Adler and Seymour Cain, *Ethics: The Study of Moral Values* (Chicago: Encyclopedia Britannica, Inc., 1962), 262–263.
21. For a cogent discussion of this debate, see, for example, Thomas L. Friedman, *Hot, Flat, and Crowded: Why We Need a Green Revolution—And How It Can Renew America* (New York: Farrar, Straus and Giroux, 2008).
22. John Rawls, *A Theory of Justice* (Cambridge, Mass.: Belknap Press, 1971), 182–183.
23. G. E. Moore, *Principia Ethica* (Mineola, N.Y.: Courier Dover Publications, 2004). See also Michael Smith, "Neutral and Relative Value after Moore," *Ethics* 113, 3 (April 2003): 576–598.
24. David Hume, *An Enquiry Concerning Human Understanding* (Indianapolis, Ind.: Hackett Publishing, 1977), 111–112. See also Robert S. Hill, "David Hume," in *History of Political Philosophy*, 2nd ed., Leo Strauss and Josephy S. Cropsey, editors (Chicago: University of Chicago Press, 1972), 509–531.
25. Immanuel Kant, *Fundamental Principles of the Metaphysics of Morals* (New York: Prometheus Books, 1987), 49. See also Pierre Hassner, "Immanuel Kant," in *History of Political Philosophy*, 2nd ed., Leo Strauss and Josephy S. Cropsey, editors (Chicago: University of Chicago Press, 1972), 554–593.

26. Nietzsche, *Beyond Good and Evil*, 13, 17–19. The Emerson quote can be found in Robert B. Cialdini, "Commitment and Consistency: Hobgoblins of the Mind," in *Readings in Managerial Psychology*, 4th ed., Harold J. Leavitt, Louis R. Pondy, and David M. Boje, editors (Chicago: University of Chicago Press, 1989), 177.
27. Benjamin Constant, "On a Supposed Right to Lie Because of Philanthropic Concerns," in Immanuel Kant, *Fundamental Principles of the Metaphysics of Morals*, 3rd ed., James W. Ellington, translator (Indianapolis, Ind.: Hackett Publishing, 1993), 63–68.
28. VanDeVeer and Christine Pierce, "An Introduction to Ethical Theory," 15–17.
29. Social Darwinism has been the subject of numerous books. See, for example, Richard Hofstadter, *Social Darwinism in American Thought* (Boston: Beacon Press, 1992).
30. On non-human rights, see, for example, Paola Cavalieri, *The Animal Question: Why Nonhuman Animals Deserve Human Rights*, Catherine Woollard, translator (Oxford and New York: Oxford University Press, 2001); Tom Regan and Peter Singer, *Animal Rights and Human Obligations*, 2nd ed. (Upper Saddle River, N.J.: Pearson Education, 1989); Paul Taylor, *Respect for Nature: A Theory of Environmental Ethics*, 25th Anniversary Edition (Princeton, N.J.: Princeton University Press, 2011).
31. VanDeVeer and Christine Pierce, "An Introduction to Ethical Theory," 34–36.
32. Rawls, *A Theory of Justice*, 36–38, 182–183.
33. See, for example, John Vogler and Mark F. Imber, editors, *The Environment and International Relations* (New York: Routledge, 1996).
34. See, for example, Andrew John and Rowena Pecchenino, "An Overlapping Generations Model of Growth and the Environment," *The Economic Journal* 104, 427 (November 1994): 1393–1410.

Chapter 2

Neoclassical Economics

> Anyone who believes exponential growth can go on forever in a finite world is either a madman or an economist.
>
> **—Kenneth Boulding, "The Economics of the Coming Spaceship Earth"[1]**

The standard narrative presents a tale of environmentalists battling proponents of industrial development in a never-ending, irreconcilable, epic confrontation. Compromise is probably undesirable, perpetually unsustainable, and practically impossible. Supporters portray the committed environmentalist as *Homo reciprocans*, a dedicated public servant advancing a noble goal on behalf of the amorphous public interest in the face of shortsighted, narrow-minded business interests. A legion of detractors presents an alternative view of self-righteous, smug, naïve, impractical idealists pursuing unrealistic, costly, silly goals that fail to consider the needs of a burgeoning population seeking enhanced employment and commercial opportunities. In this accounting, humankind can be thought of as *Homo economicus*. Captains of industry are mythologized as supplying high-paying, intellectually stimulating jobs, spurring technological innovation that leads to cleaner sources of energy, and ensuring that the United States remains competitive in a ruthlessly global economy. Alternatively, industrialists can be vilified as indiscriminately fouling the environment with little concern for the long-term consequences for nature. The distinctions are explored at greater length in Chapter 3.[2]

As with any conventional wisdom, the standard narrative is standard for a reason. The central distinction among the schools of thought concerns a fundamental disagreement over the nature of man. *Homo reciprocans* views the world as interdependent. As much as an individual seeks to maximize his particular self-interest, he recognizes he must cooperate with others, even at the risk of losing some of his personal wealth, to ensure the continuation of the community that allows him to transact business. *Homo economicus* focuses on independence and the rationality of the individual who does not worry about society writ large. He maximizes his own self-interest and leaves broader normative questions for another day.[3]

Based on the *Homo economicus* model, neoclassical economics, which Thomas Carlyle disparaged as the "dismal science," explores human behavior not as it ought to be in accordance with a noble standard, but as it occurs, based on marketplace observations of man acting on his acquisitive nature. Philosophers object that an attempt to explain human behavior using a single model invariably relies on theoretical presuppositions that do not always hold true no matter how rigorously one separates the "is" from the "ought." To assume that all persons are acquisitive or motivated primarily by economics is to gloss over important differences in human behavior. The ascetic, the deeply spiritual person, and the genuinely altruistic defy easy categorization, although philosophers of egoism contend that even these noble personalities have not divorced themselves from self-interested human nature.[4]

The point is well taken that outliers exist in any data. Economists do not profess to explain the behavior of all persons, only the "typical consumer." From a presupposition that human beings are base, self-interested creatures, neoclassical economic theory seeks to explain why they act as they do. The discipline professes to be descriptive and empirical, not a normative prescription for creating a better world, although economics sometimes serves as a starting point for normative public-policy recommendations. Economists compile experiential data from which to draw inferences about how consumers behave and therefore how markets work. Their inferences influence public policy enormously, sometimes to the dismay of committed environmentalists.[5]

Rise of Neoclassical Economics

Beginning in the eighteenth century with the celebrated work of Adam Smith (Figure 2.1) and the development of his "invisible hand" theory, capitalist economic principles were transformed into a crucial complement to a democratic system of government. If the value of a good or service could be determined, measured, and calculated, decisions about resource allocation could inform virtually every aspect of public policy. For Smith's followers, appropriate public policy encouraged the development of markets free from unnecessary government controls, although the definition of "unnecessary" would prove to be endlessly vexing. The fundamental relationship between politics and economics, so radical in Adam Smith's era, became axiomatic in a later epoch.[6]

Although the concept antedated the label, the term "neoclassical economics" originated in Thorstein Veblen's famous article "The Preconceptions of Economic Science" in 1900. The label possesses no fixed meaning, but generally it is used as a

Figure 2.1 The Scottish philosopher Adam Smith championed free-market economics. (Courtesy of the Library of Congress)

shorthand designation for mainstream microeconomics that charts the behavior of individual consumers. Adherents of the neoclassical model assume that consumers develop rational preferences and that each preference can be assigned a value. Values differ among consumers, and these differences allow consumers to exchange goods and services to maximize the relative value of an individual consumer's preferences. Exactly as individuals seek to maximize the value of their preferences, firms seek to maximize profits. The decisions that individual consumers and firms reach about preferences are based on information that may or may not be timely and accurate, but which nonetheless influences the behavior of individuals and firms.[7]

The role of government in a neoclassical economic arrangement is to ensure that each party plays by the rules. Public sector representatives lurk in the background as arbiters when disputes arise, but they are not to interfere by offering unduly restrictive public-policy initiatives or assuming the role of market participant. Based on the Founders' conception of classical liberalism, the mantra is simple: Government is best which governs least. When government enters the marketplace to advance a public-sector goal, it undermines efficiencies and skews the data on consumer preferences. For a committed free-market economist, robust intervention increases government power at the expense of human freedom.[8]

Equating market economics and human freedom can be criticized as a convenient rationale for opportunistic investors to capitalize on opportunities unavailable to the less well-to-do, but the nexus enjoys a long, hallowed pedigree hearkening back to Adam Smith. The generally acknowledged father of modern economic theory, Smith, a Scottish philosopher who came of age during the Enlightenment when the feudal guilds were dissolving, produced a masterwork, *The Wealth of Nations*, reflecting the era's emphasis on the importance of human freedom. In earlier epochs, large organized forces—a strong monarch, a feudal lord, or an elite conglomerate of wealthy landowners—controlled what subsequent thinkers such as Karl Marx would derisively label the "forces of production." Smith championed the concept of free markets as the epitome of the rational life. To unchain humanity from the shackles of elites and allow consumers the freedom to choose was a core value for capitalism. Competition among suppliers eventually would drive prices lower and increase quality as each rational actor in the marketplace sought to improve on a competitor's performance and capture market share. The results would lead to iterative change: An innovator would develop a product or a service and offer it for sale. If consumers indicated a desire for the good or service, other innovators would enter the marketplace. To entice customers to move away from the original supplier, the new innovator would have to offer either an enhanced version of the good or service or the same good or service at a lower price (or both). To compete effectively, subsequent market entrants would be compelled to experiment by adding features and improving production efficiencies. If the original supplier wished to remain competitive, he would need to match the enhancement and perhaps improve on it. Each succession would result in an improved product at a lower price, thereby meeting consumer desires and spurring innovation.[9]

The economic model assumes that government policies do not interfere with market performance. Because governments do not respond to the profit motive that drives private sector firms, when public-sector entities regulate markets or act as market participants, they skew the data on price and performance and add costs to the delicate balance between suppliers and customers. Markets have never been completely free, but Smith argued that the greater the freedom, the greater the return for everyone. Innovation is grounded in self-interest, but self-interest can benefit the whole. The self-interested entrepreneur invests heavily in the economic and political systems, which in turn promote stability and opportunities for economic advancement that benefit even the less fortunate who were unwilling or unable to produce innovations of their own. As later economists suggested, a rising tide lifts all boats.[10]

The difficulty with the concept of an unfettered market is that it is based on rational "self-interest, properly understood," as Alexis de Tocqueville wrote. Self-interest, properly understood requires that market participants do not foul the marketplace by driving away all competitors or exhausting a resource. Self-interest can serve as the core value for a regime, but sometimes self-interest is not "properly understood." A firm may act in its short-term interests without considering negative long-term consequences. In such cases, government intervention is desirable to protect the integrity of a free market. The ability to participate on a so-called "level playing field" has led to a concern about monopolies crowding out market participants. Late in the nineteenth century, a class of opportunistic businessmen labeled "robber barons" acquired sufficient wealth to manipulate markets so they could create artificial monopolies. The robber baron would offer a good or service at a loss to drive competitors out of business. When no competitors existed, the successful supplier could raise prices to whatever level he deemed desirable, especially if the good or service was thought to be integral to modern life. As citizens were forced to pay extravagant prices, they complained to elected officials who eventually responded by enacting antitrust measures to prevent wealthy financiers from manipulating the marketplace (Figure 2.2).[11]

A democratic government subscribes to multiple values aside from economic self-interest. Regime values in the United States include a belief in the beneficial aspects of pluralism and the creative possibilities found in diversity and conflict, the importance of protecting liberty and property interests, the need to safeguard minority rights through "mitigated democracy" created and maintained by institutional controls and adherence to the Bill of Rights, and the relative sovereignty of citizens who participate in consensual self-government. Governments attempt to balance these often competing values, but the neoclassical economic model sometimes is set aside. With the growth of the administrative state beginning in the middle of the twentieth century, government has become more involved in the marketplace. That involvement remains contentious among free-market economists. As government officials enter into the marketplace, efficiency is no longer the paramount objective. When government officials believe market participants have behaved in ways that undercut competition, endanger workers, or harm the natural environment, they impose regulations mandating that market participants undertake actions and incur costs that reduce efficiencies. The battle lines are clear even if a plan for compromise is not. For economists, public policy that interferes with markets is anathema.[12]

Figure 2.2 Fears about powerful economic interests ("trusts") manipulating markets and bribing U.S. senators were reflected in an 1889 cartoon titled "The Bosses of the Senate," which indicated, "This is the Senate of the Monopolists, by the Monopolists, for the Monopolists." (Courtesy of the Granger Collection, New York)

A fundamental precept of the neoclassical model as it evolved after Adam Smith's time is the notion of a general equilibrium where income distribution, the forces of production, and the prices and quantities of goods intersect. Supply and demand curves demonstrate the reciprocal relationship among these factors. When one variable changes from a point of equilibrium, other variables are affected and the marketplace changes until a new equilibrium point is reached. The closeness with which variables are tied to one another is expressed as elasticity; that is, when one variable changes and triggers a concomitant change in another variable, the relationship is said to be elastic. If the change is not direct or does not move in tandem, the relationship is inelastic. In economists' parlance, price elasticity of demand refers to a percentage change in quantity demanded caused by a percentage change in price. The price elasticity of supply measures how the quantity or amount of a good changes in response to price changes.[13]

Elasticity explains how markets operate through a push-pull system and provides policymakers with guidance on the leeway they have to enact legislation affecting the marketplace. If policymakers enact legislation that imposes a surtax on a product, the tax will change the market for that product—and not always in the way lawmakers intended. Environmental policies that interfere with markets may introduce economic inefficiencies into the marketplace that upset general equilibrium. If the policy is intended to achieve this goal, the consequences may be foreseeable and desirable but the elasticity needs to be understood or the policy can be counterproductive. Legislation might inadvertently undermine the market for a product and cause consumers to switch their preferences for a product that is less desirable and more damaging to the environment.[14]

Markets are rarely static owing to increasing or stagnating wages, changes in external events that increase or decrease production, and substitutability among products. Consumers seek to satisfy needs and desires but they can achieve satisfaction through multiple channels. If one good is unavailable or prohibitively expensive, another good may satisfy the need or desire. As consumers change their behavior in response to constantly shifting circumstances, these changes ripple across the marketplace and affect everyone involved in the supply chain, from raw material producers to suppliers to distributors to brokers to consumers.

Ideally, a market transaction requires rational participants to consider all relevant factors that affect a given exchange. A raw material supplier, for example, factors in the costs associated with supplying the material—labor, insurance, transportation, and so forth—to determine whether he is willing to participate in the market. The rational supplier attempts to limit his variable costs even as he recognizes that sunk costs, which he incurs regardless of the circumstances involved in a single transaction, should not be considered when he calculates the costs and benefits of a particular decision.[15]

Proponents of government controls contend that market participants, if left to their own devices, invariably maximize their own self-interest even if it imposes negative externalities on society. Externalities are consequences of individual behavior that parties to a transaction do not factor into their decision-making or cost calculations. Externalities can be positive, such as the benefits that accrue when someone is educated. It is difficult to know how much good will occur owing to a school instructing a pupil in exchange for tuition or tax monies, but the likelihood is that external benefits will accrue because an educated person almost always contributes more to society than an uneducated person contributes. Externalities can be negative, too. The quintessential example is pollution. If a small amount of a hazardous air pollutant is below regulatory concern and therefore is allowed, the costs of generating the pollution do not have to be taken into consideration by an individual polluter; however, when hundreds or thousands of firms pollute, the aggregate level of pollution may lead to harmful health effects for everyone.[16]

In cases where negative externalities exist, government sometimes is better suited to address the issue than private parties because where a single instance of pollution may be below a firm's threshold of concern, the aggregate level captures government attention. Regulators can impose a tax on the negative externality—a so-called "Pigouvian" tax, named for an influential Cambridge economist, A. C. Pigou—or it can develop public programs to regulate the negative externality. Programs can be financed partially by private sector funds and partially through government funding as is the case with the Comprehensive Environmental Response, Compensation, and Liability Act of 1980 (CERCLA). The goal is to ensure that the costs of externalities are accounted for so that the marginal social costs are equal to the marginal social benefits and a kind of balance is achieved.[17]

Neoclassical economists recognize that government exercises a legitimate role in regulating negative externalities, hence the notion of a Pigouvian tax, but they reject policies and programs that allow government to interfere with markets beyond a minimal level. Public sector agencies do not operate on a strict profit motive and they can enjoy advantages such as monopolistic power over markets, the ability to impose regulations, and the authority to engage in deficit spending, in some cases. These features place private firms at a distinct disadvantage and manipulate consumer preferences in often unforeseen ways. The type of public policy determines whether a neoclassical economist supports at least a minimal level of intervention.[18]

Economics, Public Policy, and the Freedom–Authority Divide

As government intervenes in the marketplace, regulators rely on different types of public policies to achieve their desired ends. Distributive policies are policy actions that deliver widespread benefits to individuals or groups that do not directly bear the costs of the policy. Tax deductions on home mortgage interest, loan guarantees provided by the federal government, agricultural price supports, and federally insured loans for college students are a few instances in which government promotes policy goals without requiring participants to bear the full costs. Home-mortgage interest deductions illustrate the point. Federal policymakers have decided that it is in the best interests of Americans to own homes because home ownership presumably gives the property owner a stake in the community and provides equity over time, both of which lead to improved neighborhoods and a higher standard of living. By allowing homeowners to deduct the interest paid on mortgages, the federal government has created an incentive for Americans to purchase homes. Government policy has distributed a benefit to a large group of people, although the wisdom of this long-standing policy became controversial in the wake of the housing crisis of 2008 and 2009.[19]

Redistributive policies involve efforts by government to shift the allocation of valued goods from one group to another. Because of shifting resources, redistributive policies often are among the most contentious issues addressed by governments. A substantial number of people contend that government efforts to collect resources from one group of citizens and transfer them to another are inequitable. They argue that redistribution represents a frontal attack on private property rights and an erosion of due process of law. Due process requires government to follow clearly established legal procedures that include an opportunity to be heard in a suitable forum using known and knowable principles of positivist law.[20]

When redistributive policies involve environmental issues, governments intervene to ensure that the costs of natural-resource degradation are not visited disproportionately on poor citizens. Environmental justice supporters contend that government is justified in intervening to delegate costs among parties as opposed to allowing environmental costs to be imposed on poor neighborhoods. Free-market economists argue that such a redistribution results in poor resource allocation. If the well-to-do are forced to accept a diminution in property values as landfills, incinerators, or other waste facilities are located nearby or if taxes are adjusted to require property owners to bear a heavier *ad valorem* burden, the parties who bear the burden may flee the jurisdiction, thereby reducing the overall tax base and saddling impoverished citizens with a more onerous problem than they faced initially. Redistributive policies influence the behavior of free-market participants, and not always in anticipated ways.[21]

Self-regulatory, or constituent, policies, while perhaps not as controversial as redistributive policies, can become politically sensitive. Governments do not possess the personnel or resources to regulate every type of behavior that policymakers believe must be regulated; therefore, they sometimes allow groups to regulate themselves subject to government oversight. The most obvious example is the licensing of professions and occupations such as physicians, lawyers, and real estate agents. The legal profession, as an illustration, develops standards for admission into the profession. In most states, this requires law school graduates to sit for a bar examination and attest to their good character. The state bar also develops codes of professional ethics for practicing attorneys and prescribes sanctions for ethical lapses or noncompliance.[22]

Self-regulatory policies are controversial because they suggest an inherent conflict of interest in allowing lawyers to police lawyers or physicians to police physicians. The mistrust of "too much power in too few hands" makes the public wary of entrusting power over a profession to members of the profession because the incentive to assist one's fellow professionals at the expense of the public interest is undeniable. Professionals argue that members of the public are not qualified to oversee professionals because the specialized knowledge and training necessary to understand and apply professional standards are not available to the layman. According to this perspective, professionals should be left alone to regulate themselves and governments or members of the public should become involved only in situations where malfeasance can be demonstrated clearly and convincingly.[23]

Environmentalists worry that when corporate interests are unregulated, business executives protect their own narrow interests and pass along the cost of negative externalities to the public sector. When the public sector enacts regulations to ensure the costs are imposed on the parties that generated them, members of the regulated community cry foul. They rail against government interference and denounce over-regulation as destroying economic competitiveness in the global marketplace. The confrontation over appropriate government interference and the need to impose costs on the parties is a wedge issue between supporters of neoclassical economics and environmentalists.[24]

The crucial divide can be stated as a question involving the tradeoff between efficiency and equity. Should the goal be to develop and implement policies using the fewest resources in the shortest amount of time (one definition of efficiency) or should the goal be to treat like cases alike (one notion of equity), even if such treatment is economically inefficient? Scholars and academics have long debated whether a tradeoff exists between efficiency and equity and, if so, what implications the tradeoff holds for public policy. Typically, the debate casts these concepts as antithetical values so that an increase in one value necessarily requires a corresponding decrease in the other. Although in many cases such a tradeoff exists, efficiency and equity need not be competing values in every instance.[25]

The tradeoff can be expressed as a debate along the freedom-authority divide. At one end of the spectrum is absolute freedom, which can be defined as a complete absence of constraints on individual behavior. In a state of absolute freedom, no government legitimately exists. A person can do whatever he or she desires without interference from the state because the state is absent. Although some extreme libertarians—William Godwin is a famous example—have argued that a state of complete liberty is tantamount to existing in a virtual Garden of Eden free from government intrusion, most mainstream western thinkers, at least since the Enlightenment, have viewed a state of absolute liberty with horror—the ultimate laboratory for licentiousness. If no external authority exists, life becomes a Darwinian struggle where the strong consume the weak until the strong succumb to a still stronger force.[26]

The seventeenth-century English philosopher Thomas Hobbes, a supporter of the social contract theory of government, famously wrote in his 1651 tome *Leviathan* that life in a "state of nature" was "nasty, brutish, and short."[27] A pre-state existence is a horrific place to be avoided at almost all costs. Clearly, rational men would gladly surrender a measure of their individual liberty in exchange for the authority of a state to protect them and meet basic needs that were not met in a state of absolute liberty.[28]

After individuals have agreed to surrender a measure of liberty to create a state, the question is how much personal liberty should be surrendered in exchange for how much state authority. Personal liberty and state authority are competing ideals. An increase in one decreases the other. There is indeed a tradeoff. With the birth of an entity called the state, decisions have to be made and a balance between freedom and authority must be struck.

At the extreme end of the spectrum, a totalitarian state can be created so that there is no individual liberty whatsoever. In such a state, government makes all political and economic decisions. The authoritarian regime creates an equality of outcome enforced by the state. Everyone is treated exactly the same regardless of the person's individual talents or industry. Unlike a state of absolute freedom, where people may starve to death because there is no one to provide for them and life becomes a Hobbesian war of all-against-all or a brutal test of natural selection where only the strong and cunning survive, a state with absolute authority can ensure that no one starves. The problem, of course, is that there is no check on the power of the state or the behavior of the people who administer its laws. A totalitarian regime risks being exploited by its leaders and oppressing its citizens. This insight, if nothing else, was dramatically illustrated by the former Soviet Union and in many totalitarian states throughout history.[29]

Most mainstream theorists have rejected a state possessing absolute authority as too extreme. There must be room for personal liberty or life is not worth living. The totalitarian state becomes a place where life is nasty, brutish, and short, but the selection of who lives and dies is not natural. It is manmade. Tell the man languishing in a gulag and awaiting a slow, horrible death how much better off he is than if he had perished in the wilds of nature, and the distinction will seem nonsensical. This conclusion is reminiscent of Patrick Henry's famous cry, "give me liberty or give me death." In Publius's words from "Federalist Number 10," when a government chooses to quash liberty to "control the mischief of factions," the "cure is worse than the disease."[30]

Fearing an omnipotent central authority, the American Founders created a weak nation-state when they enacted the Articles of Confederation in 1781. They would surrender a minuscule level of freedom in exchange for a bare-bones government, for they subscribed to the libertarian creed of limited government answerable to the populace. Don't tread on me: Government should serve as a night watchman and essentially not interfere in people's lives. Yet the Articles proved to be disastrous. Because a state that has very little authority must continually surrender to the will of the masses, and the masses are constantly in turmoil, the state simply cannot function effectively. After only eight years, American statesmen rejected the classical liberal state encapsulated by the Articles of Confederation and replaced it with a stronger system—the federal government found in the new U.S. Constitution.[31]

If the totalitarian state is too strong and the classical liberal state is too weak, the socialist state might offer a viable compromise. Socialism, according to the theorist Sidney Webb, is not a well-developed set of doctrines or clearly delineated precepts. At its core, socialism is the idea that people should enjoy a small measure of freedom, but the state will step in and provide essential goods and services when they are necessary. Private enterprise is not prohibited, but it is strictly regulated, and large companies can be nationalized if it is in the public interest to do so. Different socialist regimes provide different levels of personal liberty; in fact, some western European nations are labeled "democratic-socialist" states. In some cultures and traditions where personal freedom is not as highly valued as it is in the United States, a democratic-socialist state might be greeted with open arms. In the United States, however, the idea that government controls most of the means of production and makes major economic decisions in the public sector as part of the planned economy is unpopular. Anything that smacks of socialism, even if it inaccurately carries the socialist label, will be met with scorn and derision in the American political landscape. Politicians in the modern slash-and-burn arena of electoral politics denounce government programs as "socialist" if they offer a promise of government assistance, but even the most committed ideologue would be hard-pressed to justify such claims in the American republic of the twenty-first century.[32]

This process of elimination leaves the U.S. political system. A mixed polity—that is, a regime where people, acting through elected representatives, make some political decisions but unelected experts also make decisions within their areas of expertise—is the preferred form of government for people who profess to love individual freedom but also desire a strong state that meets many needs, especially for national defense, protecting free markets, and ensuring equality of opportunity (although the definition of the latter term is highly contentious). The U.S. political system calls to mind Winston Churchill's famous comment that a democracy "is the worst form of government except all those other forms that have been tried from time to time."[33]

The freedom-authority divide has existed throughout American history. One theory embraces unfettered markets and the concept that personal freedom is a crucial component of life. Free-market economists accept this view. John Stuart Mill, author of the classic 1859 book *On Liberty*, argued that government should intervene to prevent "other-regarding" actions if they are negative, but "self-regarding" actions should be unregulated.[34] Theorists such as Milton Friedman, the Nobel Laureate long associated with the University of Chicago, contended that free markets are far more efficient and effective than government in meeting the needs and desires of citizens, in most cases.[35] Irving Kristol and Robert Nozick are two politically conservative theorists who presented persuasive and influential arguments about the importance of private property and the liberty of the marketplace.[36]

Efficiency versus Equity: An Unbridgeable Gulf?

In light of this framework, the standard notion of "efficiency" is conceptually simple. It is the idea that a rational person seeks to achieve his or her goals while also using as few resources as possible. To be "efficient," therefore, means that the person achieves the greatest return on an investment of his or her time and money. The colloquial expression is that a person gets "the biggest bang for the buck." Presumably, an efficient course of action is also effective, albeit effectiveness is a different concept. To be "effective," as that term generally is understood, means that a person achieves the goal, but in doing so, the person may use more resources than necessary. In such a case, the person is effective, but not efficient. (Conversely, a person could use very few resources—hence be efficient—but not be especially effective.)

"Equity," by contrast with notions of "efficiency," refers to fairness. Philosophers of jurisprudence speak of treating "like cases alike." To be equitable means that distinctions must be made according to a principle of fairness so that people are not harmed or discriminated against based on factors they cannot control such as race, ethnicity, or gender. Much of modern American law attempts to ensure that legal standards and requirements are equitable.

The desire for equity returns to a fundamental issue for mainstream economists: What is the appropriate role of government in the marketplace? The traditional model of capitalism—stretching back to Adam Smith's idea that an "invisible hand" guides the marketplace—champions free markets. Choosing efficacious public policy means basing decisions on the desires of consumers. The assumption is that people are self-interested rational maximizers. They may not always maximize their self-interest and they may not always act rationally because they may be mistaken in what they believe, they may have incomplete information (asymmetric information, in economists' parlance), or transaction costs may be too high, but they attempt to maximize their self-interest in a rational manner, to the extent possible.[37]

If markets do not fail owing to externalities, asymmetric information, or the presence of natural monopolies, the marketplace is seen as a more realistic barometer of what people seek than any measure devised by governments. Public choice theorists such as Anthony Downs and Elinor Ostrom contended that the measures used by governments to gauge what people desire—voting results, feedback through political pluralism, public opinion polls, and so forth—are not nearly as precise as free markets for measuring what people want in public policies. People vote for many reasons, not all of which are clear even to voters. One cannot determine from voting studies how people feel about specific policies, despite elected officials' claims to enjoy a mandate after they are swept into office. Pluralism—the idea that interest groups compete for political power, which leads to compromises resulting in effective public policy—helps to discern some of what people want, but not everyone participates in groups.[38] Moreover, because governments are so fragmented and policies are often complex and difficult to fathom, looking to government mechanisms for understanding the "will of the people" in the policy process is misguided.[39]

Economists David Weimer and Aidan Vining observed that the condition of markets can determine when it is appropriate for government to intervene into the marketplace. Weimer and Vining focused on "contestable markets," those markets where competition exists, but only among a few big, competing firms. A market is highly contestable when there is a large private supply of a good, and ownership is vested in the private sector. In those cases, markets can function efficiently and government should stay out of the way. In cases where markets are not highly contestable—where supplies are low and ownership is almost wholly within the public sector—government intervention is appropriate. Government should intervene when there is "market failure."[40]

According to Robert Haveman in *Economics of the Public Sector*, markets can fail when there is a goal other than efficiency in the smooth, orderly transfer of goods and services from one party to another. An example can be found in the Rural Electrification Act, a statute enacted by Congress during Franklin D. Roosevelt's administration in the 1930s. Congress and the president decided as part of the New Deal that every American should have access to residential electric service. The problem was that some impoverished people living in Appalachia could not afford to pay the market rate for electricity even if it were available. In this situation, a private electric company would not have an economic incentive to invest in new equipment for ensuring that adequate electricity-generation capacity existed. The lack of a market response to a perceived need is what economists call a "market failure." Among other things, the Roosevelt administration relied on a government corporation called the Tennessee Valley Authority (TVA). TVA supplied electricity to poor people in the geographic area at subsidized rates. The market failed, so government stepped into the marketplace to correct the deficiency.[41]

Free-market economists also see a role for government in leveling the playing field. Ronald Coase, a 1991 Nobel Laureate in economics, developed a famous theorem bearing his name to explain the appropriate role of government. The Coase Theorem posited that property resources, absent transaction costs, wind up in the hands of the party that places the highest value on the resource. This phenomenon is one of the reasons why private enterprise is so efficient. People cut a deal to acquire the resources they desire most. Transaction costs skew the process so that a party is dissuaded from acquiring a resource owing to the total costs. The role of government, according to Coase, is to adopt public policies that minimize transaction costs. If railroad freight rates impede the ability of buyers and sellers to transact business, government is justified in regulating freight rates through the Interstate Commerce Commission (ICC). Coase and free-market economists harbor no love for the ICC or other government agencies, but if agencies can level the playing field, participants in the marketplace can become more efficient. Coase argued against Pigouvian taxes for exactly this reason—government-imposed fees, while they provide economic incentives, do not rely on efficient market mechanisms to change behavior. Taxes introduce more transaction costs when fewer costs are preferred.[42]

The traditional view of the efficiency-equity tradeoff associated with public policies and programs is not a universally accepted perspective. Deborah Stone offered a stinging critique of the tradeoff in her book *The Policy Paradox*. Stone viewed the traditional approach to public policy—which she derisively labeled the "rationality project"—as seriously deficient because it assumes that a linear process exists. The assumption is faulty. Mainstream economists and students of the policy process act as though people live independently from one another and make decisions as rational maximizers based on clearly defined, specific, individual goals. Stone argued that this conception of how policies are derived is simplistic and myopic.[43]

We live in a constantly evolving political community. To think that the policy process or the system of adopting public policies is somehow static and fixed with stable, ongoing issues and procedures in place is to mistake what happens. As our world changes—new policies are adopted, new leaders come into power, new technologies emerge, people's interests and desires change—the process therefore changes. Adopting a policy today may or may not involve the same process it will tomorrow. This fluid understanding of the policy process contravenes the antiquated iron-triangle model. Political scientists in past eras talked about the triangular relationship among and between Congress, the bureaucracy, and outside interest groups as though it were a fixed framework. With Hugh Heclo's concept of issue networks—where relationships constantly change and alliances form and dissolve with lightning rapidity—the understanding of iron-triangle relationships became obsolete.[44] Similarly, Stone's argument suggested that an inherent paradox exists when we assume a fixed position, such as the standard economic position on efficiency and equity, because as positions and relationships change, the idea of a fixed position must change.[45]

In the absence of fixed positions, it is possible to adopt a public policy that has the potential to increase both efficiency and equity. Unlike the standard economic perspective, which assumes that a tradeoff between efficiency and equity is unavoidable, a policy can embrace both values, although the issue requires its architects to walk a fine line. A new policy requires parties to agree to a principle known as Pareto Optimality, named for the twentieth-century economist Vilfredo Pareto. Pareto sought to understand how parties in an economic system can reach equilibrium. If the parties possess goods but at least one party can be better off without making another party worse off, the situation is inefficient. In this situation, all other things being equal, government should stand aside and allow the parties to transact business, for voluntary exchanges allow for the greatest economic efficiency. Rather than rely on an authoritarian government to dictate economic policy, the parties will decide when they have reached equilibrium owing to the satisfaction of their own needs and desires. A transaction or series of transactions ideally should lead to a Pareto improvement, which means that at least one party benefits from the transaction as long as no one else is worse off as a result. Eventually, when it is impossible for any party to be better off without worsening another party's situation, the result is Pareto Optimality. That is, the parties have reached an equilibrium that can be described as maximally efficient.[46]

Anthropocentric environmentalists might subscribe to this economic principle except for several disturbing features. First, Pareto Optimality presupposes that all other things are equal. Efficiency is considered equitable because all parties are locked in a situation where their needs and desires have been maximized. Yet the desires of the parties may be skewed because of mistaken perceptions or asymmetric information. In the case of natural resource use, a party might be satisfied with an outcome because he always wanted to build a house in a wetlands area. His desire is met, as is the desire of the banking officer who provided the loan for the mortgage and the builder who was paid to construct the house. The relationship among the parties fails to consider the broader concerns about intruding on wetlands to construct an artificial structure. The loss of wetlands may trigger a slew of negative environmental consequences such as habitat destruction, poor drainage prospects for the neighborhood, and an increase in surface heat as ever more forest acreage and wetlands give way to asphalt roads, concrete driveways, and unshaded, arid lawns.[47]

A tradeoff between efficiency and equity leads to an uncomfortable proposition: In some instances, "equity" requires an unequal distribution of resources. For persons who cannot participate in market transactions because they are poor or lack reliable information, the ability to engage in voluntary exchanges is not a valuable right. For proponents of environmental protection, an economically efficient use of resources typically means that parties consider the return on investment from a position of narrow self-interest without appreciating the scale of environmental considerations. The tragedy of the commons is that individuals hold an interest in exhausting a broad, common resource to satisfy a narrow personal interest. Garrett Hardin discussed this dilemma in a now-famous 1968 article.

According to Hardin, individual users of a publicly owned resource (a "common pool resource," in the language of neoclassical economists) have an incentive to use as much of the resource as possible, even to the point of exhausting it, to satisfy their individual desires with little regard for future uses. By taking a short-term view, these individuals risk destroying the resource for everyone. Hardin's article is a useful explanation of why people engage in behavior that harms the natural environment. With its individualistic approach to issues that can be addressed in short election cycles, the American political system emphasizes the importance of short-term personal rights, often at the expense of the long-term collective welfare.[48]

This is the rub: The neoclassical economist views the tradeoff between efficiency and equity as the point at which all the parties to a transaction are satisfied that their self-interest has been fulfilled to the extent possible without harming another party. However, what does it mean to harm another party, and how is self-interest properly understood? The environmentalist contends that the respective parties to a transaction must broaden their scope to consider the larger impact on the natural environment. To protect an ecosystem and regulate negative externalities, it may be necessary for a neutral arbiter—in this case, government—to step into the marketplace and reallocate costs and benefits among the parties, even if this means some parties will be dissatisfied with the result and even if this interference redistributes resources in an unequal manner. As long as divergent goals exist—freedom versus authority, the individual versus the collective—neoclassical economists and environmentalists will be at odds. We turn to the precise nature of their disparate views in the context of sustainability in Chapter 3.

Notes

1. Quoted in David K. Levine, *Is Behavioral Economics Doomed? The Ordinary versus the Extraordinary* (Cambridge, U.K.: Open Book Publishers, 2012), 11.

2. See, for example, Ernst Fehr and Simon Gachter, "Reciprocity and Economics: The Economic Implications of *Homo Reciprocans*," *European Economic Review* 42, 3 (May 1998): 845–859; Richard H. Thaler, "From *Homo Economicus* to *Homo Sapiens*," *Journal of Economic Perspectives* 14, 1 (Winter 2000): 133–141.
3. Fehr and Gachter, "Reciprocity and Economics," 845–848.
4. David Levy, *How the Dismal Science Got Its Name: Classical Economics and the Ur-Text of Racial Politics* (Ann Arbor: University of Michigan Press, 2001), xiii.
5. E. K. Hunt and Mark Lautzenheiser, *History of Economic Thought: A Critical Perspective*, 3rd ed. (Armonk, N.Y.: M. E. Sharpe, Inc., 2011), 380–390; Robert H. Nelson, "Economics and Environmentalism: Belief Systems at Odds," *The Independent Review* 17, 1 (Summer 2012): 5–17.
6. Hunt and Lautzenheiser, *History of Economic Thought*, 382–385; L. D. Keita, *Science, Rationality, and Neoclassical Economics* (Cranbury, N.J.: Associated University Presses, 1992), 41–44.
7. John Patrick Diggins, *Thorstein Veblen: Theorist of the Leisure Class* (Princeton, N.J.: Princeton University Press, 1999), 14–15. See also Hunt and Lautzenheiser, *History of Economic Thought*, 38–90; Keita, *Science, Rationality, and Neoclassical Economics*, 77.
8. Patrick M. Garry, *Liberalism and American Identity* (Kent, Ohio: Kent State University Press, 1992), 47–50.
9. Adam Smith, *An Inquiry into the Nature and Causes of the Wealth of Nations* (London: J. M. Dent & Sons, Ltd., 1921), 104–109, 137–156. See also Elias L. Khalil, "Beyond Self-Interest and Altruism: A Reconstruction of Adam Smith's Theory of Human Conduct," *Economics and Philosophy* 6, 2 (October 1990): 255–273.
10. The conclusion that a rising tide lifts all boats has been the subject of much debate. See, for example, Sheldon Danziger, "Do Rising Tides Lift All Boats? The Impact of Secular and Cyclical Changes on Poverty," *The American Economic Review* 76, 2 (May 1986): 405–410; L. Randall Wray and Marc-Andre Pigeon, "Can a Rising Tide Raise All Boats? Evidence from the Clinton-Era Expansion," *Journal of Economic Issues* 34, 4 (December 2000): 811–845.
11. See, for example, John Patrick Diggins, *The Lost Soul of American Politics: Virtue, Self-Interest, and the Foundations of Liberalism* (Chicago and London: University of Chicago Press, 1986), 98.
12. See, for example, Marshall E. Dimock, *Free Enterprise and the Administrative State* (Westport, Conn.: Greenwood Press, 1972).
13. Yves Nievergelt, "The Concept of Elasticity in Economics," *SIAM Review* 25, 2 (April 1983): 261–265.
14. See, for example, Bernard Salanie, *The Economics of Taxation* (Cambridge, Mass.: MIT Press, 2003), 3, 23–30; John B. Shoven and John Whalley, "Applied General-Equilibrium Models of Taxation and International Trade: An Introduction and Survey," *Journal of Economic Literature* 22, 3 (September 1984): 1007–1051.
15. Douglas R. Johnson and David G. Hoopes, "Managerial Cognition, Sunk Costs, and the Evolution of Industry Structure," *Strategic Management Journal* 24, 10 (October 2003): 1057–1068.
16. Nicholas Mercuro, "Interdisciplinary Paradigms for Environmental Policy: Interrelationships among Ecology, Law, and Economics," in *Property Rights, Economics, and the Environment: The Economics of Legal Relationships*, Michael D. Kaplowitz, editor (New York: Routledge, 2000), 247–279.

17. James Andreoni, "Warm-Glow versus Cold-Pickle: The Effects of Positive and Negative Framing on Cooperation in Experiments," *Quarterly Journal of Economics* 110, 1 (February 1995): 1–21; A. H. Barnett, "The Pigouvian Tax Rule under Monopoly," *The American Economic Review* 70, 5 (December 1980): 1037–1041.
18. Hunt and Lautzenheiser, *History of Economic Thought*, 38–90.
19. Christopher L. Foote, Kristopher Gerardi, Lorenz Goette, and Paul S. Willen, "Just the Facts: An Initial Analysis of Subprime's Role in the Housing Crisis," *Journal of Housing Economics* 17, 4 (December 2008): 291–305; Anne Schneider and Helen Ingram, "Social Construction of Target Populations: Implications for Politics and Policy," *American Political Science Review* 87, 2 (June 1993): 334–347.
20. Hans Peter Gruner, "Redistributive Policy, Inequality, and Growth," *Journal of Economics* 62, 1 (February 1995): 1–23. See also Carol M. Rose, "Liberty, Property, Environmentalism," *Social Philosophy & Policy* 26, 2 (July 2009): 7–8.
21. Richard Stillman II, *American Bureaucracy: The Core of Modern Government*, 3rd ed. (Belmont, Calif.: Wadsworth/Thomson, 2004), 171, 173.
22. See, for example, Virginia Haufler, *A Public Role for the Private Sector: Industry Self-Regulation in a Global Economy* (Washington, D.C.: Carnegie Endowment for International Peace, 2001); Andrew A. King and Michael J. Lenox, "Industry Self-Regulation without Sanctions: The Chemical Industry's Responsible Care Program," *The Academy of Management Journal* 43, 4 (August 2000): 698–716.
23. M. Christine Cagle, J. Michael Martinez, and William D. Richardson, "Privatizing Professional Licensing Boards: Self-Governance or Self-Interest?" *Administration & Society* 30, 6 (January 1999): 734–770.
24. See, for example, Daniel J. Fiorino, "Rethinking Environmental Regulation: Perspectives on Law and Governance," *Harvard Environmental Law Review* 23, 2 (1999): 441–468; Andrew J. Hoffman, "Institutional Evolution and Change: Environmentalism and the U.S. Chemical Industry," *The Academy of Management Journal* 42, 4 (August 1999): 351–371.
25. See, for example, Sudhir Anand and Kara Hanson, "DALYs: Efficiency versus Equity," *World Development* 26, 2 (February 1998): 307–310; Robert A. Becker, "Intergenerational Equity: The Capital-Environment Tradeoff," *Journal of Environmental Economics and Management* 9, 2 (June 1982): 165–185; Arthur M. Okun, *Equality and Efficiency: The Big Tradeoff* (Washington, D.C.: The Brookings Institution, 1975).
26. See, for example, Norman P. Barry, *On Classical Liberalism and Libertarianism* (New York: MacMillan, 1986).
27. Thomas Hobbes, *Leviathan* (Indianapolis: Bobbs-Merrill, 1958), 107.
28. Andrew Heywood, *Political Ideas and Concepts: An Introduction* (New York: St. Martin's Press, 1994), 27–28.
29. Heywood, *Political Ideas and Concepts*, 45–46; Kenneth R. Hoover, *Ideology and Political Life*, 2nd ed. (Belmont, Calif.: Wadsworth, 1994), 191.
30. Alexander Hamilton, James Madison, and John Jay, *The Federalist Papers*, Clinton Rossiter, editor (New York: New American Library, 1961), 78.
31. Frank Donovan, *Mr. Madison's Constitution: The Story behind the Constitutional Convention* (New York: Dodd Mead & Company, 1965), 10–12; Merrill Jensen, *The Articles of Confederation: An Interpretation of the Social-Constitutional History of the American Revolution, 1774–1781* (Madison: The University of Wisconsin Press, 1963), 244.

32. G. D. H. Cole, *Fabian Socialism* (Oxford: Routledge, 1971); Jim Tomlinson, "The Limits of Tawney's Ethical Socialism: A Historical Perspective on the Labour Party and Market," *Contemporary British History* 16, 4 (Winter 2002): 1–16.
33. Quoted in Winston Churchill, *Quotations* (Norwich, England: Jarrold Publishing, 1997), 31. See also John P. Roche, "The Founding Fathers: A Reform Caucus in Action," in *American Government: Readings and Cases*, 3rd ed., Peter Woll, editor (Boston: Little, Brown & Company, 1969), 55–78; Fred Rodell, *Fifty-Five Men* (New York: The Telegraph Press, 1936).
34. John Stuart Mill, *On Liberty* (Arlington Heights, Ill.: AHM Publishing, 1947), 9–11.
35. See, for example, Milton Friedman, *Capitalism and Freedom* (Chicago, Ill.: University of Chicago Press, 1962).
36. See, for example, Irving Kristol, *Reflections of a Neoconservative: Looking Back, Looking Ahead* (New York: Basic Books, 1983); Robert Nozick, *Anarchy, State, and Utopia* (New York: Basic Books, 1974).
37. Anand and Hanson, "DALYs: Efficiency versus Equity," 307–310; Okun, *Equality and Efficiency*, 8, 50.
38. See, for example, Anthony Downs, *An Economic Theory of Democracy* (New York: Harper & Row, 1957); Elinor Ostrom, *Governing the Commons: The Evolution of Institutions for Collective Action* (New York and Cambridge, U.K.: Cambridge University Press, 1991). See also Chapter 8 in Robert Putnam, *Bowling Alone: The Collapse and Revival of American Community* (New York: Simon & Schuster, 2000), which discusses the controversial argument about declining group participation.
39. See, for example, Vincent Ostrom, *The Intellectual Crisis in American Public Administration*, 3rd ed. (Tuscaloosa: University of Alabama Press), 2007.
40. David L. Weimer and Aidan R. Vining, *Policy Analysis: Concepts and Practice*, 2nd ed. (Englewood Cliffs, N.J.: Prentice Hall, 1992).
41. Robert H. Haveman, *Economics of the Public Sector* (Santa Barbara, Calif.: Wiley, 1976).
42. Ronald H. Coase, "The Institutional Structure of Production," *The American Economic Review* 82, 4 (September 1992): 713–719; Joseph Farrell, "Information and the Coase Theorem," *The Journal of Economic Perspectives* 1, 2 (Autumn 1987): 113–129.
43. Deborah Stone, *Policy Paradox: The Art of Political Decision Making*, rev. ed. (New York: W. W. Norton, 2002): 376–383.
44. Hugh Heclo, "Issue Networks and the Executive Establishment," in *The New American Political System*, Anthony King, editor (Washington, D.C.: American Enterprise Institute, 1978), 87–124.
45. Stone, *Policy Paradox*, 412–414.
46. See, for example, Harold M. Hochman and James D. Rodgers, "Pareto Optimal Distribution," *The American Economic Review* 59, 4 (September 1969): 542–557; A. P. Xepapadeas, "Environmental Policy under Imperfect Information: Incentives and Moral Hazard," *Journal of Environmental Economics and Management* 20, 2 (March 1991): 113–126.
47. See, for example, Daniel W. Bromley, "Environmental Regulations and the Problem of Sustainability: Moving beyond 'Market Failure,'" *Ecological Economics* 63, 4 (15 September 2007): 676–683.

48. The leading article is Garrett Hardin, "The Tragedy of the Commons," *Science* 162, 3859 (December 13, 1968): 1243–1248. See also Beryl L. Crowe, "The Tragedy of the Commons Revisited," *Science*, New Series, 166, 3909 (November 28, 1969): 1103–1107; Rose, "Liberty, Property, Environmentalism," 3–4.

Chapter 3

Environmental Ethics: An Overview

> Why should I care about posterity? What has posterity ever done for me?
>
> **—Groucho Marx, comedian, actor, talk-show host[1]**

"Sustainability" has become a cliché in American environmentalism, and an imprecise one at that. From the ivory towers of academe to corporate boardrooms to the outdoor pavilions of Keep America Beautiful affiliates, concerned citizens wistfully speak of their fervent desire to chase down the elusive phantom traveling under the *nom de guerre* of "sustainability." The less analytically inclined do not speak in specifics when they profess their undying allegiance to sustainability; they pontificate using the language of synonyms and complements, employing vague descriptions such as "environmentally friendly," "eco-friendly," "earth-friendly," "socially responsible," "intergenerational equity," or the all-encompassing fuzzy-wuzzy appellation, "green." The sobriquets sound soothing and oh-so-desirable to persons fearful of environmental degradation, but the lack of precision dooms the utility of buzzwords and trendy concepts, especially among neoclassical economists and proponents of the scientific method. "Sustainability" is a term that can mean everything or nothing; it is an assault on the status quo, a plea to change behavior even if the practicality of the change is unknown or unknowable. In a move reminiscent of Justice Potter Stewart's observations on obscenity, dilettantes cannot define sustainability, but they know it when they see it.

However it is defined, sustainability can be thought of as a normative rather than a descriptive concept, although the goal is to move eventually from "ought" to "is" behaviors. A sustainable future requires citizens and decision-makers to establish a new way of thinking about, and acting on, natural-resource preservation. The difficulty arises because reasonable minds can, and do, differ on what "sustainable" thinking and acting entail. Thus, if we cannot corner our elusive prey for a first-hand dissection, let us examine its trail in this chapter. Along the way, we can pause to visit with forbears who have walked the path and staked out a claim—if not to the truth of sustainability, perhaps to partial insight on what the concept involves. Even if we cannot bring down the evasive beast at the conclusion of the hunt, we may glimpse its essential features. Because of this glimpse, we can generalize on common characteristics so anyone can recognize sustainability if they catch sight of it in the wild.

Recall from Chapter 2 that neoclassical economics has served as the foundation for most public policy in the United States. As a descriptive summary, this declaration is indisputably true and appallingly trite. The normative question addressed in this chapter is whether calculations made pursuant to a capitalist economic system ought to consider values such as environmental sustainability. From an economist's perspective of "weak" sustainability—that is, using the tenets of mainstream, neoclassical economics, such as supply and demand—one method for reconciling environmentalists and industrial developers might be to consider sustainable development as an obligation to conduct ourselves so that we leave to the future the option or the capacity to be as well off as we are. This perspective requires adherents to accept neoclassical economic precepts whereby public policy issues are translated into units that can be measured by comparing costs and benefits to calculate utility based on the "invisible hand" of the marketplace (momentarily setting aside problems associated with a market failure). Philosophical Utilitarianism meets economic pragmatism.

The difficulty in considering sustainability from the neoclassical economists' position is that the analysis assumes *arguendo* that all claims of moral worth can be expressed as measurable units with meaningful valuations. By calculating costs and benefits without weighing preferences or differentiating between the characteristics of resources, we beg the question of evaluating what is being preserved for future generations. Yet if we seek precision by assigning weights to preferences and distinguishing among and between types of resources, problems do not dissolve; they proliferate. The endeavor guarantees no small measure of strife and dissension. If the current generation has a duty to preserve the natural environment for future generations, the nature and scope of that duty are unclear.

Timing is critical. It does not require a leap of faith to conclude that the current generation owes a duty to its children or even its grandchildren. Because many people live to see and enjoy the lives of their children and grandchildren, they may be persuaded to forgo benefits today for the promise of conserving resources for their immediate progeny. The "distance" problem of how far into the future the duty extends is compounded when we consider the lives of three, four, five, ten, or twenty generations hence. This long-term scale also raises the problem of ignorance, which increases as we extend the time of our future obligations. Who knows what people of the future will need? Perhaps their value systems will differ from our own. Nobel Laureate Robert M. Solow, a well-known commentator in this field, criticized standard definitions of sustainability because they are so amorphous in defining the nature of the duty and the length of time involved, but he offered no clear alternative. Certainly it is wise not to deplete all resources today—even the most entrenched, hard-nosed neoclassical economist would not recoil at this presupposition—but the nature of our obligation to future generations frequently is expressed as an admonition to preserve the capacity of subsequent peoples to be as well off as we are with no further explanation of the scope of the duty. This chapter explores a range of possible explanations.[2]

Criticism of the initial marriage between sustainability and neoclassical economics leads one to ask whether alternative perspectives on the current generation's obligations to the future offer greater depth of vision and clarity of expression. Throwing off the shackles of bromides and bumper-sticker buzzwords requires the researcher to dig deep into the academic literature concerning environmental ethics and search for a conceptual framework to discuss intergenerational equity far beyond the proverbs of neoclassical economic theory. Accordingly, it is helpful to examine the range of positions available to the ethicist who seeks to understand environmental sustainability.

Anthropocentrism, Neoclassical Economics, and the Scientific Method

Recall the philosophical ideas and concepts discussed in Chapter 1. At the risk of over-simplification, neoclassical economics can be analogized to Utilitarianism in the sense that the outcomes are the crucial factor in decision-making. If the benefits exceed the costs, an economic venture can be justified according to the neoclassical economic position because the results lead to the greatest good for the greatest number. By contrast, some environmentally conscious decision-makers seek to move beyond the traditional economic model. They contend that a broader understanding of the "good" is necessary. The broader conception can be characterized as a concern for sustainability, which can be thought of as factoring the ecosystem into the decision-making process. For ecologists, human beings have a duty to protect nature. The ecological position can be analogized to the Kantian perspective where duty, not consequences, is the crucial determinant. Yet even these generalizations do not shed light on the range of environmental positions, for not all environmentalists are united in their understanding of duty.[3]

Environmental ethicists identify values positioned along a continuum. At one end, the values most closely akin to neoclassical economics can be labeled "anthropocentric," literally "human-centered." The conservative anthropocentrist believes that only human beings possess intrinsic value and therefore moral standing. This perspective is reflected in numerous literary works from the Platonic dialogues, the Christian Bible, and the works of the great Enlightenment playwrights such as William Shakespeare to the philosophical rebels, including Friedrich Nietzsche, who urged human beings to transcend the limitations of their environment and become, in Nietzsche's words, "the Übermensch." Because humans are transcendent, all other living creatures as well as plants and inanimate objects are subservient to the fulfillment of human goals. Their value is derivative from human value and contingent on humanity's changing needs; non-humans are only worthy of protection if they contribute to humanity's well-being. The environmentally conscious anthropocentrist preserves nature because it is in humanity's interests to do so, not because nature or other animals possess any special characteristics or moral worth.[4]

In addition to its adherence to neoclassical economics, the standard anthropocentric worldview is grounded in a Cartesian/Newtonian understanding of science. Cartesian science, named for the seventeenth-century French philosopher and mathematician René Descartes, posited that the world is mechanistic, comprised of matter and motion, similar to a machine wound up to run without an exalted purpose or a higher meaning. The clockmaker, having perfected his invention and set it in motion, has left the building (if he was ever present in the first place). Isaac Newton's subsequent attempt to explain the physical laws of science augmented the Cartesian insight and led to a search for replicable data in all human endeavors. Newtonian mechanics delve into the specific workings of the clockmaker's invention. Whether the artisan checks in occasionally to ensure the continued operation of his or her device is open to interpretation and debate. As Professor William E. Rees, a pioneer in the field of human bio-ecology, wrote, "Modern economics owes much to this scientific world view. The founders of the neo-classical school, impressed with the spectacular successes of Newtonian physics, strove to create economics as a sister science; 'the mechanics of utility and self-interest.'"[5]

Yet for all the rigors of Cartesian science as well as economists' desires to import Cartesian/Newtonian principles into the social science realm, practitioners embraced their own mythos. According to the traditional anthropocentric narrative, logical positivists championed the scientific method, with a capital "M." The Method based its claim for supremacy on its clear, logical, intellectually rigorous, verifiable, and ultimately replicable characteristics. The Method supposedly was immune to contextual influences that infected lesser disciplines. If the Scientific Method was perfected through repeated experimentation, man could solve the great problems that had plagued human beings since time immemorial. The rise of the Method led to the glorification of data analysis and quantitative measures for ensuring precise, rigorous accountability. Uncertainties existed, but they could be reduced or eliminated by conducting additional experiments, compiling more data, or reframing confounding questions.[6]

Method once trod the earth as an unassailable king, but late in the twentieth century a group of detractors, the eclectic postmodernists, sought to overthrow the unreflective monarch. As an example, cognitive scientist Larry Laudan explored the "postmodern conundrum" of the ascendant Method in his book *Progress and Its Problems*. The Enlightenment's view of the Method allowed rational thinkers to make enormous strides in the seventeenth and eighteenth centuries, but eventually a troubling realization arose: Science was yet another byproduct of its context. The attempt to demythologize epistemology and discover the unvarnished essence of empirical "truth" became another myth. Faith in the rigors of science and economics replaced faith in philosophy and religion, but it still required a conceptual leap. Unbeknownst even to practitioners of the Method, logical positivists did not destroy the kings and priests of mythos; they took their place. If Diderot was correct that "man will never be free until the last king is strangled with the entrails of the last priest," postmodernists applied the same criticism to scientists and economists, respectively.[7]

The insight was especially worrisome because scientists viewed the Scientific Method as the most effective means for rationally discovering the world. When the scientific method was questioned, practitioners assumed that rationality was under attack. During the latter part of the twentieth century, this dawning realization led to an epistemological crisis because some anthropocentrists lost confidence in the "Method," and what commentator Philip Kitcher called the "Legend" of idealized science. "Since the late 1950s, the mists have begun to fall," Kitcher explained. "Legend's lustre is dimmed. While it may continue to figure in textbooks and journalistic expositions, numerous intelligent critics now view Legend as smug, uninformed, unhistorical, and analytically shallow."[8]

Logical positivism and empiricism were attacked on many fronts during the twentieth century, notably by Thomas Kuhn, author of the influential 1962 book *The Structure of Scientific Revolutions*. According to Kuhn, scientific practice is not understood as the inexorable march of progress, each insight necessarily building upon its predecessors as a kind of iterative, logical, empirical, puzzle-solving exercise. Instead, in Kuhn's view, useful scientific practice develops after scientists construct a series of propositions—a paradigm—about how the world works. Whenever enough new evidence contradicts existing propositions over a long enough time, scientists develop a new paradigm that displaces the previous paradigm. Kuhn called this process when one paradigm shifts in favor of another a "scientific revolution," hence the title of his book.

Kuhn's legion of critics responded that his concept of paradigm shifting ultimately leads to relativism because it undermines the notion that logically certain, generally recognizable standards can be developed, empirically tested, and unequivocally verified. Instead, scientific practice seems dependent on the time and place in which theories are developed and tested. If this is the case, science ultimately is relative and has no meaning beyond the scope of its contextual boundaries, a disturbing prospect for defenders of mainstream, idealized science.[9]

Cognitive scientists spent much of the 50 years after Kuhn's work first appeared trying to rehabilitate the scientific method without concomitantly reviving all the features of logical positivism. In his book *Explaining Science: A Cognitive Approach*,

University of Minnesota philosophy professor emeritus Ronald N. Giere suggested that science, properly understood, cannot be divorced from its practice; it is best described as a complex blend of conceptual models and scientists' interactions with the world. He argued that two competing theories of scientific practice—logical empiricism, on one hand, and relativism, on the other—are mistaken in their approaches. In the former, empiricists contend that developing theories based on "objective," logically defensible foundations such as those employed in mathematics leads to the practice of "good" science. In Giere's words, logical positivists see their primary objective as an effort to use immutable principles "to justify, or legitimate, science, not merely...explain how it works." Giere adopted a naturalistic middle ground between empiricism and relativism, which he termed a "cognitive view of science." "This is to focus on *cognitive processes*, such as those involved in representation and judgment, which are shared by all scientists," he wrote. According to this perspective, scientific practice involves the development of models (cognitive maps or mental models) that represent scientists' experience with the world, but do not mirror it directly. Indeed, a model is a shorthand way of representing the world—a correlation—but it is not an exact, one-to-one reflection of the world. Using their judgment about the efficacy and persuasiveness of a given model, scientists develop strategies about how experiments should be designed and how their results should be interpreted.[10]

In her book *Science as Social Knowledge*, Stanford University philosophy professor Helen E. Longino adopted a slightly different approach. She agreed with Giere that the traditional view of science as a value-neutral, logically grounded, empirically verifiable process is too simplistic. In her opinion, science does not consist of isolated individuals applying scientific principles to problems by conducting empirical experiments in laboratories around the world, as the idealized view suggests. Instead, she argued that science must be viewed as a practice that is conducted by groups of scientists who interact with each other through a series of social processes. In effect, science is context-specific. Every major decision—which studies to conduct, the design of the studies, and the interpretation and communication of results—depends on the generally accepted practices of the scientific community. According to Longino, this conception of science as social knowledge has far-reaching implications. "A consequence of embracing the social character of knowledge is the abandonment of the ideals of certainty and of the permanence of knowledge," she observed.[11]

This abandonment of certainty worried supporters of the traditional view of science. Empiricists argued that the perspectives advocated by Kuhn, Giere, Longino, and others undermined the primacy of the scientific enterprise. If scientific principles cannot be established and tested with a reasonable degree of certainty, the entire quest for human knowledge has no meaning and is little more than a relativistic search for ideology or values that vary according to time and place. The changing perspective on science holds profound repercussions for the anthropocentric approach to environmentalism. Anthropocentrists implicitly or explicitly assume that economic growth is theoretically unlimited and the Scientific Method, regardless of its flaws, remains the most effective means of legitimizing human endeavors, but the more environmentally conscious postmodern ecologist contends that adherence to the Method is untenable. In short, the tools of science and economics are useful, albeit limited. The creative, evolving use of those tools becomes crucial in determining how the needs of human beings can be met without depleting natural resources.

Traditional anthropocentrists refuse to go gentle into that good night. Science and economics, they insist, are powerful explanatory tools. The tools may not function properly in all instances, but they explain human behavior more accurately than alternatives. To illustrate this insight, they point to the central thesis of anthropocentrism: the fungibility of natural resources. When a particular resource is exhausted, the principle of substitutability indicates that another good can supplant the discarded resource. From this perspective, natural resources can be evaluated just as any other resource is analyzed. The underlying assumption is that total capital stock will never be depleted. Even as human beings confront a limited number of substitutes, technological advances will serve as a *deus ex machina*, heroically saving the day by providing heretofore-unknown resources in quantities sufficient to satisfy human beings indefinitely. The iterative nature of scientific progress promises to break through walls of finite possibilities and create another dimension that we cannot envision. As Hamlet famously remarked, "There are more things in heaven and earth, Horatio, than are dreamt of in your philosophy."[12]

The standard anthropocentric definition of sustainable development appeared in a 1987 report of the World Commission on the Environment and Development (WCED), the "Brundtland Report," named for the chairperson, the prime minister of Norway. According to the Brundtland Report, sustainable development is defined as development that "meets the needs of the present without compromising the ability of future generations to meet their own needs." Similarly, in *Agenda 21*, a major document adopted by almost all countries of the world at the famous 1992 United Nations Conference on the Environment and Development (UNCED) held in Rio de Janeiro, Brazil, the goal of sustainability is "to ensure socially responsible economic development while protecting the resources base and the environment for future generations."[13]

In contrast to the traditional anthropocentric view, "biocentrists"—literally "earth-centered" advocates—argue that all living creatures possess moral standing. Because all organisms inhabit the ecosphere and follow a similar path from birth to maturity to death to decay, no species can claim moral superiority and a right to dominate nature or other species. Living things can justify using natural resources to satisfy basic needs such as food, clothing, shelter, and protection from predators, but humankind's claim to exercise dominion over the earth and its creatures is fallacious. According to biocentrists, anthropocentrism implicitly adopts a "might makes right" ethic.

The world-centered position is most closely associated with Paul Taylor, whose book *Respect for Nature* explained that "the biocentric outlook, we might say, provides a general 'map' of the natural world, enabling us to see where we are and how we fit into the total scheme of things. It presents the realm of nature and life as a setting for human existence." Instead of existing apart from nature and trying to conquer it as an "other" while justifying our actions using science and economics, biocentrism posits that "we see ourselves as biological creatures." Humankind does not exist separate and above the earth; human beings are of the earth.[14]

The bifurcation between anthropocentrism and biocentrism represents a fundamental distinction among schools of thought on sustainability. It would be overly simplistic, however, to suggest that environmentalism is an "either/or" position: Either you favor economic health for human beings or you favor the ecosystem and biodiversity. In fact, some theorists contend that focusing on the anthropocentrism-biocentrism division deflects attention from profound, nuanced environmental issues. Despite this criticism (or perhaps because of it), American environmentalism traditionally has encompassed a range of perspectives extending from weak sustainability to strong sustainability, ecological sustainability, normative sustainability, and intrinsic value theory.[15]

Weak Sustainability

Long-time Yale University economist William D. Nordhaus, a prolific author, has been a renowned defender of "weak sustainability." One of his best-known works is *Managing the Global Commons: The Economics of Climate Change*. In that work, Nordhaus examined the economic impacts of climate change and concluded that much of the hysteria surrounding the issue has been misplaced. Only minor abatement programs are necessary to curb greenhouse gases. In subsequent works, he examined a variety of environmental issues using the standard tools employed by neoclassical economists. In each instance, Nordhaus contended that "price-type control mechanisms" are more effective influences on global environmental policies than "quantity-oriented control mechanisms" such as international agreements and targeted reductions, which often impose unrealistic, arbitrary goals that are difficult to enforce and fail to consider constantly changing variables such as population growth, evolving individual preferences, and unexpected global developments.[16]

Critics have charged that this faith in the substitutability of resources virtually *ad infinitum* reveals a fundamental misunderstanding of natural resources by mainstream capitalists. Owing to the delicate natural balance among species and habitats, what looks to be an acceptable loss of natural capital can be deceptive. Short-term substitutions can mask long-term crises. The sudden loss of a habitat, for example, can trigger the extinction of a species that supports a multitude of other endangered species. The loss of the species can accelerate environmental degradation in unforeseen ways that spread as a contagion and upset the careful calculations of even the most conscientious neoclassical economist. A sudden, unexpected environmental catastrophe could require hundreds of substitutes available only at an exorbitant cost or possibly altogether unavailable. In his assessment of *Managing the Global Commons*, Eric Neumayer of the London School of Economics and Political Science remarked, "The real problem of Nordhaus's methodology is his implicit underlying assumption of perfect substitutability between natural and other forms of capital." This criticism of general economic principles extends far beyond Nordhaus's work.[17]

The most environmentally conscious anthropocentrists distinguish between this traditional view and a broader perspective sometimes labeled "intergenerational anthropocentrism." Rather than focus merely on the instrumental value of nature today, persons interested in intergenerational equity contend that human behavior toward the natural environment should be evaluated based on expected resource use in the future. To deplete a resource in the short term steals from future generations on behalf of the current generation. The tragedy of the commons may appear to be an acceptable neoclassical response to resource management in the near term, but it fails to appreciate the need to preserve nature for the current generation and generations to come.

In addition, natural resource valuation is not a straightforward problem of substitutes and complements. Persons concerned with intergenerational equity understand the necessity of a robust ecological movement that accounts for differences in types of natural capital. Notice, however, that the valuation is a matter of degree, not kind. Natural capital must be measured differently from standard property and monetary capital, but a measurement with a reasonable degree of precision is theoretically possible if the parties can come to terms on acceptable values. Supporters speak of sustainable development as a means of moving beyond the traditional neoclassical economists' perspective as long as the movement does not undermine the basic tenets of western capitalism. A fundamental premise is that the global economy operates as a closed system; that is, goods are finite, but they are exchanged as though they were infinite without developing a plan for renewing the goods.

The Second Law of Thermodynamics states that a closed system eventually degrades to an unavailable state as resources are depleted. If neoclassical economics is to remain a viable global system, market participants and governments must implement environmental management systems and sustainable public policies immediately. The closed system must become an open system. Unlike radical environmentalists, however, defenders of weak sustainability contend that nature is a management problem that can be resolved with minor retooling.

Retooling requires a reorientation among policymakers. Neoclassical economic theory assumes that equilibrium is reached where income distribution, the forces of production, and the prices and quantities of goods intersect. Intergenerational anthropocentrism acknowledges the myopia of this presupposition. Equilibrium can never be achieved when human beings view their affairs as independent from the biosphere. As William Rees explained, "Functionally speaking, there is only a single entity, the biosphere, and humanity has always been part of it."[18]

Many anthropocentrists seek to preserve resources for the future while not sacrificing their current economic system or their high standard of living. The "environmental" anthropocentrist decries the frivolous use of plastic products, for example, but he indignantly refuses to eschew the latest electronic game, computer, tablet, or device constructed from plastic. He welcomes an age of smart resource use but is only dimly aware of the burdens that must be borne by persons less affluent than he as low-wage workers produce those goods in overseas factories, discarded waste is shipped to those same nations for disposal, and alternative products and processes are avoided because they would add to the cost of the desired consumer good.

Theodore Roosevelt is an example of the nature-loving anthropocentrist. As a well-known champion of conservation policies in an age not known for its environmental sensitivity, Roosevelt was the first American president to protect U.S. forests from timber interests using his bully pulpit. His efforts to establish national parks and wildlife sanctuaries are widely, and rightly, lauded. His numerous forays into the wild and his love of his country's natural treasures also are well documented. Indeed, the man's desire to preserve resources using the instrument of government is beyond dispute. The cartoon of the vigorous trustbuster confronting the lumber trust in Figure 3.1 is a typical illustration of the president's zeal for action against special interests on behalf of U.S. citizens and, by extension, on behalf of unborn Americans.

Yet for all of his environmental zeal, Theodore Roosevelt viewed nature as instrumental to human happiness. He enjoyed his adventures in the wilderness in large measure because he could carry a gun and slaughter all manner of beast as trophies, mementos of his treks through primitive lands. He did not object to treating nature as a means to an end, namely the end of humankind's contentment. No doubt, the famous conservationist would have blanched at the postmodern environmentalist's criticisms that traipsing into the woods while wielding a gun despoils the earth. Taking from the rich bounty of the earth to sustain human life can be justified, but killing animals as a blood sport is thoughtless, cruel, and wasteful.[19]

The eco-sensitive anthropocentrist eventually is faced with a harsh realization. The ecological crisis of the twenty-first century is caused because the productivity of the ecosystem is limited while the growing needs and desires of neoclassical economies are infinite. While a growing population seeks to maximize production by consuming resources, preferably in ever-increasing quantities, the ecosystem remains static. With the passage of time, the insatiable human appetite consumes an ever-greater percentage of the earth's finite resources. Even after human beings satisfy base needs, their appetites do not diminish; they expand. The typical consumer amasses a large enough fortune to purchase a home and enjoy a decent standard of living. Rather than shrug and pronounce himself satiated, the consumer immediately considers moving into a larger house or purchasing a second home. His goals change. Initially, he sought to provide for his family and enjoy

"WHOA!"

Figure 3.1 Theodore Roosevelt, 26th president of the United States, is sometimes viewed as the quintessential example of the anthropocentric environmentalist. In this 1906 cartoon, "TR" confronts lumber interests that were intent on exploiting public lands. (Courtesy of the Granger Collection, New York)

the "American dream." After accumulating wealth, he might be induced to give away part of his fortune to assist persons in need, but he must also impress the neighbors with ostentatious spending. Nothing demonstrates to one's rivals that he has achieved affluence and power as much as conspicuous consumption. To the enormously wealthy, gas-guzzling sport utility vehicles, multiple residences, and the ultimate status symbol, a private jet, flaunt the ability to waste resources at a prodigious rate. At some point—and the precise point is a matter of ferocious debate—consumption exceeds the capacity of the natural environment to replenish the dwindling resource. Especially since the advent of the Industrial Revolution, the global economy has begun to cannibalize the biosphere.[20]

Human appetites are inexhaustible, but natural resources are not. This stark reality has forced some environmentally conscious souls to move beyond the well-meaning but ultimately self-defeating world of the intergenerational anthropocentrist. Herman E. Daly, a well-known critic of weak sustainability, articulated the problem well in his arguments touting steady-state economics, the idea that an economic system should function at a stable rate that does not outgrow the capacity of natural resources to support life. He observed, "It is widely believed by persons of diverse religious convictions that there is something fundamentally wrong in treating the earth as if it were a business in liquidation."[21]

As discussed in Chapter 2, neoclassical economists view the world in terms of the rational individual consumer who accepts the Utilitarian principle of maximizing self-interest. The problem can be captured in terms of scale. Mainstream economic and scientific theories assume that the world can be understood in small increments scaled to fit human affairs. Individuals measure significance in years that correspond to election cycles, school programs, generations, or human lifespans, but an ecosystem extends across hundreds of thousands or millions of years. The divergent scales require committed environmentalists to tout a dynamic, hierarchically organized, multiscalar world.[22]

To develop effective environmental policy, the objective must be to ensure the health of the ecosystem on a large geographic and temporal scale. If the focus is left on the scale of human affairs, rational consumers are encouraged to perpetuate the tragedy of the commons. Exhausting a resource can be rationalized in the short run, but if public values are recast, the rational choice is to protect the processes that support the health of larger-scaled ecological systems.

Some ecologists refer to ecosystem management using the term autopoiesis, a Greek concept that literally means "self-making" or "production." In using this term, they mean that an ecosystem consists of self-organizing processes that constantly interact through a network with other processes. As a metaphysical concept, such a system cannot be separated from its constituents. That is, an autopoietic system operates as a self-generating feedback system. Human beings may label component parts of the system to understand and discuss the ecosystem, but they cannot step beyond the system and examine it from the outside looking in. This concept can be distinguished from an allopoietic system, which is a closed loop.

The quintessential example of an allopoietic system is a manufacturing plant that uses raw materials to produce a separate product. The product exists outside of the manufacturing plant. The system can be examined by outside forces looking inside the plant to adjust the types of raw materials, the rate of production, the quality of the finished product, and so forth.[23]

These contrasting systems shed light on the division between anthropocentrists who embrace neoclassical economics as well as Cartesian/Newtonian science and ecologists who argue for a broader understanding of the important variables. The former treat environmental management as an allopoietic system that can be tweaked from time to time to improve efficiencies. If a species is threatened with extinction or pollution levels grow unbearable, a law or regulation can be implemented to protect the species or reduce the level of negative externalities imposed by private parties on the aggregated public. The latter group contends that piecemeal environmental policies fail to appreciate that environmental problems are systemic. Incremental public policy addresses the symptoms but not the causes of ecological degradation. As Bryan Norton observed, "*Homo economicus* 'discounts' values across time, preferring to enjoy benefits sooner rather than later, and to delay losses. Economic analyses therefore discount values that will be experienced in the future; in business, costs or benefits expected in future years are discounted at a rate roughly equivalent to the rate of real interest."[24]

Strong Sustainability

To move beyond weak sustainability, some environmentally conscious persons champion the concept of ecological economics. Rather than becoming mired down in determining an appropriate discount rate for monetary evaluations, ecological economists refer to the incommensurability of values. Environmental resources are not necessarily amenable to the tools employed by neoclassical economists to evaluate and trade goods and services. Social costs associated with environmental destruction have to be valued as "extra-market phenomena" because they are borne by society as a whole. A new form of accounting and evaluation is needed. "In the framework of ecological economics," authors Joan Martinez-Alier, Giuseppe Munda, and John O'Neill observed, "the use of a multidimensional approach seems desirable. This implies that the strong comparability assumptions of neoclassical economics have to be abandoned." In lieu of the singular, mechanistic, static approach adopted by mainstream economists, a dynamic, evolutionary approach is desirable.[25]

Desirability is one consideration, but feasibility remains a concern. Ecology and economics are separate disciplines, although they both assist individuals and organizations in understanding and advancing their goals. They also use similar theoretical concepts. Despite some similarities, though, as ecologists Robert Costanza, Cutler Cleveland, and Charles Perrings noted, the disciplines "began with different first principles, addressed separate issues, utilized different assumptions to reach answers, and supported different interests in the policy process." In an attempt to reconcile the differences, several groups of prominent ecologists and economists met numerous times during the late 1980s. In 1988, they created the International Society for Ecological Economics (ISEE). The following year, group affiliates began publishing *Ecological Economics*, an academic journal dedicated to the proposition that "bringing these domains of thought together and attempting to reintegrate the natural and social sciences" in a cross-pollinated discipline called "ecological economics" is a worthwhile endeavor.[26]

The major problem for disciples of the field lies in employing concepts and tools from both ecology and economics to resolve public policy issues. Ecologists frequently argue for a systems approach that considers factors across a span of centuries, millennia, and beyond while economists generally refer to transactions among parties that occur in a short time span. For ecological economics to work, adherents of the discipline necessarily must embrace a "conceptually pluralistic" approach. Practitioners are warned that the enterprise is riddled with inconsistencies and is subject to attack by detractors for being unfocused. Ecological economists shrug off such criticism, opting instead to argue the virtues of encouraging a "completely new, hopefully more integrated, way to think about the linkages between ecological and economic systems." An integrated manner of establishing linkages requires the use of "non-linear mathematics, general systems theory, non-equilibrium thermodynamics, and ecosystem ecology." Yet ecological economics is not a genuinely new discipline. Instead, it "represents a commitment among natural and social scientists, and practitioners, to develop a new understanding of the way in which different living systems interact with one another, and to draw lessons from this for both analysis and policy."[27]

Herman E. Daly explained the mechanics of that commitment as threefold. First, ecological economics requires a broader understanding of scale, by which he meant the process of using low-entropy raw materials to create a finished product that eventually becomes a high-entropy waste material. The ecological economist seeks to recast the debate so that human activities conform to the larger long-term needs of the biosphere. Second, the relative distribution of goods and services is accomplished in an equitable manner so that the resources of today are not depleted, thereby robbing future generations of access to a reasonable standard of living. Finally, Daly contended that ecological economics is concerned with an efficient allocation of scarce, finite resources so consumer needs are met without violating the first two stated principles.[28]

As Robert Costanza explained in his book *Ecological Economics: The Science and Management of Sustainability*, this new understanding changes traditional economics, which sees individual tastes and preferences as the primary consideration in a static system. Ecological economics, by contrast, is dynamic, adopts a broader focus, and considers individual preferences as but one feature among many. Instead of a short-term scale, ecological economics evaluates preferences from days to eons along a multiscale continuum. Human beings, while important, are only one of many species that possess moral worth. Ecological economists are suspicious of unbridled economic growth—Costanza confessed that he was "prudently skeptical"—because they understand that resources are limited and no system can be overloaded indefinitely. Rather than focusing on economic or scientific measures alone, ecological economists employ multiple approaches to conceptualizing a problem and seeking solutions.[29]

John B. Cobb, Jr., another supporter of strong sustainability, authored several works arguing that citizens and their leaders must make immediate, systemic changes. In *Economics, Ecology and Justice*, he wrote,

> The concern here is for the healthy survival of humanity [on] a healthy planet, and the conclusion is that to attain this we must shift drastically from our role as masters of creation to one of participants within it. We need to see ourselves as simply one species among others, rather than as standing outside of nature and dominating it. We should move from a stance of moral responsibility for nature to one of fitting into it.[30]

Ecological Sustainability

Ecological sustainability is similar to the idea of strong sustainability except that it emphasizes the need for maintaining resilience in the ecosystem as the crucial feature in decision-making. The classically anthropocentric preservation movement of Theodore Roosevelt and John Muir early in the twentieth century promoted public policies that set aside lands for human use and enjoyment. Even environmentally conscious policymakers who sought to move beyond this classic standard often focused on finding ways of maximizing sustainable yield. C. S. Holling, an ecologist who spent much of his career at the University of Florida before he retired, wrote extensively on the reasons why maximizing the yield is not an ideal method for thinking about ecology.

In a groundbreaking 1973 article, he explained that some scholars seek system stability as a kind of equilibrium where an ecosystem can be preserved and maintained. Holling believed that a complex, dynamic system is constantly changing and is never stable. The search for equilibrium is misguided. Instead, resilience is required for the ecosystem to respond to a multitude of changes. The 1973 essay reflected Holling's belief that populations interact similar to the way predators and prey react to each other—somebody wins and somebody loses.[31] Thirty years later, Holling clarified the goal. "Single equilibria and global stability have made ecology focus on near equilibria behavior," he wrote. Yet he argued that ecology is fluid and dynamic. Reflecting on the repercussions of his 1973 essay, he observed, "The multi-stable state reality opened an entirely different focus on behavior far from equilibrium and on stability boundaries. High variability became an attribute to maintain existence and learning. Surprise and inherent unpredictability was the inevitable consequence for ecological systems." In his view, the focus should be on variability, not stability. In other words, the study of ecology involves an analysis of "the amount of disturbance a system can take before its controls shift to another set of variables and relationships that dominate another stability region."[32]

In another article, Holling and a co-author discussed the steps necessary to adopt ecological sustainability. In choosing whether to undertake an economic venture, decision-makers must factor in systemic environmental concerns. This insight means that an endeavor "should be deemed unworthy of undertaking not only if its costs exceed its benefits, but if it will compromise the health of the (relatively) macroscale ecosystems in which it is embedded and the (relatively) microscale ecosystems on which it is imposed." Ecosystem health, in short, is a critical factor in any economic decision.[33]

Elsewhere he and several co-authors explained the need for "adaptive management" to forestall the loss of ecosystem resilience. When an ecosystem is ailing, it loses biological productivity and can jeopardize the ability to support all forms of life. By irreversibly changing the options available to human decision-makers, the loss of resilience imperils the ability of future generations to enjoy a decent standard of living as environmental catastrophes such as desertification, soil erosion, and the depletion of groundwater sources occur. Finally, as an ecosystem loses capacity, it creates uncertainties about how additional changes occur. These uncertainties pressure decision-makers to enact policies that are not well thought-out or tested and thereby exacerbate the deleterious effects of change. Holling and his colleagues wrote that "economic growth is not a panacea for environmental quality; indeed, it is not even the main issue. What matters is the content of growth—the composition of inputs (including environmental resources) and outputs (including waste products)." Policymakers must adapt by modifying economic institutions to meet environmental challenges. "These institutions need to be designed so that they provide the right incentives for protecting the resilience of ecological systems." Only then will human beings "assure a sustainable scale of economic activity within the ecological life-support system."[34]

Ecological-sustainability followers embrace this notion of adaptive management. Because large, complex, dynamic systems contain numerous variables and fluid processes, they change frequently. Human beings do not always understand the processes of change, which means that human decision-making always takes place in the face of uncertainty. To ensure that the decisions are made using the best data available, ecologic systems must be monitored continuously and iterative changes implemented as circumstances require. As Carl J. Walters, an expert on fisheries, observed along with co-author Ray Hilborn, "Uncertainty is a pervasive feature of ecological management problems." Even if all the data could be captured, "we would still be faced with the uncertainty of various environmental agents." We also would have to account for problems in measuring data and communicating it clearly to third parties. Adaptive management allows ecologists to develop models that factor uncertainty and frequent changes into the decision-making process, which is a distinct improvement over deterministic prediction models.[35]

Author Kai Lee viewed adaptive management as an ongoing process—"implementing policies as experiments." This approach contravenes command-and-control management systems as well as the traditional scientific method that requires practitioners to develop a research hypothesis, test it against standard measures, and wait to modify the hypothesis until the experiment has concluded. Adaptive management, as the name implies, allows researchers to adjust the hypothesis as data are collected and anomalies are discovered. Because the decision-making process is fluid and the effort is counterintuitive for many policy analysts, adaptive management is difficult to sustain. Yet it is a critical endeavor because "all aspects of preserving the life-support systems of earth, particularly the protection of biological diversity and ecological processes, lie at or beyond the frontier of established practice or reliable knowledge." Acknowledging uncertainties and allowing researchers to alter the experiment as they collect data and learn new information are integral to the success of adaptive management.[36]

Normative Sustainability

Normative sustainability is closely related to ecological sustainability except that the emphasis is placed on maintaining options in environmental management, a notion that allows for more flexibility than simply maintaining resilience. Rather than approaching sustainability as a concept to be analyzed, normative sustainability supporters contend that it is an ethical value that underlies any discussion of human development. The core imperatives are to safeguard the environment and reduce poverty for different groups and nations and ensure political participation. The key insight is in meeting current needs, which are not restricted to economic necessities, while ensuring that future generations do not suffer from a lack of meaningful options.[37]

In his 2005 book *Sustainability: A Philosophy of Adaptive Ecosystem Management*, Bryan G. Norton of the Georgia Institute of Technology argued that an action is not sustainable if it reduces the ratio of opportunities to constraints for future generations. Every generation finds itself in an asymmetric relationship with future generations. Our actions today will have outcomes known to future generations but unknown to the current generation at the time the decision is made. Therefore, the current generation must consider myriad permutations and combinations of actions and consequences that can occur and affect people yet unborn. If adaptive management allows decision-makers to adjust variables, normative sustainability allows ethicists to state that new information and data gleaned moving forward ought to modify public policy decisions and subsequent actions.[38]

Intrinsic Value Theory and Its Discontents

Intrinsic value theory is sometimes referred to as "biocentrism" or "ecocentrism" to explain that "nature has moral consideration because it has intrinsic value, value aside from its usefulness to humans." The question is why moral standing should be limited to human beings or even to living creatures. If environmentalism is to be consistent and comprehensive, a broader ethic is required to evaluate nature using a holistic approach. Ecosystems and species rise and fall over millennia owing to interconnected relationships that human beings poorly understand because they tend to evaluate natural resources in a limited time frame using narrow criteria found in Utilitarianism and neoclassical economics. Aldo Leopold, sometimes called the "father of environmental ethics," argued that humankind's duty extends not simply to other living creatures, but to the land (Figure 3.2).[39]

"Deep ecologists" worry that much of what passes for environmentalism consists of superficial calculations reminiscent of thinly veiled neoclassical economics. They

Figure 3.2 Biocentrists contend that the needs of the planet should be central to any discussion of environmental ethics. Their holistic view is reflected in this photograph of earth that Apollo 11 astronauts took on July 17, 1969, from 113,000 miles away in space. (Courtesy of the Granger Collection, New York)

seek to probe deeper, to burrow beneath the façade created by western civilization to understand humankind's place in the biosphere. Coined by the Norwegian philosopher Arne Naess in a 1973 article, "The Shallow and the Deep, Long-Range Ecology Movements," the term "deep ecology," as opposed to "shallow ecology," is the idea that the sum is greater than the parts. The western notion of selfhood, thanks to Descartes, intimates that all that can be known for certain is the existence of the ego. All other matter may be illusion or fodder for self-gratification. The language of economics, therefore, is the language of maximizing material things in hopes that the self will be fulfilled. It is little wonder that so many westerners experience a deep sense of longing and alienation. Without a sense of spirituality and an appreciation of the interconnectedness of life, human beings will be adrift. As two deep ecologists, Bill Devall and George Sessions, explained, "We must see beyond our narrow contemporary cultural assumptions and values, and the conventional wisdom of our time and place, and this is best achieved by the deep questioning process."[40]

In addition to redefining the concept of the self so that self-interest is not narrowly construed, deep ecologists argue that all living creatures have a right to live, and they possess intrinsic worth. Human beings cannot justify dominion over the earth on moral grounds, especially in light of Charles Darwin's theory of natural selection, which posits that *Homo sapiens* is descended from lower-order animals. Rather than embracing the rampant consumerism inherent in neoclassical economics, deep ecologists insist that human beings must learn to conserve, recycle, and protect resources so the biosphere is not harmed to meet non-vital needs. Aside from Arne Naess, Bill Devall, and George Sessions, J. Baird Callicott and Laura Westra are well known for their work in this area.[41]

Some deep ecologists speak of the Gaia principle, a notion famously articulated by chemist James Lovelock. The principle suggests that all organisms are part of a holistic, complex system. Weak Gaia theory contends that organisms constantly adapt and evolve over time as part of this system. Even changes brought about by human beings can be characterized as adaptations that, while potentially destructive in the short run, can be corrected by future adaptations among different species. The concept of co-evolutionary Gaia mirrors much of the literature on evolutionary biology and has not proven to be as controversial as other aspects of the theory. The idea of homeostatic Gaia, however, has generated substantial disagreement among scholars because it posits that the presence of life on earth is an iterative process that becomes a self-correcting mechanism; changes among the species lead to stability and order in nature.

Strong Gaia theory is even more contentious because it suggests that all living organisms are purposive. According to "Optimizing Gaia," organisms seek to create favorable environmental conditions for the perpetuation of their species. Although human beings enjoy their self-perception as the only beings capable of intentionally altering nature, in reality all living creatures act on their environment, albeit their actions often are imperceptible. Similar offshoots of strong Gaia theory set forth metaphysical constructs for understanding how organisms have worked purposively with the ecosystem to preserve the planet.[42]

The Gaia principle has been frequently ridiculed as New Age silliness, a vestige of the Hippie era when amateur ecologists spoke of "Mother Earth" as though she were a conscious, active force in the world. To many critics, the concept is a reinvigoration of Aristotle's ancient concept of *telos* where living beings pass from a young stage of potentiality to a mature stage of actuality. Such metaphorical descriptions, detractors charge, hold no explanatory power and might be counterproductive because they mix scientific methods with abstract metaphysics that cannot be replicated.[43]

Deep ecology is not merely a subject for scholars. The movement's insights permeated popular culture with the publication of Daniel Quinn's successful 1992 novel, *Ishmael*. The book opened with an advertisement stating that a teacher sought a pupil who possessed "an earnest desire to save the world." The remainder of the book was a dialogue between the teacher—Ishmael, a gorilla who professed to be an expert on the subject of captivity—and his student, the human narrator of the novel. As with Plato's dialogues from the fourth century BCE, *Ishmael* introduced an interlocutor (in this case, the narrator) who posed a series of questions reflecting conventional assumptions about society while a wise figure (in this case, Ishmael assumed the role of Socrates) questioned the assumptions and nudged the student onto a different path. The purpose of the book was to puncture myths perpetrated by what Ishmael called "Mother Culture," or the systematic socialization of human beings in industrialized society to believe cultural myths about the legitimacy of human dominion over the world. By critically examining Mother Culture's myths, Ishmael educated the narrator (and the reader) on the virtues of one group of human beings (the Leavers) that respected the natural environment and the vices of another group (the Takers) that sought to overcome and dominate the natural environment. Ishmael urged his pupil to live harmoniously within the natural environment in the manner of Leavers, those citizens outside of advanced capitalist societies who adapt to their environment in lieu of making the environment adapt to them.[44]

As enjoyable as *Ishmael* is to read and as thought provoking as the teacher's insights are to ponder, the work can be criticized as an overwrought polemic. In fact, the inadvertent self-parody of mystical ecologists has worried more than a few environmentalists. Murray Bookchin, a pioneering ecologist, lamented the "exotic tendencies" of some mystics who "espouse deep ecology, biocentrism, Gaian consciousness, and eco-theology, to cite the main cults that celebrate a quasi-religious 'reverence' for 'Nature' with what is often a simultaneous denigration of human beings and their traits." By dismissing social issues among human beings as no more important or legitimate than any other species' concerns, the mystics reject anthropocentrism because it is unworthy of serious consideration. Human beings are not the proper unit of analysis in environmentalism, the fanatics assert; such greedy, rutting beings barely merit our attention even though we are those beings. Bookchin quoted *Gaia Meditations*, a mystical account penned by several well-known deep ecologists, as typical of the "drivel" and "bumper-sticker slogans" that pass for non-anthropocentric environmentalism. The authors invited the reader to "think of your next death. Will your flesh and bones back into the cycle. Surrender. Love the plump worms you will become. Launder your weary being through the foundation of life."[45]

Bookchin argued that in the quest to oppose the fiercely anthropocentric outlook championed by neoclassical economists and proud Utilitarians, deep ecologists risk an overcorrection. A wholesale rejection of human problems and concerns not only alienates potentially sympathetic audiences, but it threatens to reduce the environmental movement to irrelevancy. "The ecology movement is too important to allow itself to be taken over by mystics and reactionary misanthropes," Bookchin cautioned. Rather than be distracted by absurd red herrings, he suggested that persons interested in environmental issues must refocus on the crucial question: whether the "'grow or die' economy, lurching out of control" can be infused with workable sustainability concepts or whether the *status quo ante* will continue, thereby eroding the "fragile conditions necessary to the maintenance of advanced life-forms on this planet."[46]

Ramachandra Guha, a renowned Indian historian, offered an even more devastating assessment of deep ecology in his essay "Radical American Environmentalism and Wilderness Preservation: A Third-World Critique." Many developed nations, mostly in the West, exploited the natural environment for decades, developing a largely unrestrained appetite for resource consumption in the search for a higher standard of living. It became clear to environmentalists—especially those in the deep ecology movement—that this destructive behavior could not continue indefinitely. Thus, deep ecologists propounded a set of four principles designed to curb the desire for unrestrained economic growth. First, our way of thinking must evolve from an anthropocentric to a biocentric perspective. Second, the focus is on preserving unspoiled wilderness areas as well as restoring degraded lands to their former condition. Third, deep ecologists often cite eastern spiritualism as the philosophical impetus for this change. Finally, deep ecologists see themselves as existing on the cutting edge of environmentalism.

Guha addressed each of the four tenets in his essay. Although he welcomed a change in attitude from the neoclassical-economic model, he worried that discussing resource conservation as part of a dichotomous world view—bad guys (anthropocentrists) versus good guys (biocentrists)—was "of very little use in understanding the dynamics of environmental degradation." He suggested, "What is unacceptable are the radical conclusions drawn by deep ecology, in particular, that intervention in nature should be guided primarily by the need to preserve biotic integrity rather than the needs of humans." Human needs are important. Patterns of consumption should not be considered at the aggregate level, however, because patterns differ among nations. A pressing problem concerns over-consumption by industrialized countries at the expense of peoples living in developing nations. The anthropocentric/biocentric debate does not address this issue satisfactorily.

In addition, deep ecologists who desire to preserve large tracts of unspoiled wilderness harm the developing world disproportionately. Industrialized nations long ago degraded their wilderness to create an infrastructure for industrial development and economic growth. Now that other nations seek to emulate this pattern, deep ecologists chastise them for harming nature. Owing to this imbalance, Guha argued, "setting aside wilderness areas has resulted in a direct transfer of resources from the poor to the rich."

Guha also was disgusted by the New Age cooptation of eastern philosophy to justify deep ecology. He argued that well meaning but naïve dilettantes cherry-pick the concepts they believe will support their preconceived views of the mystical East. Yet eastern philosophy shares with its western counterparts an essential inscrutability. Many great works can be interpreted in multiple ways. Moreover, nuances and differences among spiritual traditions as diverse as Hinduism, Buddhism, and Taoism cannot be smoothed over with platitudes about the eastern respect for nature or its unitary view of the cosmos. Setting aside the misuse of spiritually and philosophically rich traditions and texts, Guha deplored the desire to misappropriate "primal" peoples as untrained, naturally inclined deep ecologists. The position may not be intentionally paternalistic, but Guha found it distasteful to assume that the natives will be in harmony with deep ecology principles by virtue of their proximity to the land. What a highly industrialized, well-fed, biocentric ethicist finds to be a charming land ethic might strike a native as an unfortunate and inescapable existence. The poor villager does not necessarily revel in his oneness with nature, although he may. Then again, he may wish to find a way out of poverty to enjoy the "good life" of the over-consuming western anthropocentrist.

"To my mind," Guha wrote as his final point, "deep ecology is best viewed as a radical trend within the wilderness preservation movement." It is trendy, and its proponents seem to take comfort in their deliberate alienation from mainstream neoclassical economic thought. Yet deep ecologists are not as chic and cutting edge as they believe. The movement "runs parallel to the consumer society without seriously questioning its ecological and socio-political basis." Equating environmental protection with wilderness preservation is a "distinctively American notion, borne out of a unique social and environmental history," but it is not relevant to many global environmental concerns. In Guha's opinion, a genuinely radical ecology ought to seek a synthesis of technology and lifestyles in the framework of the American political and economic systems. Otherwise, its supporters risk being marginalized—"out there" ex-hippie types who wallow in their irrelevance even as serious environmental issues go unaddressed.[47]

The Ethical Dilemma in a Nutshell

In each school of thought outlined in this chapter, environmental-ethicists seek to understand and explain humanity's relationship with nature. The anthropocentric perspective accords human beings moral worth; all other living things are instrumentally valuable if and only if they meet human needs. The biocentric perspective sees humankind's existence as only one part of a larger biosphere, and certainly not a being that can claim moral superiority over other creatures. Proponents of the positions between these extreme views struggle to balance, and not always successfully, human needs with environmental concerns.

No ecologist seriously contends that a consensus on what constitutes sustainability can be developed to the satisfaction of all parties, but perhaps the myriad ethical positions can be appreciated by examining changing views of nature throughout history. Positions are developed and nurtured at distinct historical periods and for specific contextual reasons. Therefore, Section II of this book examines the historical foundations of American environmentalism to understand how and why competing schools of thought originated.

Notes

1. Quoted in Ulrich Steger, Wouter Achterberg, Kornelis Blok, Henning Bode, Walter Frenz, Corinna Gather, Gerd Hanekamp, Dieter Imboden, Matthias Jahnke, Michael Kost, Rudi Kurz, Hans Gottfried Nutzinger, and Thomas Ziesemer, *Sustainable Development in the Energy Sector* (Berlin, Heidelberg, and New York: Springer, 2005), 55.
2. Robert M. Solow, "Sustainability: An Economist's Perspective," in *The Environmental Ethics & Policy Book*, 2nd ed., Donald VanDeVeer and Christine Pierce, editors (Belmont, Calif.: Wadsworth, 1998), 450–455.
3. For a good discussion of this issue, see, for example, Bryan G. Norton, *Toward Unity Among Environmentalists* (Oxford and New York: Oxford University Press, 1991).
4. The Übermensch is discussed *in* Friedrich Nietzsche, *Thus Spoke Zarathustra*, Graham Parkes, translator (New York and Oxford: Oxford University Press, 2005). See also Friedrich Nietzsche, *Beyond Good and Evil*, Walter Kaufmann, translator (New York: Vintage, 1966). As for the anthropocentric position, see, for example, Emily Northrop, "Normative Foundations of Introductory Economics," *The American Economist* 44, 1 (Spring 2000): 53–61; John Passmore, *Man's Responsibility for Nature: Ecological Problems and Western Traditions* (New York: Scribner, 1974). As discussed in Chapter 1, not everyone agrees that the Judeo-Christian tradition requires dominion over the earth or presents an antagonistic environmental ethic. See, for example, Calvin DeWitt, *The Environment and the Christian: What Does the New Testament Say about the Environment?* (Grand Rapids, Mich.: Baker Academic, 1991); J. Patrick Dobel, "Stewards of the Earth's Resources: A Christian Response to Ecology," *Christian Century* 94, 2 (October 12, 1977): 906–909; Peter J. Hill, "Environmental Theology: A Judeo-Christian Defense," *Journal of Markets and Morality* 3, 2 (Fall 2000): 158–172.
5. Quoted in William E. Rees, "The Ecology of Sustainable Development," *The Ecologist* 20, 1 (January/February 1990): 18. See also Athanasse Raftopoulos, "Was Cartesian Science Ever Meant to Be *A Priori*?" *Philosophy of Science* 62, 1 (March 1995): 150–160.
6. Theodore M. Porter, *Trust in Numbers: The Pursuit of Objectivity in Science and Public Life* (Princeton, N.J.: Princeton University Press, 1995), 11–21.

7. Larry Laudan, *Progress and Its Problems: Towards a Theory of Scientific Growth* (Berkeley: University of California Press, 1977). For a discussion of Diderot's famous comment, see, for example, Pat Duffy Hutcheon, *Leaving the Cave: Evolutionary Naturalism in Social-Scientific Thought* (Waterloo, Ontario, Canada: Wilfrid Laurier Press, 1996), 48.
8. Philip Kitcher, *The Advancement of Science: Science without Legend, Objectivity without Illusions* (Oxford and New York: Oxford University Press, 1993), 5.
9. Thomas S. Kuhn, *The Structure of Scientific Revolutions*, 2nd ed. (Chicago: The University of Chicago Press, 1970). The literature on Kuhn and his effect on science is voluminous. See, for example, Barry Barnes, *T. S. Kuhn and Social Sciences* (New York: Columbia University Press, 1982); Alexander Bird, "Kuhn, Naturalism, and the Positivist Legacy," *Studies in History & Philosophy of Science Part A* 35, 2 (June 2004): 337–356; Alisa Bokulich, "Heisenberg Meets Kuhn: Closed Theories and Paradigms," *Philosophy of Science* 73, 1 (January 2006): 90–107; Ipek Demir, "Incommensurabilities in the Work of Thomas Kuhn," *Studies in History & Philosophy of Science Part A* 39, 1 (March 2008): 133–142; Michael Friedman, "Ernst Cassirer and Thomas Kuhn: The Neo-Kantian Tradition in History and Philosophy of Science," *Philosophical Forum* 39, 2 (Summer 2008): 239–252; Michel Ghins, "Thomas Kuhn on the Existence of the World," *International Studies in the Philosophy of Science* 17, 3 (October 2003): 265–279; Alan W. Richardson, "Narrating the History of Reason Itself: Friedman, Kuhn, and a Constitutive *A Priori* for the Twenty-first Century," *Perspectives on Science* 10, 3 (Fall 2002): 253–274; David Weinberger, "Shift Happens," *The Chronicle Review: A Weekly Magazine of Ideas* 58, 34 (27 April 2012): B7–B9.
10. Ronald N. Giere, *Explaining Science: A Cognitive Approach* (Chicago: The University of Chicago Press, 1988), especially 4–7, 16–19, 23, 37.
11. Helen E. Longino, *Science as Social Knowledge: Values and Objectivity in Scientific Inquiry* (Princeton, N.J.: Princeton University Press, 1990), 232.
12. William Shakespeare, *The Tragedy of Hamlet: King of Denmark* (New York: Airmont Books, 1965), Act I, Scene v, 40.
13. World Commission on Environment and Development (WCED), *Our Common Future: From One Earth to One World* (Oxford and New York: Oxford University Press, 1987), 8; Wilfred Beckerman, "'Sustainable Development': Is It a Useful Concept?" in *The Environmental Ethics & Policy Book*, 2nd ed., Donald VanDeVeer and Christine Pierce, editors (Belmont, Calif.: Wadsworth, 1998), 463.
14. Paul Taylor, *Respect for Nature: A Theory of Environmental Ethics*, 25th Anniversary Edition (Princeton, N.J.: Princeton University Press, 2011), 156. See also James Sterba, "A Biocentrist Strikes Back," *Environmental Ethics* 20, 4 (Winter 1998): 361–376.
15. See, for example, Nicholas Agar, "Biocentrism and the Concept of Life," *Ethics* 108, 1 (October 1997): 147–168; Bob Pepperman Taylor, "Environmental Ethics and Public Policy," *Polity* 23, 4 (Summer 1991): 567–583.
16. William D. Nordhaus, *Managing the Global Commons: The Economics of Climate Change* (Cambridge, Mass.: MIT Press, 1994), 94. See also William D. Nordhaus, "After Kyoto: Alternative Mechanisms to Control Global Warming," *The American Economic Review* 94, 2 (May 2006): 31–34.
17. Eric Neumayer, "Global Warming: Discounting Is Not the Issue but Substitutability Is," *Energy Policy* 27, 1 (January 1999): 33–43. See also John Gowdy and Sabine O'Hara, "Weak Sustainability and Viable Technologies," *Ecological Economics* 22, 3 (September 1997): 240–242.
18. Rees, "The Ecology of Sustainable Development," 19.

19. The best book on Theodore Roosevelt's views and policies on the environment is Douglas Brinkley, *The Wilderness Warrior: Theodore Roosevelt and the Crusade for America* (New York: HarperCollins, 2009). See also Ethan Fishman, "The Quality of Theodore Roosevelt's Environmentalism," in *A Companion to Theodore Roosevelt*, Serge Ricard, editor (West Sussex, England: Wiley-Blackwell, 2011), 173–185.
20. Rees, "The Ecology of Sustainable Development," 20.
21. Herman E. Daly, "Economics, Environment, and Community," *Earth Ethics* 20, 1 (Fall 1989): 10.
22. Bryan G. Norton and Robert E. Ulanowicz, "Scale and Biodiversity Policy: A Hierarchical Approach," *Ambio: A Journal of the Human Environment* 21, 3 (May 1992): 244–249; Carol M. Rose, "Liberty, Property, Environmentalism," *Social Philosophy & Policy* 26, 2 (July 2009): 1–25.
23. Norton and Ulanowicz, "Scale and Biodiversity Policy," 246.
24. Bryan G. Norton, "Evaluating Ecosystem States: Two Competing Paradigms," *Ecological Economics* 14, 2 (August 1995): 115–116.
25. Joan Martinez-Alier, Giuseppe Munda, and John O'Neill, "Weak Comparability of Values as a Foundation for Ecological Economics," *Ecological Economics* 26, 3 (September 1998): 277–286.
26. Robert Costanza, Cutler Cleveland, and Charles Perrings, *The Development of Ecological Economics* (Cheltenham, U.K.: Edward Elgar Publishing, 1997), 45.
27. Ibid., 46–47.
28. Herman E Daly, "Allocation, Distribution, and Scale: Towards an Economics That is Efficient, Just, and Sustainable," *Ecological Economics* 6, 3 (December 1992): 186.
29. Robert Costanza, *Ecological Economics: The Science and Management of Sustainability* (New York: Columbia University Press, 1991), 3–6.
30. John B. Cobb, Jr., *Economics, Ecology and Justice* (Maryknoll, N.Y.: Orbis, 1992), 97–98.
31. C. S. Holling, "Resilience and Stability of Ecological Systems," *Annual Review of Ecology and Systematics* 4, 1 (1973): 1–23.
32. Quoted in Carl Folke, "Resilience: The Emergence of a Perspective for Social-Ecological Systems Analyses," *Global Environmental Change* 16, 3 (August 2006): 254.
33. J. Baird Callicott and Karen Mumford, "Ecological Sustainability as a Conservation Concept," *Conservation Biology* 11, 1 (February 1997): 36.
34. Kenneth Arrow, Bert Bolin, Robert Costanza, Partha Dasgupta, Carl Folke, C. S. Holling, Bengt-Owe Jansson, Simon Levin, Karl-Goran Maler, Charles Perrings, and David Pimentel, "Economic Growth, Carrying Capacity, and the Environment," *Ecological Applications* 6, 1 (February 1996): 15.
35. Carl J. Walters and Ray Hilborn, "Ecological Optimization and Adaptive Management," *Annual Review of Ecology and Systematics* 9, 1 (1978): 157. See also Ussif R. Sumaila and Carl Walters, "Intergenerational Discounting: A New Intuitive Approach," *Ecological Economics* 52, 2 (January 2005): 135–142.
36. Kai Lee, "Appraising Adaptive Management," *Ecology and Society* 3, 2 (September 1999): 14.
37. Joachim H. Spangenberg, "Will the Information Society Be Sustainable? Towards Criteria and Indicators for a Sustainable Knowledge Society," *International Journal of Sustainable Development* 1, 1/2 (November 2005): 89–90.
38. Bryan G. Norton, *Sustainability: A Philosophy of Adaptive Ecosystem Management* (Chicago: University of Chicago Press, 2005), 98–99.

39. See, for example, Katherine V. Kortenkamp and Colleen F. Moore, "Ecocentrism and Anthropocentrism: Moral Reasoning about Ecological Commons Dilemmas," *Journal of Environmental Psychology* 21, 3 (September 2001): 261–262. See also Aldo Leopold, *A Sand County Almanac; with Essays on Conservation from Round River* (New York: Ballantine Books, 1990 [1949]).
40. Quoted in Bill Devall and George Sessions, "Deep Ecology," in *The Environmental Ethics & Policy Book*, 2nd ed., Donald VanDeVeer and Christine Pierce, editors (Belmont, Calif.: Wadsworth, 1998), 222. See also Peter B. Corcoran and Eric Sievers, "Reconceptualizing Environmental Education: Five Possibilities," *Journal of Environmental Education* 25, 4 (Summer 1994): 4–8; Charles T. Rubin, *The Green Crusade: Rethinking the Roots of Environmentalism* (Lanham, Md.: Rowman & Littlefield, 1998), 179; George Sessions, "The Deep Ecology Movement: A Review," *Environmental Review: ER* 1, 2 (Summer 1987): 105–125; Arne Naess, "The Shallow and the Deep, Long-Range Ecology Movements," in *Deep Ecology for the 21st Century*, George Sessions, editor (Boston: Shambala, 1995), 151–155.
41. See, for example, J. Baird Callicott, *In Defense of the Land Ethic: Essays in Environmental Philosophy* (Albany, N.Y.: State University of New York Press, 1989), 129–148; J. Baird Callicott, "Non-Anthropocentric Value Theory and Environmental Ethics," *American Philosophical Quarterly* 21, 4 (October 1984): 299–309; and Laura Westra, *Living in Integrity: A Global Ethic to Restore a Fragmented Earth* (Lanham, Md.: Rowman & Littlefield, 1998).
42. James Lovelock, *Gaia: A New Look at Life on Earth* (New York and Oxford: Oxford University Press, 2000). See also Andrew Free and Nicholas H. Barton, "Do Evolution and Ecology Need the Gaia Hypothesis?" *Trends in Ecology and Evolution* 22, 11 (November 2007): 611–619; James W. Kirchner, "The Gaia Hypothesis: Can It Be Tested?," *Reviews of Geophysics* 27, 2 (May 1989): 223–235; James W. Kirchner, "The Gaia Hypothesis: Fact, Theory, and Wishful Thinking," *Climatic Change* 52, 4 (2002): 391–408; David M. Wilkinson, "Is Gaia Really Conventional Ecology?," *Oikos* 84, 3 (March 1999): 533–536.
43. Charles H. Eccleston and Frederic March. *Global Environmental Policy: Concepts, Principles, and Practice* (Boca Raton, Fla.: CRC Press, 2011), 30–33.
44. Daniel Quinn, *Ishmael: An Adventure of the Mind and Spirit* (New York: Bantam, 1992).
45. The quote is reprinted in John Seed, Joanna Macy, Pat Fleming, and Arne Naess, *Thinking Like a Mountain: Towards a Council of All Beings* (Gabriola Island, B.C., Canada: New Society Publishers, 1988), 43.
46. Murray Bookchin, "Will Ecology Become 'The Dismal Science'?" in *The Environmental Ethics & Policy Book*, 2nd ed., Donald VanDeVeer and Christine Pierce, editors (Belmont, Calif.: Wadsworth, 1998), 230–235.
47. Ramachandra Guha, "Radical American Environmentalism and Wilderness Preservation: A Third World Critique," in *The Environmental Ethics & Policy Book*, 2nd ed., Donald VanDeVeer and Christine Pierce, editors (Belmont, Calif.: Wadsworth, 1998), 515–522. See also Ramachandra Guha, *Environmentalism: A Global History* (New York: Longman, 2000).

II

HISTORICAL FOUNDATIONS OF AMERICAN ENVIRONMENTALISM

Chapter 4

Pre-American Attitudes about the Natural Environment

> We do not inherit the earth from our ancestors; we borrow it from our children.
>
> **—Native American Proverb**[1]

Historians typically describe the publication of Rachel Carson's classic 1962 book *Silent Spring*, the first Earth Day celebration in 1970, and the multitude of laws and regulations enacted during the 1970s and 1980s as precursors to the twenty-first century American environmental movement. Undoubtedly, these seminal milestones were important developments that influenced subsequent attitudes on nature. Accordingly, they will be described in due course. Yet the environmentalism of the late twentieth and early twenty-first centuries was the culmination of many ideas and concepts that evolved over centuries, a long chain linking present and past epochs. Environmental concerns predated the American republic as well as the advent of the Industrial Revolution, although developments after 1800 exacerbated the consequences of negative externalities. To understand how this chain of logic was forged, it is necessary to trace the evolution of attitudes about man's relationship with nature. The story begins in antiquity.

Ancient Greeks

Alfred North Whitehead once suggested that all western philosophy is one long footnote to Plato. Although this widely cited comment is an unvarnished exaggeration, it is clear that the Ancient Greeks provided the foundation for much of what followed in the western intellectual tradition. If the ancients could not provide definitive answers to profound philosophical problems, they could at least pose questions that would confound thinkers throughout the centuries. The philosophical inquiry commenced with the pre-Socratics in the fifth and sixth centuries BCE.[2]

These all-but-forgotten thinkers predated Socrates by more than a century, hence their modern designation as "pre-Socratic." They were among the first western philosophers to seek a non-mystical explanation for the ur-stuff of the world, a remarkable feat considering the prevalence of mystical cults and fantastical religious beliefs at the time. Little is known of Thales, the earliest of the pre-Socratics and a resident of Miletus, a Greek colony in Ionia, except that he reputedly predicted an eclipse that occurred in 585 BCE. This keen observer of natural phenomena sometimes is called the father of philosophy because he posited that water is the source of all things. From a later perspective, this simplistic observation can be dismissed as clumsy and amorphous, but for a thinker anxious to move beyond the religious traditions of ancient Greece, the assertion was not farfetched or absurd. Water is necessary for all forms of life to exist. In addition, because water can be found in a solid, liquid, or gaseous state, it seems to possess semi-magical properties.

Thales is revered today not so much for the content of his propositions as for his attempt to explain the world in concrete, measurable, secular terms. Unwilling to express blind faith in unseen, mystical, abstract forces, he probed the natural world through observation and critical thinking. He suggested that one thing—in this case, water—was the starting point for all other things that flowed from it; in so doing, he sought a "scientific" explanation for substances and processes that once had been considered the sole province of the gods. This concept, labeled "monism," described the diversity of all physical objects in the world as deriving from unity; that is, the multitude of objects today called "matter" originated as one thing, "water." A principle of change was implicit in this concept: Thales asserted that water is "full of gods," presumably his shorthand method of explaining how matter changes. We know from Homer that the Greeks viewed the gods as active agents motivated by the same emotions—jealously, greed, lust, anger, and love—that move human beings to act upon the world. For Thales, water filled with gods contains a natural interior force that propagates change as opposed to an external force forcing change.[3]

Despite his attempt to describe the natural world using concrete terminology and rational processes, Thales lapsed into mysticism with his assertion that water is full of gods. Unable to explain specifically how all things flowed from water, he left the task of developing a more intellectually rigorous explanation to his progeny. Fortunately, younger Milesians were equal to the task.

More is known about Anaximander, Thales's pupil in Miletus, because he wrote a book outlining his views. His contemporaries cited the book numerous times, although only one sentence has survived the centuries: "From what source things arise, to that they return of necessity when they are destroyed; for they suffer punishment and make reparation to one another for their injustice according to the order of the time." This statement initially strikes the modern reader as opaque, but Anaximander raised an important point. The world is indeed composed of ur-stuff, as Thales contended, and the world is formed from this ur-stuff. Yet Anaximander developed the concept of the undifferentiated thing—the idea of "matter," to use current terminology. He also believed that natural law determines how the properties of matter operate. The ur-stuff is "boundless," or unlimited, and earth, air, water, and fire issue from the boundless ur-stuff. This refinement of Thales's observations allowed Anaximander to distinguish between opposites such as hot and cold, day and night, and so forth.

Anticipating evolution by almost 25 centuries, Anaximander suggested that the world as human beings experience it is separated from the boundless ur-stuff through a series of eddies that form and eventually dissolve. As an example, hot and cold are properties found in the world. Heat circles cold as bark circles a tree. Applying this principle to evolutionary biology yields a crucial insight: Over long periods of time, land developed and the seas were distinguished from the land. Building on Thales's idea that life needs water to exist, Anaximander saw that in moist areas of the world, life developed. Human beings therefore developed from lower creatures, such as fish, that rose from the oceans.[4]

Anaximander's thinking represented an enormous step forward in explaining the origins of the natural world, but he left many questions unanswered. The third pre-Socratic, Anaximenes, also hailed from Miletus, and he took up the questions that Anaximander did not address. Younger than Thales and Anaximander, Anaximenes wrote a book that contemporaries often cited. As with his predecessors, Anaximenes believed that the world could be traced back to an essential ur-stuff. Yet he was perplexed by Anaximander's concept of "boundless" matter. The abstract notion suggested a world without limits, and yet the things that derive from that limitless ur-stuff were themselves limited. For change to occur, Anaximenes argued that air must be the primary stuff of life because all other things exist because of air. Although this position seemed to echo Thales's monism (except in substituting air for water), Anaximenes advanced philosophy because he identified a crucial problem: How can permanence and change simultaneously exist? This question was to influence the development of western philosophy from his day until the present.[5]

Another Ionian, Heraclitus, posited that the fundamental element is fire because the material world is always changing. Human beings are born, grow to maturity, become old, and perish. The cycle of change never ends. To explain how human beings persist as a species even as particular human beings die, he concluded that ur-stuff is not a physical property; it is an underlying pattern. This conclusion was an attempt to explain how one thing can become many things, how the ur-stuff could translate into other properties. If the "oneness" of the world is an abstract pattern, the particularity of specific matter makes sense. "Change according to the measures," in Heraclitean terms, sets the stage for the world of reality versus the world of appearance that Plato later would discuss. Heraclitus stopped short of explaining the interaction between permanence and change. Although he identified fire as the symbol of physical change based on a permanent pattern, he failed to explore the full implications of his insight.[6]

Parmenides of the Eleatic school took issue with Heraclitus, arguing that "being" either exists or it does not exist. Things that exist must have some form of permanence. The idea that reality is ever changing is therefore illusory. He asked, "How could what is perish? How could it have come to be? For if it came into being, it is not; nor is it if ever it is going to be. Thus coming into being is extinguished, and destruction unknown." This concept of static monism stressed the permanence of matter in direct contrast to Heraclitus. Permanence versus change, dynamism versus static monism, appearance versus reality—these concepts would endure across the centuries, outliving the Ancient Greeks and capturing the attention of a multitude of modern and postmodern philosophers.[7]

As the Milesians and Eleatics debated permanence and change, another Ancient Greek, Pythagoras, taught a natural philosophy that relied on the underlying pattern found in numbers. Little is known about the historical figure of Pythagoras, and most of his work has been filtered through his followers. He seems to have founded a school whose adherents expressed a quasi-religious faith in the primacy of mathematics. According to the Pythagoreans, nature consists of numerical harmonies that can be appreciated using mathematics. The Pythagorean Theorem—in a right triangle, the two sides squared equal the hypotenuse squared—is exactly the kind of numerical precision that appealed to the Pythagoreans' sense of order and balance in the universe.[8]

The desire for precision in explaining the natural world was perhaps best expressed through Atomism, a concept at odds with Parmenides as well as monism. Leucippus, Democritus, Epicurus, and Lucretius, the most celebrated of the Atomists, argued that the world is composed of atoms, or indivisible units of undifferentiated matter whirling in a void. Because they could not be seen or empirically verified, at least not using the primitive tools available to the ancients, their size, shape, and relative location could be expressed only through mathematics. By arguing that matter is comprised of smaller units, the Atomists provided a blueprint for the physical sciences that would follow centuries later.[9]

In their own lurching, tentative ways, each of the pre-Socratics sought to understand the natural world through its mechanisms and operations. They were not satisfied with the traditional explanations offered by religion or mysticism that unseen forces controlled by angry or jealous gods initiated changes in the world. They insisted on a deeper explanation for the laws of nature. Their radical, new approach formed the foundation of the western scientific tradition.[10]

When modern students of philosophy speak of the "Ancient Greeks," however, they usually do not refer to the pre-Socratics. They think of the great triad of Socrates, Plato, and Aristotle. These three monumental thinkers advanced the pre-Socratic movement away from mythos toward a more rationally based discourse, so much so that we often speak of the school of Athens as distinguished from the school of Jerusalem that represents religious thought. Socrates, Plato, and Aristotle, citizens of the Greek city-state of Athens at one time or another, synthesized the elements of pre-Socratic rationality to form a more comprehensive, coherent narrative.

Most of what the world knows of Socrates comes from the great teacher's most famous pupil, Plato. Socrates, we are told, was a gadfly renowned for his temerity and willingness to ask uncomfortable questions challenging the ruling elite of Athens. The maxim that the unexamined life is not worth living is attributed to this intriguing figure who ultimately ran afoul of the city elders. Pronounced guilty of corrupting the city's youth and defying the gods, Socrates was forced to drink hemlock as punishment for his crime. Plato and other admirers were left to carry on the Socratic tradition, and carry on they did.[11]

To bridge the gap between Heraclitus and the concept of change as well as Parmenides and the idea of permanence, Plato developed his celebrated theory of "forms," or blueprints for the world of appearance. This concept was part of his plan for drawing upon the knowledge developed by the pre-Socratics and synthesizing it into a coherent cosmological theory. The "forms," Plato explained in several dialogues, consist of unseen, nonphysical, nonspatial, and nontemporal ur-stuff that serves as the foundation for all matter. Adopting the Parmenidan concept of permanence, he viewed the forms as the unchanging, eternal, objective blueprints that underlie everything. Without the generality allowed by the existence of an underlying, invisible ur-stuff, the particularity of the world encountered by humankind through its senses—the constant changes that Heraclitus observed—could not exist. Without the forms, the universe would not be an orderly, coherent place. With the forms, a universal construct existed for the particularity of the world, whether the thing was concrete and physical, such as a horse, or an abstract concept such as justice, love, or ethics.[12]

Although Plato developed the concept of the forms in multiple dialogues, *The Timaeus* presents perhaps the clearest perspective for understanding environmentalism. The dialogue recounts a creation myth, common in many cultures, but in this case explaining how appearance and reality coexist. Timaeus, the protagonist, posits that the eternal world (reality) exists and can be understood only through reason. By contrast, the physical world (appearance) can be appreciated through sensation, which leads to the formation of opinion about the meaning of that sensation. Because changes cannot be occasioned without a fundamental, or prime, cause, the cause must be a godhead, which sometimes is translated from Greek as a "demiurge." The demiurge does not possess the characteristics of the Judeo-Christian God, but the concept involves an eternal maker, the architect that took the disorder of existence and fashioned it into a coherent universe with patterns (or forms).[13]

The divine creator relied on the four elements—fire and earth, augmented by water and air—the pre-Socratics had wrestled with in their works. Each element was necessary to create a proportional balance in the world. The demiurge fashioned that world into the form of a globe because a circle is a perfect figure without beginning or end. The circular motion of the world represents a uniform movement, hence harmony and balance. Within this world, two varieties of Sameness exist, namely indivisible sameness and divisible sameness. The world contains indivisible and divisible Differences as well as indivisible and divisible Being. The mixture of these elements leads to compound substances such as intermediate Sameness, intermediate Difference, and intermediate Being. According to Plato, this blueprint explains how permanence and change seem to exist side-by-side; they are mixtures of different substances just as colors can be mixed to form a new, distinct color.[14]

Perhaps the most confusing portion of *The Timaeus*, at least to modern sensibilities, involved the World Soul. The demiurge created matter according to what is good or fair. It created a balance among the elements of fire, earth, water, and air, which allowed for harmonious existence. Afterward, the demiurge connected the body and soul of the universe. From this perspective, the world is a living, conscious creature. According to Plato, "When he was framing the universe, he put intelligence in soul, and soul in body, that he might be the creator of a work which was by nature fairest and best. Wherefore, using the language of probability, we may say that the world became a living creature truly endowed with soul and intelligence by the providence of God."[15]

Plato's work was influential in the western tradition because it offered a compelling explanation for unseen patterns behind the everyday, mundane existence of the world without denying a human being's day-to-day experiences. Analogizing this insight to American environmentalism, the implications are that short-term decision-making can occur without rejecting its legitimacy, but long-term patterns, unseen through everyday processes, must be appreciated. The bifurcated Platonic worldview can be used to justify anthropocentric decisions while concomitantly carving out a place for ongoing, long-term biocentric patterns. The concept of a World Soul also lent support to biocentrists who later argued that Mother Earth possesses consciousness.

The philosopher Karl Popper contended that all scientific theories originate with myths.[16] If so, Aristotle, Plato's most famous student, was well situated to apply a rudimentary scientific inquiry to Platonic myth. As compelling as the theory of forms was as an explanation for the world of appearance and the world of reality, the precise nature of the concept was never clear. The process by which an unseen blueprint for matter could create matter or exercise any causal function in relation to matter was bizarre and never adequately explained. The theory collapsed into a replica of the mysticism it was designed to replace. Descartes would suffer from the same duality problem centuries later when he fumbled his way through an explanation of how mind and body could coexist when one concept required belief in the unseen while the other was rooted in the particularity of matter, of the here and now. The unseen and the seen, the thought-of and the experienced, the *a priori* and the *a posteriori*—these dualistic concepts would haunt generations of thinkers.[17]

Aristotle discounted Plato's theory of forms for precisely these reasons: Accepting the unseen world as somehow more genuine than the world encountered through the day-to-day experiences of human beings seems not only factually suspect and counterintuitive, but also downright perverse. To understand the world, Aristotle emphasized inductive reasoning. Rather than reason deductively from the general to the particular, Aristotle began with the specific and, based on observation, he reasoned to the general. A valid scientific process requires the theory to fit the facts, not vice versa. For twenty-first-century environmental scientists, the goal is to collect specimens and ruminate on their commonalities, which tell the observer something about how the world works and humanity's place in the schemata. From these common features, a diligent observer can develop a theory of general processes.[18]

The Ancient Greek contribution to the history of American environmentalism was to create a framework for understanding nature. Close observation of natural processes provides data on matter and man's relationship to his environment. Despite contradictions among the various thinkers and their doctrines, the Greeks established the tradition of questioning assumptions and testing alternative theories of matter. This intellectual pedigree ensured that western thinkers would not be content to accept conventional wisdom. They would strive to refine their understanding of the world through deductive and inductive processes. Logos would triumph over mythos.[19]

Biblical Tradition

Historians sometimes speak of the division between Athens and Jerusalem in the western intellectual tradition. While the Ancient Greeks established a framework for rational discourse, Christianity emphasized the need for faith in a higher spiritual being. The division was manifested in myriad ways across the broad canvas of philosophy and theology—appearance versus reality; reason versus faith; materiality versus spirituality; Caesar versus God; earth versus heaven.[20]

Recall from Chapter 1 that the Christian Bible recounted a creation story where God created lower animals to be tame and useful to human beings. After human beings appeared, God instructed His highest creation to exercise dominion over all other creatures of the kingdom. This arrangement was so pleasing that God rested on the seventh day of creation. Later, God sent the lower animals to man to be named.

The Garden of Eden (Figure 4.1), man's first home, was an idyllic land where scarcity did not exist. Nature was benign. All human needs were satisfied as long as the first two human beings, Adam and Eve, did not eat the fruit from the tree of knowledge of good and evil. Even at this early date, the Christian tradition recognized the primacy of human needs. The earth's bounty was available and plentiful for *Homo sapiens*. The first view of a specific non-human animal in this oasis was a

Figure 4.1 For many Judeo-Christians, the Garden of Eden was an idyllic land where scarcity did not exist. (Courtesy of the Library of Congress)

serpent, an embodiment of Satan, the wicked spirit that infected men's hearts and tempted humans to sin.

After Eve listened to the evil serpent and consumed the forbidden fruit, enticing her husband to join her in the sin, humanity faced an angry deity. God drove Adam and Eve from the Garden of Eden and made them live in an inhospitable landscape overgrown with thorns and thistles. He punished the serpent by making him crawl in the dust. A comfortable existence would no longer be assured, even for the highest creature on earth. Adam faced scarcity for the first time; he was forced to labor to raise enough food to sustain himself and his family.

The Judeo-Christian God was vengeful, but He was not merciless. In addition to promising a path for redemption from earthly sin, God made coats for Adam and Eve from the skins of animals. He allowed them to procreate, blessing them with two sons, Cain and Abel. Both sons possessed wicked hearts like their parents, but the younger son, Abel, pleased God by offering up a lamb for sacrifice. Cain refused to repent; his offering consisted of grain taken from the field. By not offering up a lamb or another innocent creature in tribute, Cain angered God. Rather than repent, a petulant Cain, upset that God was angry with him, killed his brother in a jealous fit.[21]

This well-known tale from the Book of Genesis illustrates the standard view of environmentalism as explained by Lynn White, Jr. In every instance when nature and lower animals are referenced, they are given life only in the context of the human story. Animals gain a name, but only when human beings deign to anoint them with one. An innocent lamb has value because he is slaughtered in a deliberate sacrifice to appease God and redeem human beings. The Garden of Eden is desirable because it meets human needs—not because the garden possesses intrinsic value. When human beings misbehave, they are denied access to the plentiful storehouse and forced to labor for sustenance. Even God Himself is characterized as possessing human traits. He rests. He is pleased. He is angry. He is resentful.[22]

For all of the divisions between the competing traditions of Athens and Jerusalem, they share a common view of humankind as the unit of analysis. In the Athenian tradition, reason requires a being with sufficient brainpower to work through complex logical puzzles to determine the good. Jerusalem's followers look to God for salvation, but they understand that even a sinful, imperfect, fallen being stands atop the hierarchy of living creatures. Both traditions adopt an anthropocentric worldview even if some strains of Greek and Judeo-Christian thought can be interpreted as heralding a biocentric outlook.[23]

Europeans in the Middle Ages and the Renaissance

The fall of the Roman Empire in the fifth century CE was the end of one world and the beginning of another. The break between the classical world and a new order was traumatic in every sense of the word. For all of its defects, the Roman world had been orderly and predicated on centuries of learning in the Greco-Roman tradition. To be overtaken by hordes of Visigoths and barbarian tribes that depended on a simpler dynamic to govern a complex society was an affront to men of learning and refinement. The centuries that followed have been called the Dark Ages precisely because the medieval mind rejected much of what had been accomplished and archived in the great libraries of antiquity. Centuries would pass before the Renaissance ushered in a second phase of intellectual refinement.

It is not strictly accurate to say that advanced learning died during the thousand years between the fall of Rome and the rebirth of the arts in Western Europe, but clearly, the collapse of a recognized, well-defended, authoritative institutional structure drove learned men underground. Where government once had been the prime innovator of technological advances and a patron of the arts, it was no more. Individuals who sought to figure their equations, write their poems, or play their songs were left to their own devices.

One major institution arose in the West during those long years to provide, among other things, solace to men of learning. In one of history's supreme ironies, Christianity, the fringe cult that once had seemed so threatening to the Romans, became the means by which the western intellectual tradition survived in the centuries to come. With the fall of Rome, the Catholic Church transformed itself into one of the few remaining institutional vestiges of a civilization lost to most citizens. The works of the Ancient Greek and Roman thinkers were consigned to the ash heap of history, eclipsed for hundreds of years by an oral tradition of myths that had so incensed rational thinkers. The masses were not educated and were governed, to the extent they were governed at all, by small enclaves that depended on family and tradition as the unifying factors. The city-state was dead, and the nation-state had not yet been born.[24]

The evolution of the Catholic Church into an instrument of earthly power and progress was hardly a seamless transition. Early church leaders, still wary and suspicious from the weight of governmental persecution, hesitated to debase themselves with worldly matters. Rome fell, they contended, because the citizenry rotted from within; wickedness infected the hearts of men who engaged in all manner of debauchery and decadence. If Christians lowered their gaze to the city of men in lieu of raising their eyes up toward the city of God, they risked undermining the integrity of the religion. It was only a matter of time before they, too, would suffer from the same hubris that had toppled proud, arrogant Caesar. Had Constantine the Great not converted to Christianity during the fourth century, a rapprochement might have been beyond reach. His conversion initiated an incremental process of acceptance among differing classes of men concerned with earthly affairs and heavenly beatitudes.[25]

It was left to St. Augustine of Hippo, the great theologian and philosopher of the late fourth and early fifth centuries, to wrestle with reconciling the traditions of Athens and Jerusalem. These years were uncertain for the Christian religion. Despite its spread, Christianity might yet sputter and fail as so many mystical cults had risen and fallen in the medieval era. A towering intellect of his or any age, Augustine understood the magnitude of his task. Never a reticent fellow, he spent much of his career exploring the uneasy relationship between earthly and heavenly duties. In the final analysis, he reluctantly admitted that the Christian church must not be entirely removed from politics. The concept of a religious institution engaging in human governance was relatively novel, but in time the church would become a powerful political force that rivaled European monarchies in power and influence. In the century following Augustine's death, the papacy began its long ascendancy until it grew powerful enough to challenge the political legitimacy of all but the most stalwart of earthly monarchs.[26]

As the church grew in power and influence, Christianity's emphasis on man's dominion over all creatures, human and non-human, increased, assuming that one accepts the standard narrative. The church procured large tracts of land and instructed followers to clear a path for elaborate cathedrals reaching to the sky. Woe to man or beast that dared to obstruct the progress of the righteous. The Crusades were Christian soldiers' way of taking up arms against supposed obstructionists who refused to recognize the dominion of the church over the citizenry. The Catholic Church eventually became so strong and invested in worldly political matters that popes engaged in mischief that mimicked the intrigue surrounding long-dead Roman emperors.[27]

By the thirteenth century, St. Francis of Assisi felt compelled to plead for a return to the monastic lifestyle of an ascetic. In its early days, Christianity had been a simple religion calling on the poor and downtrodden to accept Christ the savior into their lives. In the ensuing years, the church had become a large bureaucracy, an edifice so high and mighty that the peasants, of whom there were multitudes during the Dark Ages, were increasingly closed off from their God. Francis yearned to strip away the accoutrements and return to the early days. Despite this potentially heretical notion, he was not viewed as a threatening figure. His simple, humble appeal to common people pleased the church so greatly that Pope Innocent III endorsed the Franciscan order in 1210. Acceptance by the church fathers was by no means certain. Francis might have been denounced as a dangerously unbalanced iconoclast, defrocked, and driven into exile or executed. That he was allowed to build a monastic order with the blessing of the church was no mean accomplishment.

What survive of the Franciscan order are tales of the saint's gentle nature and his love of animals. Here was a model for western man's benevolence toward nonhuman creatures. Among the more famous *Fioretti* (literally, "little flowers"), a series of legends and folk tales, is the story of Francis traveling with companions on a road lined with trees where a flock of birds nested in the limbs. Legend holds that Francis turned to his companions and said, "Wait for me while I go preach to my sisters the birds." So soothing was his voice that the creatures descended from the branches and surrounded him as he spoke. Not a single bird flew away.

Another often-told vignette concerns a vicious wolf terrorizing the inhabitants of Gubbio, a small city where Francis resided in Italy. When he learned the rough beast had devoured men and animals, Francis departed in search of the animal in the company of civic-minded townsfolk. Eventually, fear got the better of his companions and Francis was left alone to pursue his quarry. When man and beast found themselves face to face, legend has it the man of God made the sign of the cross and instructed the wolf to come to him. The animal placidly approached and collapsed at Francis's feet, whereupon the holy man told him, "Brother Wolf, you do much harm in these parts and you have done great evil. All these people accuse you and curse you.... But brother wolf, I would like to make peace between you and the people."

Francis led the wolf into Gubbio and presented the animal to the citizens. Explaining that the creature of God possessed no evil intent—he had been hungry—Francis reached an agreement with all parties. The wolf would not attack shepherds or their flocks and, in return, the townspeople would feed the wolf. Dogs that roamed through the town would not harass the wolf nor would the wolf threaten them. To seal the compromise, Francis blessed the wolf.

These myths reinforce the saint's image as a man who preached that God recognized the moral worth of all His creatures as well as man's duty to protect nature from despoliation. Rather than emphasize dominion through a heavy hand, Francis viewed the church's responsibility through a kind heart. God intended for human beings and lower animals to coexist and praise His name. So great was Francis's love of living creatures that on his deathbed he thanked the donkey that had carried him on his journeys. After the kind man died, the donkey wept.

Gentle St. Francis remains the quintessential environmentalist of the Middle Ages, the figure on whom nature lovers can focus with a sigh—"if only, if only." His was a lone voice, an exception to the trend toward brutal western attitudes on nature. Within the Judeo-Christian tradition, most men of the cloth preferred to consider humankind's sin and redemption in lieu of pondering the puzzles involving other forms of life. Outside of the church tradition, persons forced to eke out a subsistence living in the muck of the feudal system had little time or patience for abstract considerations. Similarly, their political leaders were practical men who looked on a vicious, dichotomous world—conquer or be conquered. To analogize from Machiavelli, nature, like virtue, is a weak woman who will be ravished by stronger parties that care little for her fine qualities. The march of human progress under force of arms wielded by the papacy and European monarchs would permit no regard for abstract niceties on protecting nature and her creatures.[28]

With the dawn of a new epoch, Europeans established nation-states headed by a monarch that sought to prop up the regime through elaborate fortifications and vast armies. Taking a lesson from the Crusades, a king could ensure longevity through financial stability, which could be maximized by entering into lucrative trade agreements with neighboring countries or engaging in territorial conquests. Over time, the kings grew rich, their empires grew strong, and their sense of entitlement grew large. What better way to stave off internal and external dissent than to unite the kingdom behind foreign adventures in search of land and riches? Thus did the European experiment in colonization begin as a means of perpetuating earthly rule—if not in perpetuity, perhaps for the length of the monarch's life and the lives of his immediate progeny.

As Europeans searched the known world for commercial opportunities, some enterprising souls looked to the horizon and dreamed of untamed lands that might offer up untold riches. The prospect seemed to be limited only by the imagination of the would-be adventurer. European governments eventually invested in entrepreneurial quests to discover a shorter route to the Orient in support of the spice trade and to subdue brown and yellow peoples around the globe.[29]

With the increasing confidence and sophistication of earthly governments at the dawn of the Renaissance beginning in the fourteenth and fifteenth centuries, the church found itself assailed from all quarters. European monarchs quarreled with popes about which institution—church or state—should dominate public affairs. It was an argument, violent at times, that would extend for centuries. By the early sixteenth century, the Catholic Church faced an even more disturbing challenge, namely dissension within the ranks. When Martin Luther nailed the ninety-five theses to the church door in Wittenberg, Germany, in 1517, he ushered in the Protestant Reformation that undermined the foundations of Catholicism unlike any threats the church had faced in a millennium. Luther had been incensed by the inaccessibility of the church to the masses. The Catholic focus on good works, he believed, did not sufficiently accentuate the need for faith. Luther threatened the legitimacy of the Catholic Church because he argued that individuals could have a direct relationship with God through the Bible in lieu of intercession from church leaders. These Protestants were a dangerous lot, and more was yet to come.[30]

In the sixteenth century, John Calvin set forth a doctrine that bore his name. Calvinism presupposed that human beings were tainted with original sin, yet not all hope was lost. Through God's grace, some individuals were predestined for forgiveness and redemption, although it was impossible to identify the chosen few. The faithful were well advised to live a life worthy of redemption by industriously working to improve their lot in life. Calvin's suggestion that idleness is the devil's playground fit perfectly with the European quest to export its civilization into other lands. Coupled with technological developments in weaponry and transportation as well as the beginning of mercantilism—an economic system that replaced feudalism and gradually gave way to capitalism—the new openness in religion eventually ensured that the Anglo-American culture would extend into far-flung corners of the globe.[31]

In his book *Gold and Gold: Britain, America & the Making of the Modern World*, Walter Russell Mead explained the conditions that led to the ascendancy of the English-speaking world following the Glorious Revolution of 1688. In his view, western Europeans, especially the English, bested competitors owing to the presence of freedom of thought, expression, and action present in English society, features that were more prevalent in England than in other countries. The English also recognized the advantages of exporting their political and economic systems so that commerce could expand among trading partners. Having constructed a global system of trade and commerce, the English invited other countries into the order, cutting off access to the global system if a would-be participant declared war. The English discovered that cultivating trade partners was ultimately more profitable than engaging in warfare. To that end, they promoted the development of liberal institutions in the countries where they conducted business to ensure a common understanding of the rules of trade.

This "maritime order," as Mead labeled it, arose in the English-speaking world owing to a variety of favorable conditions that were not replicated elsewhere. England, he observed, is not too large, such as Russia, and not too small, such as Holland. It is located not too close to continental Europe, such as France and Germany, or too far away, such as countries in Asia or North Africa. The accidents of geography proved to be historically advantageous. The Reformation in Britain was not too hot, such as in Germany, where the Reformation triggered internecine conflicts, or too cold, such as in Italy, where the counter-Reformation largely suppressed freedom of thought. In later centuries, the United States embraced the English system and improved on it. Blessed with abundant natural resources and a large land mass, committed to an open society as well as a dynamic political and economic system, the new nation adapted the maritime order to its unique circumstances.

According to the Whig interpretation of history, England was a land where the "union of order and liberty" occurred. Liberal democratic institutions and processes—a long tradition of the "rights of Englishmen" dating to the signing of the Magna Carta in the thirteenth century, the development of English common law, the longevity of the English constitution, and the rise of the Anglican Church in the wake of Luther's Protestant Reformation—led to a dynamic and yet ordered society. English colonies enjoyed the benefits of the maritime system as well.[32]

The system was successful enough that its champions, having established order and liberty at home, could seek out new opportunities abroad. Following the abortive attempts of Spanish conquistadors to subdue indigenous peoples during the sixteenth century, English explorers finally established an ongoing colonial enterprise in North America in the following centuries. When they arrived on foreign shores, they encountered a culture that was alien to the English tradition. In time, the natives would be shunted to one side, victimized repeatedly through a series of episodes that cast indelible shame on American history, but in the early years, the North American tribes and their European visitors attempted to reach an uneasy accommodation.[33]

Native Americans

Ah, the Native Americans of the North American continent—has any group been so simultaneously revered and maligned? On one hand, they have been praised as the original American environmentalists—peace-loving, gentle, cognizant of the human footprint on the land, careful to preserve the fragile balance between human beings, lower animals, and the earth. The standard myth portrays wise, benevolent Indians as stewards of the Garden of Eden. To the everlasting detriment of Mother Earth, these environmentally conscious peoples eventually succumbed to the superior firepower and chicanery of destructive Europeans who transported death and ethnic hatred to the New World. On the other hand, when frontier children played cowboys and Indians, invariably the Indians were cast as bad guys. The natives were primitive savages who preyed on helpless women and children, obstreperous malcontents that refused to accept the benefits of progress. They were childlike and docile when isolated in their villages, but ferocious and bloodthirsty when under the influence of alcohol or while slaying white pioneers in the name of a primitive, pagan religion. Tales abounded of merciless natives on the warpath, desirous of scalping white men, abducting and enslaving their children, and gang-raping their women. Perhaps it is the fate of aboriginal peoples to serve as a *tabula rasa* for the hopes, fears, myths, and rationalizations of advanced societies. If so, the plight of Native Americans was predictable when compared with European encounters with the original inhabitants of other continents.

Setting aside the myths to understand Native American environmental attitudes is no easy matter. The evidence is spotty and not always reliable, but it appears that Native Americans during the time of Christopher Columbus's visit were advanced in numbers as well as sophistication. They had cleared forestland, cultivated crops, built roads, and established a flourishing commercial trade when white Europeans first encountered the great tribes. They were diverse, with some tribes living in one place as hunter-gatherers while plains Indians lived as hunters pursuing the bison and other animals across the vast expanse of the interior prairies. Numbers from this era are highly suspect, but scholars place them at between 40 million and 100 million Native Americans at the end of the fifteenth century, the figures representing the cumulative effect of 15,000 years of rich cultural history.[34]

Archaeologists and anthropologists have excavated Indian mounds and other indicia of native lives dating from 500 CE to 1500 CE and found evidence of the destructive effects that seem to accompany all human settlements. Forestlands were cleared to make room for human dwellings during this period. Soils were depleted of precious nutrients owing to overplanting as well as poor agricultural practices. Some species of animals were damaged because of overhunting, although it seems unlikely that Native Americans drove any animal populations to extinction. One especially repellant practice, according to some early accounts written by white settlers, was to drive herds of bison toward a precipice and wait until the hapless beasts crashed into a huge pile of carcasses in the ravine below. Although typically Native Americans used every part of the animal for food and clothing, such mass slaughter seems counterintuitive until one recognizes the religious rituals involved in the practice. The spirits of the departed creatures were thought to be responsible for the limitless abundance represented by the bison. Even in the face of these wasteful practices, the scale of negative environmental effects triggered by Native Americans did not approach the scale found in western European societies. It would be a mistake to insist that Indians left no footprint in their settlements, but the footprint should not be exaggerated.[35]

Native American life did not remain static over time. During the 250 years after Columbus's arrival in the New World, native populations declined precipitously. Estimates suggest that by 1750, the North American wilderness was more pristine than it had been when the conquistadores arrived owing to a dramatic loss of population. In some estimations, 90 percent of the indigenous peoples perished from 1500 until 1750, mostly from diseases spread by white Europeans, yet only 1.3 million Europeans and their slaves inhabited the continent at the time. The white European population would not deliberately displace Native Americans for many decades.[36]

As for the environment, Native Americans were not as technologically advanced as their Europeans counterparts with one exception. When Europeans, beginning with Columbus, transported horses to North America and the animals escaped and multiplied, Indians learned to capture and domesticate the useful animals. Riding on horseback significantly improved a man's ability to travel and hunt. With this new development, Native Americans compensated for their declining numbers by improving the efficiency with which they pursued their prey.[37]

Despite this advance in transportation and hunting, natives lacked the means and the will to wreak havoc on pristine lands to the same extent as Europeans.

Circumstances forced Native Americans into a practical conservation alien to whites hailing from London, Paris, Madrid, and Lisbon. Because they depended on forests for timber, soil for agriculture, and animals for food, Native Americans developed a culture that preserved the resources necessary to sustain life. They made a virtue of necessity. Unlike the standard narrative of the Judeo-Christian tradition, which stressed man's dominion over nature and the fear of Paganism presumed to exist in the wild, and the western European tradition of moving man away from a state of nature into an artificially constructed city that minimized contacts with pristine lands, Native American beliefs promoted harmony with nature (Figure 4.2).[38]

Religious beliefs and cultural practices were born from a deep respect for the land and animals. People revere that which sustains life. Some tribes subscribed to animism, the idea that the spirituality of nature can be found in all things. Native Americans followed a philosophical path far removed from Plato's theory of forms, or Christianity's division between heaven and earth. Animists conceived of a monistic metaphysics somewhat akin to the pre-Socratics' beliefs. For an animist, the world is not divided into an unseen, non-temporal world of reality—whether it is a demiurge, an omnipotent God, or an otherworldly heaven—and the physical world of the here and now. Animists believe that nature infuses all things. When a person picks up a rock, cuts down a tree, kills an animal for food, or encounters a rising brook, he encounters the spirituality of nature. Each thing, whether it is animate or inanimate, has a soul.

One reason biocentrists tout Native Americans as the prototypical environmentalists is because the spirituality found in animism is reminiscent of the views of deep ecologists who insist on extending the concept of moral worth beyond human beings. If spirituality can be found in all things, all things possess moral standing. The Native American philosophy of acting as one with nature contravenes the western tradition of grinding nature and her animals under the boot heels of European progress. Neoclassical economics, Cartesian/Newtonian science, and the brutal march of the Industrial Revolution across the centuries of the Enlightenment are brought to a halt, as this new notion of progress requires human beings to take the long view of their actions and develop respect for the natural environment.[39]

Conclusion: The March of Anthropocentric Progress

Students of history look to the past for clues as to how the world arrived at its current position. They gaze through a dark glass to understand the march of anthropocentric progress through the epochs of humankind's development. What are they to say of the trek? If human beings, especially twenty-first-century Americans, are the Takers that Daniel Quinn described in *Ishmael*, which part of the historical record can serve as an object lesson on the path not taken?

Figure 4.2 Native Americans sometimes are viewed as the original American environmentalists. (Courtesy of the Library of Congress)

History is reluctant to yield helpful answers. People of ages past faced challenges unfathomable to denizens of a later time. The pre-American attitudes on nature discussed in this chapter, for example, suggest an awareness of nature and a curiosity about the relationship between human beings and the world. Yet citizens of antiquity had not passed through the Industrial Revolution of the eighteenth through twenty-first centuries. They were less concerned about fouling the land, water, and air than they were in fighting natural disasters, diseases, and their fellow man. The roots of western anthropocentrism were visible as early as the pre-Socratic period of Greek history and extended through the biblical tradition, the Middle Ages, and the Renaissance. A solitary figure such as St. Francis of Assisi occasionally raised a discordant note amid a symphony of anthropocentrism, but an overriding fear of cacophonous nature and its chaotic discord drowned out shrill, dissenting voices.

Modern environmentalists point with pride to the animism of some Native American tribes of North America as reflective of a nascent biocentrism. It is undoubtedly true that these indigenous peoples were sympathetic to environmental concerns compared with their European brethren. Yet it would be a mistake to examine these indicia through a prism of modern ecological sensibilities. Historians caution against presentism, the tendency to apply modern customs and values to past events. Because persons in ages gone by lived with different societal customs, values, information, technology, and awareness of the natural world, their perspectives were radically different from our own. The mythic Indian was not a budding environmentalist in the same sense as a postmodern environmentalist who consciously champions public policy initiatives designed to reduce human destructiveness. Native Americans crafted a lifestyle that suited their position in time and space. This observation in no way denigrates the Native American respect for nature or disqualifies these peoples as potential models for environmentally conscious Americans, but it suggests that romanticizing the tribes for adopting a lifestyle that fit their situation imbues them with a sense of purpose they did not possess.

Twenty-first-century ecologists decry the shortsightedness of modern inhabitants in planning for the future because the typical two- to four-year cycle of economic and political decision-making ignores the long-term interests of the environment. They implore policymakers to adopt long-range environmental plans that incorporate a robust concept of sustainability into their calculations, and they express bewilderment at the lack of environmental awareness among the current generation of Americans. Yet the origins of the traditional anthropocentric perspective are readily apparent. Most human beings throughout time have focused on present difficulties and the need to eke out a living with little thought for the extended future. A trek through ancient history reminds us that human beings typically adopt a short-term, human-centered view of nature. Native Americans also were not immune to human needs—even if the fulfillment of those needs occasionally damaged the natural environment.

Homo sapiens has been on an anthropocentric march for so long that few conscientious souls are aware of the historical roots of the ecological crisis, as Lynn White,

Jr., labeled it. Ancient history reveals only a dim awareness of the natural environment among western man. It would be left to a later epoch to wrestle with the destructive effects of civilization, and even then, the full scope of the issue would remain shrouded in mystery and fraught with contention. Nonetheless, a few early environmentalists raised their voices. As Europeans moved into the Enlightenment and gradually populated the American continent during the seventeenth and eighteenth centuries, a more sophisticated, nuanced view of nature and man's relationship with it developed. The evolution of that perspective is the subject of Chapter 5.

Notes

1. Quoted in J. Rainsnow, *The Message of Rainsnow: A Spiritual and Cultural Vision for Beginning to Save the Earth* (Lincoln, Neb.: Writers Club Press, 2002), 27.
2. W. T. Jones, *The Classical Mind*, 2nd ed. (New York: Harcourt Brace Jovanovich, 1970), 108–109; Stuart Rosenbaum, "How To Be a Graceful Footnote to Plato," *Southwest Philosophy Review* 25, 1 (January 2009): 1–13.
3. Frank Egerton, "A History of the Ecological Sciences: Early Greek Origins," *Bulletin of the Ecological Society of America* 82, 1 (January 2001): 94; Curtis N. Runnels, "Environmental Degradation in Ancient Greece," *Scientific American* 272, 3 (March 1995): 96–99.
4. C. Joachim Claussen, "Anaximander and Anaximenes: The Earliest Greek Theories of Change?" *Phronesis* 22, 2 (1977): 89–102; Egerton, "A History of the Ecological Sciences: Early Greek Origins," 94; Jones, *The Classical Mind*, 11–12.
5. Claussen, "Anaximander and Anaximenes," 89–102; Jones, *The Classical Mind*, 12–14.
6. Helen Morris Cartwright, "Heraclitus and the Bath Water," *The Philosophical Review* 74, 4 (October 1965): 466–485; Charles H. Kahn, *The Art and Thought of Heraclitus: An Edition of the Fragments with Translation and Commentary* (New York and Cambridge, U.K.: Cambridge University Press, 1979).
7. Quoted in Jones, *The Classical Mind*, 21–24. See also Patricia Curd, *The Legacy of Parmenides: Eleatic Monism and Later Presocratic Thought* (Princeton, N.J.: Princeton University Press, 1998).
8. Jones, *The Classical Mind*, 31–39. See also Charles H. Kahn, *Pythagoras and the Pythagoreans: A Brief History* (Indianapolis: Hackett Publishing, 2001).
9. Erwin Schrodinger, *Nature and the Greeks and Science and Humanism* (New York and Cambridge, U.K.: Cambridge University Press, Canto edition, 1996), 50–52.
10. See, for example, Bruce Scofield, "Gaia: The Living Earth—2,500 Years of Precedents in Natural Science and Philosophy," in *Scientists Debate Gaia: The Next Century*, Stephen Henry Schneider, James R. Miller, Eileen Crist, and Penelope J. Brown, editors (Cambridge, Mass.: MIT Press, 2004), 151–152.
11. Jones, *The Classical Mind*, 108–118; Gregory Vlastos, "The Historical Socrates and Athenian Democracy," *Political Theory* 11, 4 (November 1983): 495–516.

12. The literature on the Theory of Forms is voluminous. See, for example, Jones, *The Classical Mind*, 124–146; K. W. Mills, "Some Aspects of Plato's Theory of Forms: *Timaeus* 49 c ff," *Phronesis* 13, 2 (1968): 145–170; A. K. Rogers, "Plato's Theory of Forms," *The Philosophical Review* 45, 1 (January 1936): 61–78; Schrodinger, *Nature and the Greeks and Science and Humanism*, 27.
13. For a richer discussion of the demiurge, see, for example, Francis M. Cornford, *Plato's Cosmology: The Timaeus of Plato* (Indianapolis: Hackett Publishing, 1997 [1935]); Glenn R. Morrow, "The Demiurge in Politics: The *Timaeus* and The *Laws*," *Proceedings and Addresses of the American Philosophical Association* 27 (1953–1954): 5–23; Jaroslav Pelikan, *What Has Athens to Do with Jerusalem? Timaeus and Genesis in Counterpoint* (Ann Arbor: University of Michigan Press, 1997), 39, 711.
14. Mills, "Some Aspects of Plato's Theory of Forms: *Timaeus* 49 c ff," 145–170; Alfred North Whitehead, *The Concept of Nature: Tarner Lectures Delivered in Trinity College, November 1919* (New York and Cambridge, U.K.: Cambridge University Press, 1920), 17–18.
15. Plato, *The Timaeus*, Benjamin Jowett, translator (Teddington, U.K.: Echo, 2006), 16–17. The World Soul sounds remarkably similar to the Gaia concept. See, for example, Gabriela Roxana Carone, "Plato and the Environment," *Environmental Ethics* 20, 2 (Summer 1998): 115–133; Cornford, *Plato's Cosmology*, 57–97; Scofield, "Gaia," 152.
16. Karl Popper, *The Myth of the Framework: In Defense of Science and Rationality* (New York and London: Routledge, 1994), 41–43.
17. Mario Bunge, *Matter and Mind: A Philosophical Inquiry* (London and New York: Springer, 2010), 143–157.
18. Gail Fine, *On Ideas: Aristotle's Criticism of Plato's Theory of Forms*, 2nd ed. (Oxford and New York: Oxford University Press, 2004); Jones, *The Classical Mind*, 218–220.
19. An extended discussion of ecology in the Ancient Greek world is beyond the scope of this chapter, but many sources exist. See, for example, Warwick Fox, "Deep Ecology and Virtue Ethics," *Philosophy Now* 26, 1 (April/May 2000): 21–23; Trish Glazebrook, "Art or Nature? Aristotle, Restoration Ecology, and Flowforms," *Ethics and the Environment* 8, 1 (Spring 2003): 22–36; J. Donald Hughes, "Ecology in Ancient Greece," *Inquiry: An Interdisciplinary Journal of Philosophy* 18, 2 (1975): 115–125; Robert Sallares, *The Ecology of the Ancient Greek World* (Ithaca, N.Y.: Cornell University Press, 1991).
20. Pelikan, *What Has Athens to Do with Jerusalem?* 2–3.
21. See the first five chapters of Genesis as well as Charles Foster, *The Story of the Bible* (Philadelphia: A. J. Holman Company, 1911), 9–15.
22. Lynn White, Jr., "The Historical Roots of Our Ecological Crisis," *Science* 155, 3767 (March 10, 1967): 1203–1207. For a contrary view, namely that Christianity and ecology are compatible, see the discussion in Chapter 1. See also Steven Bouma-Prediger, *For the Beauty of the Earth: A Christian Vision for Creation Care*, 2nd ed. (Grand Rapids, Mich.: Baker Academic, 2010); Jeanne Kay, "Human Dominion Over Nature in the Hebrew Bible," *Annals of the Association of American Geographers* 79, 2 (June 1989): 214–232. The distinction seems to be in the types of arguments advanced and the different interpretations of Christian ecology. White advanced a descriptive argument; he believed that Christianity historically was interpreted as a justification for earthly dominion. Bouma-Prediger and Kay set forth normative arguments. They contended that Christianity *ought* to be interpreted as an admonition by God to humanity to practice wise environmental stewardship. Moreover, White's

article appeared in 1967, before the modern environmental movement sprang into full flower, while the later works of Christian apologists appeared after the movement had gained momentum.

23. See, for example, Nina Rosenstand, "Everyone Needs a Stone: Alternative Views of Nature," in *The Environmental Ethics & Policy Book*, 2nd ed., Donald VanDeVeer and Christine Pierce, editors (Belmont, Calif.: Wadsworth, 1998), 59–61.
24. W. T. Jones, *The Medieval Mind*, 2nd ed. (New York: Harcourt Brace Jovanovich, 1969), 139–142; Theodore E. Mommsen, "Petrarch's Conception of the 'Dark Ages,'" *Speculum: A Journal of Mediaeval Studies* 17, 2 (April 1942): 226–242.
25. Peter Brown, *The Rise of Western Christendom: Triumph and Diversity AD 200-1000* (Oxford, U.K.: B. H. Blackwell, 1996); Jones, *The Medieval Mind,* 142–143; Ramsay MacMullen, *Christianizing the Roman Empire (AD 100–400)* (New Haven, Conn.: Yale University Press, 1984).
26. Herbert A. Deane, *The Political and Social Ideas of St. Augustine* (New York: Columbia University Press, 1963), 172–212; John von Heyking, *Augustine and Politics as Longing in the World* (Columbia: University of Missouri Press, 2001), 51–59.
27. The literature on this point is voluminous. See, for example, Francis Oakley, *The Western Church in the Later Middle Ages* (Ithaca, N.Y.: Cornell University Press, 1979), 15–22; John Van Engen, "The Christian Middle Ages as an Historiographical Problem," *The American Historical Review* 91, 3 (June 1986): 519–522.
28. Roger D. Sorrell, *St. Francis of Assisi and Nature: Tradition and Innovation in Western Christian Attitudes toward the Environment* (Oxford and New York: Oxford University Press, 1988), 7, 58, 133.
29. Ferdnand Braudel, *The Wheels of Commerce* (Berkeley and Los Angeles: University of California Press, 1992), 443–452; Herbert Luthy, "Colonization and the Making of Mankind," *The Journal of Economic History* 21, 4 (December 1961): 483–495.
30. Martin Brecht, *Martin Luther: His Road to Reformation, 1483–1521* (Kitchener, Ontario, Canada: Fortress Press, 1985), 190–202; Kenneth Clark, *Civilization* (New York: Harper & Row, 1969), 157–161.
31. Philip S. Gorski, *The Disciplinary Revolution: Calvinism and the Rise of the State in Early Modern Europe* (Chicago and London: University of Chicago Press, 2003), 20–23.
32. Walter Russell Mead, *God and Gold: Britain, America, and the Making of the Modern World* (New York: Knopf, 2007).
33. See, for example, Irving A. Leonard, *Books of the Brave: Being an Account of Books and of Men in the Spanish Conquest and Settlement of the Sixteenth-Century New World* (Cambridge, Mass.: Harvard University Press, 1992).
34. William M. Denevan, "The Pristine Myth: The Landscape of the Americas in 1492," in *American Environmental History*, Louis S. Warren, editor (Oxford, U.K.: B. H. Blackwell, 2004), 5–7; Benjamin Kline, *First Along the River*, 13–14.
35. Paul A. Delcourt and Hazel R. Delcourt, *Prehistoric Native Americans and Ecological Change: Human Ecosystems in Eastern North America Since the Pleistocene* (New York and Cambridge, U.K.: Cambridge University Press, 2004), 74–86; Dan Flores, "Essay: The Great Plains 'Wilderness' as a Human-Shaped Environment," *Great Plains Research* 9, 2 (Fall 1999): 343–355.
36. Denevan, "The Pristine Myth," 6–7; Michael Williams, "Dark Ages and Dark Areas: Global Deforestation in the Deep Past," *Journal of Historical Geography* 26, 1 (January 2000): 28–46.
37. Kline, *First Along the River*, 16.

38. Delcourt and Delcourt, *Prehistoric Native Americans and Ecological Change*, 3–7; Kline, *First Along the River*, 16–17.
39. Kline, *First Along the River*, 14–15; Christopher Vecsey, "American Indian Environmental Religions," in *American Indian Environments: Ecological Issues in Native American History*, Christopher Vecsey and Robert W. Venables, editors (Syracuse, N.Y.: Syracuse University Press, 1980), 1–37.

Chapter 5

The Enlightenment and the American Founders

Nature never deceives us; it is always we who deceive ourselves.

—Jean-Jacques Rousseau, *Emile, Or, On Education*[1]

The Dark Ages of the Western European intellectual tradition lasted for 1,000 years. When the Renaissance triggered a rebirth of learning and scholarly activity beginning in the fifteenth century, European man awoke to new possibilities. The ways of the past, so steeped in tradition and superstition, no longer suited the challenges of the day. Craftsmen, artisans, painters, sculptors, writers, philosophers, and religious leaders undertook new ventures that revolutionized western culture. On the heels of the initial rebirth, the Age of Enlightenment, a philosophical and political analogue to the artistic renaissance, hastened the pace of reform, wrenching open the European mind in the latter half of the seventeenth century and forever closing off the darkness of the Middle Ages. Although it appeared in slightly modified form in different countries, the movement shared common features: faith in human reason as a tonic for western ailments, a firm commitment to scientific principles and processes, the radical idea that government need not be left to despots to dictate policies to the masses, and a desire to correct the abuses of an increasingly autocratic Catholic Church.

The Enlightenment Project

Enlightenment man was determined to dust off the old works and become reacquainted with the masters. Yet it was not a retrograde movement. In rediscovering the ancients, the new class of thinker sought to build on their foundations to reach a higher plane in arts, music, literature, government, science, and even religion. Sir Isaac Newton, a symbol of Enlightenment progress, reputedly remarked, "If I have seen farther than others, it is because I have stood on the shoulders of giants," or words to that effect. He would use the past as a building block to the future, but he would not be stymied by mistakes and prejudices of long-dead masters. Whereas medieval man had presumed that inclement weather, astronomical events, and geologic processes were the inevitable consequences of an anthropomorphic deity motivated by crass human emotions, learned men of a dawning era searched for rational, scientific explanations. God was not necessarily absent from this new milieu—although some thinkers embraced atheism as part of a mechanistic perspective on the universe—but He was not as ready to intervene into human affairs as earlier generations had supposed.[2]

Newton, the English scientist and mathematician who did as much as anyone to banish the vestiges of mysticism from European thinking, personified the intellectual of the new age. Not satisfied to rely on the old ways, he pioneered work that created integral calculus and set forth the laws of motion and gravitational attraction that even today are referred to as Newtonian mechanics. He was an inquisitive, rigorous investigator, unafraid to expose deeply held beliefs to inductive scrutiny and adjust his thinking accordingly. His painstaking efforts to sift through data and articulate general principles, or laws, of natural science using the language of mathematics became a template for researchers that followed him. Alexander Pope summarized Newton's importance to the Enlightenment using metaphors of light and darkness:

> Nature and Nature's laws lay hid in night;
> God said, "Let Newton be!" and all was light.

Hyperbole aside, Pope was correct to focus on the search for nature's laws as the crucial component of enlightenment. In this new era, human progress would be judged in practical terms according to universal rules that would be known and knowable if sufficiently investigated. Thus, the Newtonian project of examining data, developing working hypotheses based on observation, and testing the hypotheses through subsequent experimentation became the primary components of accepted scientific methodology. The Dark Ages were dark because western man had forgotten the lessons he had learned through hundreds of years of careful, logical inquiry. He had set aside the works of the pre-Socratics and the Athenian Greeks, embracing ancient myth in lieu of reasoned judgment. After reaching a plateau of superstition, he had settled in for the dark nights to follow, complacent in his ignorance.

Those dark nights were banished as Newton's enlightened approach to natural laws altered western man's perspective on nature. She had been portrayed as an enigmatic, temperamental mistress, often visiting her wrath on human beings based on a whim, seemingly with little or no provocation. Yet Mother Nature need not be anthropomorphized as a stern, unfathomable goddess pursuing her own inscrutable agenda. In Newton's world, "nature" was a convenient shorthand term for a series of physical laws and chemical processes that formed a backdrop for life. These laws and processes were subject to knowable principles. The scientist's role was to discover and articulate the laws, processes, and principles of nature, explaining, if not quite taming, what once had appeared to be a wild, inexplicable beast. Humankind no longer needed to fear nature as a savage animal poised to pounce on unsuspecting humans. Darkness would not survive the bright light of Enlightenment science. The adage is that sunlight is the best disinfectant for cleaning up corruption, but the same comment can be made for ignorance, fear, and myth. Newton's greatest contribution to the Enlightenment project was to offer a path toward the light of knowledge—not only in his own field, but also in virtually every field of human study. When subsequent thinkers wrote about the Newtonian revolution—and when Alexander Pope lionized the great man in his poetry—it was this achievement, the commitment to chase down facts and ferret out the false idols of imagined gods, they meant to praise.[3]

Yet even enlightenment had its discontents. As Jean-Jacques Rousseau and other critics of Enlightenment science argued, the Newtonian view of nature was mechanistic and impersonal. Newton's universe was a sterile and inhospitable place where atoms smashed into one another, each action triggered an equal and opposite reaction, and existence chiefly consisted of deterministic cause-and-effect outcomes. In banishing primitivism and superstition from human life, Newton erred too far in favor of a coldly rational, purposeless void. Only a corrective course of action could right the scales. The solution to this sobering dose of rationality, to the extent there was one, consisted of humankind engaging in creative, artistic endeavors. For counter-Enlightenment thinkers, human beings supplied purpose and meaning through culture and "refinement." They created beautiful poems and literary masterpieces *ex nihilo*, thought up new proofs, theorems, and syllogisms, and erected majestic cathedrals reaching to the sky. In the tension between the Enlightenment and counter-Enlightenment movements were the seeds of Romanticism planted in the hard soil of rationality and empiricism.[4]

John Locke, the Social Contract, and Private Property

The Romantics were far in the future as the seventeenth century drew to a close. Until the dawn of that distant epoch, Isaac Newton was a leading intellectual figure. Contemporaries and progeny soon followed the path he blazed. The seventeenth century English thinker John Locke was one such colleague. Just as Newton revolutionized the natural sciences, Locke revitalized natural philosophy, in the process becoming a central Enlightenment figure in his own right. Locke's writings on the nature of democratic governments also greatly influenced the Founding Fathers of the American republic. He considered the concept of a state of nature in *Two Treatises of Government*, a work that brilliantly hinted at the Enlightenment's approach to nature and environmental preservation.[5]

Locke's political teachings can be summarized succinctly: All legitimate government is limited in its powers and exists only by the consent of the governed. Men are born free in a state of nature. A long ago time—perhaps metaphorically, perhaps existing so far back in ancient history that the details have been lost—nature governed all men as they lived in a peaceable condition well before anyone mused over the advantages of creating a collectivity. Order was maintained, in Locke's words, through the "executive power of the law of nature," which was exercised by every human being. The law of nature imbued humankind with tangible rights, guaranteeing people the right to own property, for example.

Before defending private property rights, Locke argued against Robert Filmer's defense of the divine right of kings. In his work *Patriarcha*, Filmer had echoed a long line of polemicists who contended that because the king was God's emissary on earth, all political rights flowed from the king. To go against the king, therefore, was to go against God and risk eternal damnation. This brutal logic rationalized a monarch's power, Locke observed in *The First Treatise of Government*, and allowed for indiscriminate authoritarian abuse. As a counter-argument, Locke suggested that God bestows rights on human beings the moment they are born. No political or religious figure stands between individuals and their God, a sentiment stretching back to Martin Luther and the Reformation. The implications were enormous: If people possessed God-given rights that antedated the establishment of a civil government, no government could justify its operation absent the consent of the governed. Rational human beings must surrender some of their natural rights to create a government, but that government must act according to the people's will or lose its legitimacy.[6]

People freely forfeited a small measure of liberty and established a social contract because a peaceful state of nature could degenerate into a state of war. In a state of war, individual lives would be lost and property rights would become unstable. To avoid the loss of peace and the possibility of an ungoverned and potentially ungovernable war, rational human beings agreed to pool their natural political virtue to establish a sovereign power. The sovereign was not to rule without limits. Indeed, the sovereign was to govern in perpetual trust with the citizenry. If the sovereign violated that trust, the people had a right, nay an obligation, to cashier their governors and install other leaders in their stead. In egregious situations, the people could abolish the entity known as the sovereign and return to a state of nature. In Locke's words, political power was "a right of making laws for the regulating and preserving of property, and of employing the force of the community, in the execution of such laws, and in defense of the commonwealth from foreign injury, and this only for the public good."[7]

Because Locke's emphasis on private property was so influential in the American conception of government, property, and nature, the concept bears thoughtful examination here. In Chapter 5, "Of Property," in *The Second Treatise of Civil Government*, Locke focused on the central dilemma concerning the acquisition of property. If God granted the earth to humankind in common, as explained in Genesis, how could a particular individual claim a private property right? As Locke explained the conundrum, "It seems to some great difficulty how any one should ever come to have property in anything."[8]

To overcome this difficulty, Locke asserted that an individual possesses the "unquestionable" ownership of his own labor. Therefore, if the person mixes what he unquestionably owns with property, he can claim ownership. This celebrated Labor Theory of Value was a core concept in Locke's thinking: "Whatsoever then he removes out of the state that nature hath provided and left it in, he hath mixed his labour with, and joined to it something that is his own, and thereby makes it his property." Otherwise, the fruits of the earth would remain in their natural state and would be rendered useless to humankind because such bounty would be unavailable for consumption. Because items remaining in their natural state spoil more rapidly than if they are used, untapped nature benefits no one. An apple hanging from a tree eventually rots and falls to the ground if it is not eaten.[9]

One caveat must be acknowledged: While the original owner has a claim right that cannot be violated by subsequent parties, no one can justify taking more than he needs. A balance must be struck between the desires of the haves and the needs of the have-nots. A property-claimant's right is valid "at least where there is enough and as good left in common for others." The speculator who would grab up as much property as possible to sell the surplus value for a handsome profit cannot justify his behavior. "As much land as a man tills, plants, cultivates, and can use the product of, so much is his property. He by his labour does, as it were, enclose it from the common." Assuming the claimant is not hoarding surplus property, his right must be protected from encroachment, and envious parties must recognize that right. "Nobody could think himself injured by the drinking of another man, though he took a good draught, who had a whole river of the same water left to him to quench his thirst," Locke observed, perhaps reflecting too much sanguinity about human nature.[10]

The issue of scarcity can never be far removed from this discussion. Examining history, especially biblical passages, Locke remarked, "But when there was not room enough in the same place for their herds to feed together, they, by consent, as Abraham and Lot did (Gen. xiii. 5), separated and enlarged their pasture where it best liked them." Consent once again appears as a crucial component of Locke's concept of property rights. The modern problem, however, is that land is a finite resource; eventually, enlarging a pasture is not possible. When human beings congregate in cities, often they own little or no property other than their personal effects.[11]

Recognizing this concern, Locke argued that "unassisted nature" was sufficient to meet primitive man's needs, but man in civil society must construct artificial devices holding a generally recognized value. The concept of monetary currency was born. Men agreed to assign a value to currency and trade among each other based on the agreement. According to Locke, the use of money does not destroy the Labor Theory of Value. In fact, money aids in preventing spoilage because it allows men to exchange items of value that might otherwise lie fallow. "And thus came in the use of money," he wrote, "some lasting thing that men might keep without spoiling, and that by mutual consent men would take in exchange for the truly useful but perishable supports of life."[12]

Despite his antipathy for persons who would deprive the needy of resources by hoarding surplus property, Locke was not naïve. He understood that some individuals would be more industrious than their neighbors would be. Such industry would drive civil society to greater heights and must necessarily be supported within reason. An industrious person would be encouraged to engage in enterprises that resulted in the "disproportionate and unequal possession of the earth" as long as his behavior was not exploitative. Civil governments would serve as a neutral arbiter of disputes involving property and would determine what constituted exploitation. Assuming the person who used his industry to acquire more property did not violate the rules of civil government and as long as scarcity did not drive other citizens into dire poverty, the arbiter would protect the unequal distribution. Equality was a concept that applied to human beings at birth—an equal chance in the competition—but egalitarian tendencies did not guarantee the outcome of the race.[13]

Locke's philosophy influenced the Enlightenment perspective on the natural environment: His decidedly anthropocentric focus reflected the mainstream narrative that the Christian Bible allows human beings to exercise dominion over the earth and its creatures. The Labor Theory of Value was predicated on the idea that the fruits of the earth are valuable only insofar as they benefit humanity. Nature in a pristine state has no value to speak of; it is fruit figuratively, and sometimes literally, rotting on the vine.[14]

The use of money when men gather in a collectivity allows for greater flexibility and encourages economic expansion, but it also concludes a simple existence of men in a state of nature. Locke was not troubled by the loss of natural man, but subsequent environmentalists and counter-Enlightenment thinkers, notably Rousseau, were deeply disturbed by this change. Locke also was not worried about treating natural resources as a fungible resource that could be traded as though it were any other commodity. This concept, too, would prove to be controversial in a later age. Finally, Locke's facile treatment of the scarcity issue would become a point of contention when the Industrial Revolution of the eighteenth century revealed the fragility of nature and triggered shortages that could never be alleviated by separating and enlarging "their pasture where it best liked them."[15]

American Enlightenment Figures

Despite his shortcomings as a philosopher of nature, Locke's theories of government, and by extension his views on the environment, influenced the founders of the American republic, especially Thomas Jefferson of Virginia, principal author of the Declaration of Independence, who borrowed heavily from Locke's *Second Treatise of Civil Government*. It is no exaggeration to say that Jefferson and the other Enlightenment-era Americans were thoroughgoing Lockeans in their understanding of government, property, and the natural environment. Yet their views were not homogeneous or static. Although all of the Founders possessed an instrumental view of nature—that is, the environment is valuable if it benefits human beings, but it is not valuable in and of itself—they were not united in their views on how natural resources should be used in service of humankind. To understand the major differences, it is instructive to focus on Jefferson's views of nature and contrast them with his political rival Alexander Hamilton.[16]

Jefferson and Hamilton represented distinct philosophical schools of thought on the appropriate role of government in American life. The fissure they symbolized persists to this day. The two great Founders also epitomized two perspectives on the instrumental value of nature. Thomas Jefferson, a Renaissance man owing to accomplishments in government, political philosophy, science, and letters, captured the hearts of the many generations of Americans that followed him, but Alexander Hamilton, the savvy, practical, unsentimental symbol of *Homo economicus*, captured their heads—and their wallets.[17]

Jefferson was a champion of small, limited government. He believed that rural townships where citizens were more or less self-governing would cultivate individual virtue and allow it to thrive and prosper. As with so much of his thinking, though, his actions did not always match his philosophy. He vehemently supported freedom and individual rights, famously remarking, "I have sworn upon the altar of God eternal hostility against every form of tyranny over the mind of man," and yet he owned slaves and probably fathered children by his slave mistress, Sally Hemings. He proselytized for limited government, and yet when he served as president of the United States, he purchased the Louisiana Territory from Napoleon—certainly one of the greatest land deals in American history and the single most enduring event of the Jefferson presidency—thereby increasing the power of the federal government and serving as an exemplar of the activist chief executive for his successors to emulate.[18]

In keeping with the frequent inconsistencies between his thoughts and his deeds, Jefferson's philosophy of nature was anything but straightforward. He was not a systematic thinker nor did he pen an all-encompassing book on his thinking. The renowned historian Henry Adams remarked, "Jefferson's writings may be searched from beginning to end without revealing the whole measure of the man."[19] Another historian, Joseph J. Ellis, described the man as an "American Sphinx" owing to the multitude of guises and pretensions that he adopted throughout his long life.[20]

Despite the occasional meanderings and opacity in his writings, Jefferson influenced generations of Americans with his vision of the new nation, although that vision remains highly contested. Reflecting Locke's influence, Jefferson argued for decentralized political authority as a means of protecting both individual liberty and property. By exercising local control, yeoman farmers—God's chosen people, as he described them in Query XIX of his one published book, *Notes on the State of Virginia*—can best promote citizen virtue. "Those who labour in the earth are the chosen people of God, if ever he had a chosen people, whose breasts he has made his peculiar deposit for substantial and genuine virtue," he wrote. He distinguished the virtuous farmer from persons engaged in commerce and crowded into large cities without room to live or space to roam, soon to suffer from "a degeneracy" that he deemed a "canker."[21]

In a letter he wrote around 1808, near the end of his second term as president, Jefferson remarked, "Nothing is so easy as to learn to cultivate the earth; all your women understand it, and to make it easier, we are always ready to teach you how to make plows, hoes, and necessary utensils." After a virtuous man has established himself as a country farmer, he then expects government to assist him, but only in a limited capacity. "When once you have property," Jefferson explained, "you will want laws and magistrates to protect your property and persons, and to punish those among you who commit crimes."[22]

One commentator has noted, "Nature was Jefferson's myth for all purposes, a flexible idea that gathered together his deepest beliefs."[23] Yet tilling the soil and harvesting the land were not metaphors for Jefferson. Although he believed the commonly held myth of virtuous farmers who threw down their plows and reluctantly marched off with a sense of *noblesse oblige* to perform their civic duty, à la Cincinnatus, more importantly, he believed that agricultural pursuits kept citizens in touch with their communities. Democracy requires such a connection or all is lost, in Jefferson's opinion. In *Notes on the State of Virginia*, he contended that "corruption of morals" is a natural consequence of moving away from an agrarian lifestyle. Corruption necessarily occurs as a "mark set on those, who not looking up to heaven, to their own soil and industry, as does the husbandman, for their subsistence, depend on it for the casualties and caprice of customers. Dependence begets subservience and venality, suffocates the germ of virtue, and prepares fit tools for the designs of ambition."[24]

Jefferson seized on a point that later writers, especially Rousseau, would develop to greater effect: Loss of virtue occurs when an individual moves away from a natural life into a city, where "civilizing" forces, such as the need to earn a living, require a man to make compromises that invariably compromise his character. When a man labors to satisfy customers, he sets aside his principles to cater to the desires of the moneyed men. This relationship creates a dependency that robs the laborer of his virtue because he has fewer options for deciding his own destiny. For Jefferson, it was the "customers"—which could be thought of in his day as the rich New England and New York merchants who supported Hamilton and the Federalist

Party—that corrupted the masses of honest, hard-working tillers of the soil. By contrast, men who worked their own land and lived off its bounty could remain self-sufficient and virtuous, free from pernicious influences found in America's large, already crowded cities (at least, crowded by eighteenth century standards).

Jefferson's instrumental view of nature and the centrality of agricultural pursuits as the basis for civic virtue were reflected in many letters he wrote to his friends and admirers over the years. For example, he succinctly expounded on this view in correspondence he sent to James Madison on the eve of the constitutional convention. "This reliance [on manufacturing] cannot deceive us as long as we remain virtuous; and I think we shall be so, as long as agriculture is our principal object, which will be the case, while there remain vacant lands in any part of America."[25] He reiterated this point numerous times in his long correspondence with Madison. "I think our governments will remain virtuous for many centuries; *as long as they are chiefly agricultural*," he remarked on one occasion.[26] Lest the reasons be obscured, he explained elsewhere that in America "the immense extent of uncultivated and fertile lands enables everyone who will labor to marry young, and to raise a family of any size."[27]

The connection between moral virtue and agriculture was a common theme in Jefferson's work. When men farmed their own lands and kept political decisions in their own hands, they fulfilled the promise of the Revolutionary generation. To the extent that nature yielded a bounty sufficient to sustain the yeoman farmer, man and nature harmoniously coexisted. When nature did not yield a sufficient bounty, the agrarian life and moral virtue were threatened.[28]

According to Jefferson, a proper civic education prepared citizens to defend their rights against unscrupulous would-be despots who would seize power and oppress the people. Although Jefferson generally refrained from mentioning Hamilton and the Federalists by name when he championed civic education, his concerns about Hamilton as a new American Napoleon figured into Jefferson's support for a virtuous life close to nature. To counteract the corrupting influences of the Federalists and elites, Jefferson developed a plan for universal public education that featured primary and secondary schools. In these schools, young minds could be molded into model citizens—the voters and watchdogs of a future generation that would inherit the mantle from the Founders.[29]

From the outset, Jefferson assumed that educated yeoman farmers would take their rightful place as masters of the earth, harvesting its fruits to create fully functioning farms and plantations. Yet the Jeffersonian project encountered two problems that he never satisfactorily resolved. First, not everyone would possess the ability or the interest in acquiring the kind of liberal, rigorous, high-quality education he envisioned. A "natural aristocracy" was not objectionable to Jefferson because the capable student would take advantage of the educational opportunities presented to him. Yet Jefferson failed to see that the leisured class—that is, the well-to-do plantation owners and other "moneyed men"— would be well positioned to receive a sound education while poor, itinerant laborers and subsistence farmers who barely eked out a living would be forced to forgo an education. Jefferson's championing of egalitarian ideals was not as democratic as it initially appeared.[30]

Second, young minds taught the most "scientific" advances of the day could hardly be expected to labor for themselves. The notion of undertaking backbreaking labor was unseemly to a gentleman farmer of the Virginia aristocracy. Despite Jefferson's numerous writings singing the praises of the simple landowner working his fields and cows, the well-educated man of agriculture was not a virtuous soul who dirtied his hands in the soil of his native Virginia. He was the head of a large plantation that relied on slave labor to increase the yields from planting cash crops such as tobacco and cotton. All the talk of simple yeomen farmers who cherished an "empire of liberty" was mostly a myth—and, in light of the hypocrisy over slavery, an unpalatable one as well.[31]

Jefferson bitterly denounced slavery in *Notes on the State of Virginia*, concluding that it debased both master and slave because the institution reflected "a perpetual exercise of the most boisterous passions, the most unremitting despotism on the one part, and degrading submissions on the other."[32] Despite this firm, seemingly unequivocal, statement of principles, the reality of American plantation life was far different. The vision of Jeffersonian America ultimately became a devotion to white liberty built on a cornerstone of black slavery. According to one commentator, "Jefferson's...devotion to human freedom (especially the freedom of the mind) led him to adopt in whole or in part the Lockean natural law teaching, but it was a classical Roman taste and temperament that led to his very un-Lockean agrarianism."[33] Another commentator remarked that Jefferson "was a romantic with an almost bewildering inability to see the gap between his political ideas and American realities. In his own Virginia, the happy, independent, small farmer was a down-at-the-heel myth in a state dominated by large slaveholders like Jefferson."[34]

Because Jefferson refused to subscribe to Alexander Hamilton's view that agricultural wealth could be enhanced through commerce and because he insisted that manufacturing should be rejected, he was forced to argue for the sufficiency of agricultural yields. Just as Locke stumbled in his discussion of how scarcity could be ameliorated, Jefferson did not face an easy task. *Notes on the State of Virginia* devoted many pages to describing natural phenomena in Virginia, probably to demonstrate the plentiful bounty supplied by available natural resources. If the educated yeoman farmer could be made to see that agriculture would preserve his virtue and still not rob him of the good life, perhaps the corrupting Hamiltonians who sought to tempt him with promises of a flashier, more exciting livelihood would not succeed in their nefarious labors.

Jefferson wrote most of *Notes on the State of Virginia* as a disinterested scientist would write, providing only brief commentaries and conclusions about the state's natural resources. The passion and eloquence so prominent in the Declaration of Independence and in his private correspondence was missing from the work. Reading *Notes* is much like reading an early American almanac, and that was Jefferson's intention. He sought to record the available data on his home state with a precision and accuracy missing from his eloquent polemical works.[35]

In one telling passage, he referred to difficulties with nature in an uncultivated state owing to "the spontaneous productions of the forests and waste fields," which were barely sufficient "to support indifferently the domestic animals of the farmer." Farmers must labor to increase their agricultural production. Even so, most farmers had little incentive to improve their lands because scarcity of resources was not a problem. "He therefore finds it more convenient to receive them from the hand of nature in that indifferent state, than to keep up their size by a care and nourishment which would cause him much labour." Despite Hamilton's arguments against the inefficiency of farming absent manufacturing, Jefferson contended that natural resources were so abundant in Virginia that farmers did not need to exert themselves to meet their material needs. The land easily could meet those needs. Moreover, in the event that a farmer sought to improve his lot, he could undertake additional labors, as necessary, and thereby extract even greater wealth from his land.[36]

The emphasis on an individual private landowner laboring to produce wealth captured the major theme in Jefferson's agrarian philosophy. Building on Locke's Labor Theory of Value, Jefferson implicitly argued that nature was not valuable until man mixed his labor with it to produce something useful. The theory was not unique to Locke or Jefferson, but it did occupy a fundamental role in each man's thinking. In the typical Lockean articulation, the theory posits that because man "has a property in his own person," it is man's labor that creates value when it is mixed with something taken from the state of nature.[37]

Land that was not closed off was deemed to be part of the commons, but that idea later created a dilemma for environmentalists who did not share Jefferson's Lockean perspective on the integral role of private property rights in the American regime. If land is sold off and developed as a private right, the natural, pristine beauty of the land often is spoiled. If it is not cut into smaller parcels and sold, it remains part of the public domain. Unfortunately, public domain land often is subject to the tragedy of the commons.[38]

Jefferson did not discuss such specifics. One of the few clear points from his writings was his belief that land is valuable when it is privately owned. In a letter to Madison, he observed that the earth was created as "a common stock for men to labour and live on," but the success of the new regime required that at least a small portion of land be granted as property to small farmers. "The small landowners," in Jefferson's opinion, "are the most precious part of a state."[39]

In light of his intellectual heritage stretching back through Locke to antiquity, Jefferson simply could not envision the need to consider the environment as anything other than a means to an end. As long as men left enough material resources to share, why worry about scarcity or environmental degradation? As far as Jefferson was concerned, men who worked on farms and spent their time working in the natural environment would take steps to protect the earth. It was in their best interests to preserve their property.[40]

And what of Alexander Hamilton, Jefferson's competitor—how did he distinguish himself from the Virginian? If anything, Hamilton was less inclined than

Jefferson was to value nature and its fruits. He believed that a weak central authority such as had existed under the disastrous Articles of Confederation would lead to internal dissension and strife, ultimately destroying the American republic. Only through a strong centralizing force could the fledgling nation quell violent dissent, erect stable political institutions, create sound monetary policy with a common currency, and protect itself from foreign nations that constantly filibustered along the western edges of the nation's boundaries. Hamilton regarded Jefferson's love of nature and individual virtue as silly and absurd. Nature was useful in supplying land for harvesting crops and constructing houses and public buildings, but worshipping the earth as a mythical place for encouraging citizen virtue was a contemptible notion. Citizens would be encouraged to support the regime when a strong nation and a sound economy were established.[41]

The context in which Alexander Hamilton wrote his major works, especially the "Report on Manufactures," was far less reflective than the context in which Jefferson produced his philosophical and scientific reflections in *Notes on the State of Virginia*. Always the practical politician, Hamilton adeptly fused political ideology and argument with political rhetoric in the midst of spirited partisan politics in *The Federalist Papers* early in his political career. So, too, in the "Report on Manufactures" did he mix politics and philosophy, although in that case he set forth a solid foundation for the fiscal health of the nation.[42]

Unlike almost every other American of the eighteenth century, Hamilton understood the conditions necessary for establishing a polity in which commerce could flourish. As an early champion of unbridled capitalism and, most famously, as the first secretary of the Treasury, he constantly pushed for the adoption of public policies to marry private business and public institutions together. Government was designed to establish laws that allowed private entrepreneurs to meet market demands in a laissez-faire atmosphere that eventually generated surplus wealth for everyone, in Hamilton's view. This placed him at odds with Jefferson, who feared the loss of virtue with the growth of manufacturing enterprises.[43]

It is ironic that in his day Hamilton impressed critics such as Jefferson as a proponent of big, powerful government that threatened to impede humankind's progress and severely curtail human freedom. "Our Bonaparte," Jefferson famously called Hamilton in a letter to Jefferson's son-in-law, Thomas Mann Randolph. This characterization seems patently unfair today, or at least inaccurate. As the great capitalist of the Founding Era, today Hamilton would seem to be the quintessential free-market economist of the University of Chicago. Government plays an integral role in promoting economic concerns, Hamilton would tell us, but compared with the welfare state created by reform liberalism in the twentieth century, he appears as a libertarian of the highest order. He envisioned a regime in which public policies cleared the way for manufacturers to produce goods that could be sold with minimal government constraints. Thus, while Hamilton would seem to be an apologist for big government when compared with Jefferson's agrarian ideal, he was far more sophisticated in his understanding of the proper role of a regime than he might initially appear.[44]

Government was the means; economic progress was the end. When roads were not available to ship goods to market, government had a role in financing new road construction through taxation. When farmers wanted to block free trade policies or avoid payment of their lawfully incurred debts, government had to intervene to ensure that orderly financial transactions could occur. In the words of one commentator, "Hamilton wanted the United States to have banks, factories, a stock market...because he saw a future in which an industrialized continental-sized America would reduce England to a mere appendage. Alas, for Hamilton, it was a hard sell to farmers who were being told they were chosen of the earth, and warned that Federalist policies favored businessmen and discriminated against honest tillers of the soil." Hamilton's vision proved to be prescient. The changed understanding of his program has less to do with the substance of his original arguments than with American's evolving understanding of government's role in the 200 years since the founding.[45]

Hamilton's specific purpose in writing the "Report on Manufactures" was to address Congress and defend his support for manufacturing against the charge that he would undermine agricultural production, the "most beneficial and *productive* object of human industry," in his words. Hamilton readily conceded that the farmers' views of nature were not fundamentally mistaken. Agriculture certainly was an important method of generating wealth; in fact, it remained the one true estate—the real estate.[46]

Hamilton opened his report by praising the role of agriculture before he moved to his second point that he believed mankind to be capable of far greater yields than agriculture could produce:

> It ought to be readily conceded that the cultivation of the earth—as the primary and most certain source of national supply—as the immediate and chief source of subsistence to man—as the principal source of those materials which constitute the nutrient of other kinds of labor—as including a state most favorable to the freedom and independence of the human mind—one, perhaps, most conducive to the multiplication of the human species—has *intrinsically a strong claim to pre-eminence over every other kind of industry.*[47]

In Hamilton's view, depending only on the soil for human sustenance was a mistake because human ingenuity could accomplish far more than could be realized by depending on the vagaries of climate, weather, and soil composition. Owing to advances in technology, the creation of commercial markets, and the promise of future innovations, humanity could engineer ever-larger yields of wealth. Although he did not provide an exhaustive listing, Hamilton briefly enumerated a few reasons why human beings could produce wealth far beyond what the land could provide. He listed a number of "scientific" innovations and historical developments: the division of labor, the use of machinery, additional employment of persons not generally included in the laboring class, emigration from other countries, furnishing a greater scope for the diversity of talents among men, offering a larger area for commercial enterprise, and creating markets and thus "a more certain and steady demand for the surplus produce of the soil."[48]

These factors, when combined, could greatly increase humankind's overall wealth, and hence the benefits that accrued to it. To ensure that his point was not misconstrued, Hamilton set forth the reasons why manufacturing should enjoy primacy over agricultural pursuits. "The bowels as well as the surface of the earth are ransacked for articles which were before neglected." Animals, plants, and minerals acquire a utility that is left "unexplored" by agriculture.[49]

The "Report on Manufactures" was not Hamilton's first articulation of his central thesis that manufacturing and commercial enterprises improved citizens' lives and need not be antithetical to agricultural interests. In "Federalist 12," he had argued that the two interests were complementary in a well-ordered regime. "The often-agitated question between agriculture and commerce has from indubitable experience received a decision which has silenced the rivalship that once subsisted between them, and has proved, to the entire satisfaction of their friends, that their interests are intimately blended and interwoven," he contended. "It has been found in various countries that in proportion as commerce has flourished land has risen in value. And how could it have been otherwise?"[50]

Hamilton's statements in the "Report on Manufacturers" and "Federalist 12" explained much of the Hamiltonian enterprise. Jefferson's almost mystical faith in the bounty of the earth already was obsolete—if it ever held any truth. Manufacturing, broadly taken as the pursuit of commercial enterprises, allows men to use nature fully, to extract every ounce of utility possible from available resources. Hamilton did not dwell on virtue in his defense of commercial enterprises. Such discussions had little place in the context of pecuniary endeavors.

This insight does not suggest that Hamilton was completely immoral or amoral, although certainly that is the way that the Jeffersonians portrayed him. Hamilton believed that high-minded appeals to the "virtue" of the people were naïve and sentimental. At one point in his career, Hamilton asked, rhetorically, "Take mankind as they are, and what are they governed by? Their passions.... One great error is that we suppose mankind more honest than they are." Although he did not refer to Jefferson by name here, it is no great leap of faith to realize that he probably had the Sage of Monticello in mind. "Our prevailing passions are ambition and interest; and it will ever be the duty of a wise government to avail itself of these passions, in order to make them subservient to the public good; for these ever induce us to action."[51]

Ever the rationalist, Hamilton stripped away the masks that disguised citizens' true character. People were not the simple, honest, well-meaning individualists yearning to be left alone and free of a centralized government that Jefferson supposed them to be. They were passionate, self-interested, often greedy creatures that had to be taken as they were without pretense or illusion. Accordingly, the "Report on Manufactures" avoided discussions of virtue and opted for an appeal to self-interest. Hamilton went on to explain how manufacturing protected man's "real interests" far better than agriculture did because manufacturing advanced those interests rather than injured them. If a worker could use his labor to produce two or three times the same commodities that a farmer could produce, would he not always choose the higher yield? The answer was intuitively obvious to Hamilton. The later divisions between capital and labor—between management and workers—were not foreshadowed in his work. Hamilton left it to nineteenth-century thinkers such as Dickens, Marx and Engels, and the Utopian Socialists to expound upon the dark side of the capitalist system.[52]

Despite their ideological differences, Jefferson and Hamilton agreed that nature should be harvested to serve man. They differed in degree, but not in kind. For Jefferson, man was of nature; his task was to learn to live off the land in loosely organized, decentralized communities. Where Jefferson envisioned a community of individuals living in rural areas dependent on the soil, Hamilton wanted to separate man from nature and build centralized communities in large cities divorced from nature. In modern parlance, cities consolidate population centers and lower transaction costs, thereby increasing access to markets. After he is far away from the state of nature, man is well situated to exploit nature's bounty and experiment with increasing the yields through commercial enterprises.[53]

Aside from Jefferson and Hamilton, Benjamin Franklin was an indispensable American Enlightenment figure interested in the environment. Born into a family of modest means in Boston, Massachusetts, in 1706, by the end of his long life, Franklin, one of the great self-made men of the eighteenth century, had become, with the possible exception of George Washington, the most famous American of his age. Politician, statesman, writer, inventor, scientist, and philosopher, this Renaissance man seems to have dabbled in just about everything and known just about everybody. His wandering, restless mind epitomized the spirit of the Enlightenment.[54]

Although he was not an environmentalist in the modern sense of the term, Franklin was interested in understanding the laws of nature and their application to human life. He did not enjoy a classical education, and his understanding of formal mathematics was rudimentary at best; consequently, he never advanced into the ranks of first-rate scientists who develop theories that lead to other theories. Yet he was a practical tinkerer who nonetheless contributed to the scientific enterprise and joined the pantheon of great American inventors. Franklin's natural curiosity and his obvious intelligence suited him well as he engaged in what he wryly called his "scientific amusements." As a young man of 20, he noticed dolphins swimming around the ship on which he was transported home from a voyage to England, and he studied their behavior. He also used a lunar eclipse to calculate his location on the ship. Later in life, he was curious as to why currents in the North Atlantic Ocean differed depending on which location was selected along the east coast of the United States. He eventually developed a theory of Gulf Stream currents and produced a chart laying out his ideas. Sea captains ignored the work at their peril. Years later, after the chart was generally accepted, vessels shaved two weeks from their sailing times by riding with the currents.[55]

A wood-burning stove was among his early inventions. Recognizing both the danger and discomfort of relying on most commonplace stoves to provide heat and cook food, Franklin developed an ingenious design for a metal device to channel smoke up through the chimney. The stove reduced the level of smoke and drafts that engulfed homes of the era. Although the practical device never worked quite as well as Franklin had hoped, it served as a template for later models.[56]

His most famous scientific discovery involved electricity. Newtonian science had made great strides during the seventeenth century, but the principles and practical uses of electricity remained mysterious and poorly understood in Franklin's time. After being entertained by a traveling showman who performed parlor tricks using glass tubes to elicit sparks from static electricity, Franklin turned his attention to the subject. He eventually recognized that electricity generated positive and negative charges, a concept known as the conservation of charge or the single-fluid theory of electricity. He later "snatched lightning from the sky" when he famously flew a kite during a thunderstorm and observed lightning striking a key tied to the kite string. The experiment not only electrified the scientific community, literally and figuratively, but it captured the public imagination. In 1805, long after Franklin's death, Benjamin West painted a famous, albeit inaccurate, scene, "Franklin Drawing Electricity from the Sky," that immortalized the event.[57]

Before his later canonization and death, Franklin became a celebrated public figure because of his scientific feats. Unlike many European Enlightenment thinkers, he was not an abstract philosopher, despite his intellectual bent and his role in founding the American Philosophical Society. He was first and foremost a pragmatist in the American tradition—a tradition that would become well known during the nineteenth century. This pragmatic, anthropocentric perspective carried over into Franklin's efforts on behalf of the environment.[58]

While still in his thirties, Franklin became alarmed at Philadelphia tanneries dumping waste in the city's central business district, primarily because it lowered property values. In a petition to the Pennsylvania Assembly, Franklin argued that the public rights of citizens to be free from odors, disease, and diminution in property values outweighed the private rights of tannery owners to dispose of waste as they saw fit. The Assembly agreed. Recognizing the necessity of waste management, Franklin also helped to develop a system of waste collection in Philadelphia during the 1760s. If these efforts fell short of a full-fledged commitment to environmental protection, they nonetheless were remarkable, even visionary, for their time.[59]

Environmental visionaries such as Benjamin Franklin were in short supply in North America during the eighteenth century. Among the few who thought and wrote about such matters, one man stood above the rest. Although he was European born and bred, John Hector St. John de Crèvecœur nonetheless provided a distinctly American voice to concerns about the effects of European settlers on the natural setting of the New World. Perhaps because he straddled both the old world and the new, Crèvecœur was uniquely positioned to comment on the clash of cultures on the American frontier. He emigrated to the New World in the 1750s, and later wrote a book, *Letters from an American Farmer*, that was hailed as a pioneering work of ecology long before the subject was recognized as a field of study.

He was born Michel Guillaume Jean de Crèvecœur on the last day of 1735 in Normandy, France. At the age of 19, he came to Canada to act as an engineer and surveyor for the French Colonial Militia during the French and Indian War. Injured in battle on the Plains of Abraham in 1759, he resigned his commission as a lieutenant and moved to New York. He also changed his name to John Hector St. John de Crèvecœur, perhaps in an attempt to sound less French, and hence more desirable to his English neighbors. With his newly anglicized name, he gained citizenship, married an American woman, and settled down to the hard life of a farmer in Orange County, New York.

Crèvecœur began his extensive travels during the late 1750s. From 1759 until 1765, he journeyed to Nova Scotia, Vermont, and Virginia. In 1767, he traveled down the Ohio River to south of St. Louis and up the Mississippi River to the Great Lakes region and the area that subsequently became Chicago. These visits were to affect his later reflections on nature. In an era when few people ventured beyond their homes and communities, he was a cosmopolitan man with a deep understanding of the land.

During the 1770s, Crèvecœur began writing essays that would later appear in *Letters from an American Farmer*. He also reflected on weighty issues of the day such as the evils of slavery, what it means to pursue the American Dream of wealth created through industry, and his fears for the looming American Revolution. On this last point, Crèvecœur had good reason to be worried about an armed conflict between England and her colonies. Having been born in France, briefly lived in England, and adopted colonial citizenship, he was at once a citizen of the world and a man without a fixed home among eighteenth-century nations. His New York neighbors were suspicious of his loyalist tendencies after war erupted, and the British were suspicious of his American roots. When he tried to travel from British-occupied New York City to France in 1779, the British were convinced he was a colonial spy and imprisoned him without trial for three months. He eventually won his release and sailed for Europe. Narrowly escaping a shipwreck off the coast of Ireland with his son in tow, he made his way to London and published his first volume of *Letters from an American Farmer*. A two-volume version appeared in 1784, and yet another version in 1787. He also published other works early in the nineteenth century, but the original volume of *Letters from an American Farmer* remains his best known and most influential work.[60]

Crèvecœur embraced an undeniably anthropocentric perspective, but he also waxed eloquent on the aesthetic pleasures that nature provides for the simple farmer. Reflecting on his encounters with bees, birds, and other creatures, he explained how he initially tried to drive them away as pests but came to recognize their intrinsic value. "The pleasure I receive from the warbling of the birds in the spring, is superior to my poor description, as the continual succession of their tuneful notes is [forever] new to me," he wrote in Letter II. Humankind has much to learn from birds. They are cleaner than human beings and their sense of duty seems far superior. In addition, they add to the beauty of life. "Who can listen unmoved, to the sweet love tales of our robins, told from tree to tree?"[61]

Crèvecœur's ruminations on nature led him to explain what it meant to be an American. When he wrote his essays in the 1770s, it was not a certainty that the colonies would become a standalone nation, but Crèvecœur understood that the world of eighteenth-century America possessed virtues not found in Europe. "Here are no aristocratical families, no courts, no kings, no bishops, no ecclesiastical dominion, no invisible power giving to a few a very visible one; no great manufacturers employing thousands, no great refinements of luxury." America was a land populated by rural denizens who had escaped the shackles and stodgy traditions that stifled Europeans. Reflecting a decidedly Jeffersonian view of agrarianism, Crèvecœur observed that "we are all tillers of the earth," and so much better for it. Americans enjoyed a natural equality absent from established societies that were far removed from the natural environment. "We have no princes for whom we toil, starve, and bleed," he explained in an especially eloquent passage in Letter III. Instead, "we are the most perfect society now existing in the world. Here man is free as he ought to be; nor is this pleasing equality so transitory as many others are."[62]

"What then is the American, this new man?" he asked rhetorically. He immediately answered his own question with an extended monologue: "Americans are the western pilgrims, who are carrying along with them that great mass of arts, sciences, vigour, and industry which began long since in the east; they will finish the great circle." Echoing Locke's Labor Theory of Value, Crèvecœur praised the American experiment because family connections or social position were not determinants of success; the sweat of a man's brow was the crucial factor. "Here the rewards of his industry follow with equal steps the progress of his labour; his labour is founded on the basis of nature, *self-interest*; can it want a stronger allurement?" Sharing Rousseau's concern that civilized man becomes alienated when he moves away from nature and allows himself to grow fat and contented, Crèvecœur contrasted the hardy, fit American farmer with "a despotic prince, a rich abbot, or a mighty lord" in European society. An American, Crèvecœur contended, was a new man acting on new principles. He was not hobbled by "involuntary idleness, servile dependence, penury, and useless labour."[63]

Crèvecœur also recognized that the natural environment played a part in shaping a man's perspective. In his view, "it is natural to conceive that those who live near the sea, must be very different from those who live in the woods; the intermediate space will afford a separate and distinct class." In fact, the natural environment was so important to character development in Crèvecœur's opinion that an analogy would best illustrate his point: "Men are like plants; the goodness and flavour of the fruit proceeds from the peculiar soil and exposition in which they grow. We are nothing but what we derive from the air we breathe, the climate we inhabit, the government we obey, the system of religion we profess, and the nature of our employment."[64]

Crèvecœur was not a naïve prophet of the natural life as a cure for all the ills of western man. He realized that some forms of natural life were superior to others. He was especially worried that when individuals took to the woods as nomadic hunters, they became self-indulgent, perhaps even greedy, because they were no longer tied to the soil as a farmer is rooted to his land. A hunter must follow the game. He loses his love of place and his desire to be sociable with his neighbors. He is not a genuine member of a community. For Crèvecœur, the manners of farmers "are rendered simple and inoffensive by tilling the earth; all our wants are supplied" by farming, which leaves no time for "the commission of great misdeeds." The life of hunters, however, "is divided between the toil of the chase, the idleness of repose, or the indulgence of inebriation. Hunting is but a licentious, idle life, and if it does not always pervert good dispositions; yet, when it is united with bad luck, it leads to want: want stimulates that propensity to rapacity and injustice, too natural to needy men, which is the fatal gradation."[65]

Despite his concern that some natural environments would encourage degeneracy, Crèvecœur extolled the virtues of his adopted land. Unlike Europe, where a man "becomes a stranger as soon as he quits his own kingdom," America offered an enticing prospect: the ability to remake himself from whole cloth into a new, self-sufficient individual limited only by his own industry. "We know, properly speaking, no stranger," he wrote of this New World; "this is every person's country; the variety of our soils, situations, climates, governments, and produce hath something which must please [everybody]."[66]

He summed up his love affair with America in this stirring passage:

> There is room for [everybody] in America; has he any particular talent, or industry? [H]e exerts it in order to procure a livelihood, and it succeeds. Is he a merchant? [T]he avenues of trade are infinite; is he eminent in any respect? [H]e will be employed and respected. Does he love a country life? [P]leasant farms present themselves; he may purchase what he wants, and thereby become an American farmer. Is he a labourer, sober and industrious? [H]e need not go many miles, nor receive many informations before he will be hired, well fed at the table of his employer, and paid four or five times more than he can get in Europe. Does he want uncultivated lands? [T]housands of acres present themselves, which he may purchase cheap. Whatever be his talents or inclinations, if they are moderate, he may satisfy them. I do not mean that [everyone] who comes will grow rich in a little time; no, but he

> may procure an easy, decent maintenance, by his industry. Instead of starving he will be fed, instead of being idle he will have employment; and these are riches enough for such men as come over here.[67]

Here was an Enlightenment man who understood the interrelationship between human beings and their environment unlike any other western thinker except Rousseau. He seemed to combine St. Francis of Assisi's love of nature with Locke's Labor Theory of Value, Jefferson's dreams of a self-sufficient yeoman farmer, and the American Founders' desire to craft a liberal government. He was an anthropocentric ecologist before either concept was articulated. In some ways, Crèvecœur prefigured Alexis de Tocqueville as an admirer of the American project and yet a foreigner on the outside looking in at society. He adored America and wrote passionate, eloquent phrases championing her people and their closeness with nature.

Yet it was not an anomaly that he lived out his final days in France, settling on lands he had inherited from his father. Despite his love of the American way of life, Crèvecœur was not a permanent resident. He suffered numerous setbacks during his absence from the new nation's shores. He lived in Europe during the three years preceding the signing of the Treaty of Paris that ended the Revolutionary War and required Great Britain to recognize the new country known as the United States of America. During those years, his wife died, his farm was destroyed, and his neighbors took in his children. Although he recovered his children and went on to enjoy a successful career as a writer and consul in New York City, his time living as an American farmer had ended.[68]

Humanity and Classification

While Jefferson, Hamilton, and the other American Founders wrestled with questions of the proper role of government in a political and economic system devoted to protecting private property rights, Franklin tinkered with his "scientific amusements," and Crèvecœur ruminated on the American character and its relationship with nature, European Enlightenment thinkers explored methods for communicating new-found Enlightenment knowledge as clearly and precisely as possible. The desire for clarity and precision led natural scientists to develop classification systems as a means of laying out clear, replicable, logically consistent properties. To classify is among the most basic of human endeavors, and necessarily so. The world is filled with so many natural phenomena, so many specific, and sometimes unfamiliar, things that the task of identifying those things and communicating useful information about the vagaries of the world is almost impossible unless classification systems and standards for classifications are developed according to shared values and measurements. Without classification systems and attendant standards, the world is too complex to discuss so that one group of people living in a particular time and place can understand the discussion thread. Shorthand colloquialisms can unite a small clique for a time, but ultimately the absence of recognizable means of communicating among differing groups of people can lead to an intellectual cul-de-sac. At the risk of overgeneralization, one might argue that the Dark Ages were exactly that—an academic dead-end where those who would engage in intellectual discourse were rendered speechless and inarticulate.

According to scholars Geoffrey C. Bowker and Susan Leigh Star, two distinct terms must be defined and distinguished. The first term, "classification," can be thought of as a "spatial, temporal, or spatio-temporal segmentation of the world." A "standard" is "any set of agreed-upon rules for the production of (textual or material) objects." A standard spans across more than one community, exists across heterogeneous metrics, and is enforced by a recognized, authoritative governing body. Classifications and standards can be thought of as complementary concepts. A classification is a means for organizing information about the world, but the classification system may not be recognized as definitive by everyone. As more practitioners adopt the classification, the system eventually becomes the agreed-upon standard for communicating knowledge according to a specific set of parameters.[69]

The task for Enlightenment man was to develop appropriate classification systems as a precursor to standardizing practices within fields of knowledge. In the biological sciences, an eighteenth-century Swedish botanist and university professor, Carl Linnaeus, developed a classification system based on binomial nomenclature, the practice of assigning two names to living plants and animals, the first name being the genus of the organism and the second its species. In time, his system became the standard practice among botanists and zoologists and earned Linnaeus the sobriquet of the "father of taxonomy."

He was committed to field research, often traveling abroad to examine and collect specimens. When he was unavailable, Linnaeus, ever the university professor, encouraged his students, whom he called "apostles," to travel on their own. His system can be described as a nested hierarchy. He identified three kingdoms of organisms that could be subdivided into classes, orders, genera, and species. Modern taxonomy has been refined, but it retains the essential Linnaean classification structure based on shared traits.[70]

Despite the ease and ubiquity of his system, Linnaeus suffered from no shortage of critics. Georges Louis Leclerc, Count of Buffon, most famously found fault with the Linnaean enterprise because the taxonomy seemed fixated on developing a static classification system. Buffon believed this endeavor was rife with error because it failed to account for historical processes, geological change, and what subsequent scientists labeled "evolution." Fearing that classifications could degenerate into a rigid status quo, thereby making it difficult to change categories when they required radical modification, Buffon favored a looser approach to scientific investigation and reporting. Yet tearing down the work of another is simpler than replacing it with a separate typology. Because his work did not offer a systematic alternative to Linnaeus, Buffon did not supplant the Linnaean taxonomy with his own. Instead, he became an ambiguous, elusive Enlightenment figure, eventually falling out of favor as Linnaeus captured the attention of natural scientists. For all of his subsequent obscurity, Buffon was a brilliantly cantankerous character that did much to call attention to the pitfalls of a taxonomic program in need of refinement. He also influenced the development of biological evolutionary theory, although Charles Darwin subsequently claimed to know nothing directly of Buffon's work.[71]

As natural science grew in sophistication, new figures advanced the discipline. Gilbert White, an eighteenth-century Englishman, became a pioneering ornithologist celebrated for his belief that more could be learned from observing birds and animals in their native habitats than in collecting specimens. Best known for his work *The Natural History and Antiquities of Selborne*, a compilation of letters he wrote to leading zoologists Thomas Pennant and Daines Barrington, White presented a potent combination of elegant prose, close observation, and thoughtful analysis of the natural history surrounding his home. First published in 1789, *The Natural History and Antiquities of Selborne* has never been out of print. It has appeared in 300 editions and been one of the bestselling English-language books of the past two centuries. White's work was so impressive that some historians have dubbed him the first English ecologist.[72]

Friedrich Wilhelm Heinrich Alexander Freiherr von Humboldt (Figure 5.1) was an even more important figure in Enlightenment environmentalism than Gilbert White was. The German naturalist began his professional career in the waning days of the eighteenth century, only a few generations removed from Linnaeus. During his time, Humboldt was hailed as the "scientific discoverer of America," a towering figure revered as the "second Columbus" for his achievements in exploring and explaining geography in Europe and the Americas. In September 1869, citizens of the United States marked the centennial of his birth with a flurry of monument unveilings accompanied by speeches, parades, and concerts. Ralph Waldo Emerson remarked that Humboldt was "one of those rare wonders of the world, like Aristotle, like Julius Caesar" because he demonstrated "the possibilities of the human mind, the force and the range of the faculties—a universal man." Yet despite his prominence and prestige in the eighteenth and nineteenth centuries, today he is all but forgotten except by historians of science and students of American environmental history.[73]

Among those who remember Humboldt, he has become a symbol of two contradictory impulses of the late eighteenth century—Enlightenment rationalism and budding Romanticism. Rationalists saw him as a consummate champion of classification. An indefatigable traveler and collector, he spent much of his life observing nature and grappling with the appropriate means for categorizing and quantifying specimens. The term "Humboldtian science," coined in the 1970s, refers to the classical empiricism of observing the environment, sampling subjects, describing their common features, classifying the components of each feature, and measuring their physical parameters. Humboldt argued that a dedicated scientist who carefully observed the world and collected as many specimens as possible could, from this close contextual examination, advance human knowledge of geography. Romantics, by contrast, focused on his descriptions of nature. One commentator

Figure 5.1 Friedrich Wilhelm Heinrich Alexander Freiherr von Humboldt was a towering public figure during the late eighteenth and early nineteenth centuries, owing to his achievements in exploring and explaining geography in Europe and the Americas. (Courtesy of the Library of Congress)

has remarked that Humboldt "adopted a pose of theoretical abstinence." In lieu of designing an experimental technique to test research hypotheses with the goal of developing theories, Humboldt described what he had witnessed in distinctly "non-scientific" prose. At heart, Romantics contended, he was an environmentalist, believing that unity existed in nature. The challenge was to understand that unity from close examination of the variety of plants and animals in the world. In light of these seemingly contradictory strains, perhaps Humboldt could be characterized as a bridge between rationalism and Romanticism, "being neither a naïve empiricist, nor a Romantic idealist, but engaging in an Enlightenment redefinition of the authority of the philosopher."[74]

He always lived the inquisitive life of an Enlightenment figure, even from his earliest days. Born into an accomplished Prussian family in Berlin, he collected plants, insects, and shells as a child. Although trained for a career in government and finance, Humboldt never lost his fascination with field research. While still in his early twenties, he indulged in a love of travel and began studying geology, botany, and anatomy. He achieved his first taste of fame when he explored Spanish colonies in Latin America from 1799 through 1804. Because he self-financed the trip, he was not beholden to a patron or university that might have insisted on a specific research methodology or scientific focus. Freed from such encumbrances, Humboldt chose not to limit his observations to a single field, but allowed his restless mind to take in the plants, animals, minerals, and climate of each place he visited.

Following his Latin American adventures, Humboldt traveled to the United States and visited with President Jefferson and his cabinet several times, a fortuitous series of meetings. The American president, worried about the intentions of the Spanish colonists and curious about the latest scientific discoveries of the day, was delighted to receive a visitor who could provide him with information on Spanish territories south of the U.S. border. In addition, Jefferson had recently dispatched Meriwether Lewis and William Clark on their epic voyage to explore the western lands of the continent. No doubt, he welcomed a visitor who could enlighten him on the topography and life forms of distant areas. Jefferson and Humboldt enjoyed a warm relationship that spanned many years.

Humboldt settled in France for more than two decades before returning to Berlin after exhausting his personal fortune. Later in life, he visited sites in Europe for extended periods, including an eight-month sojourn to Russia. With his observations of isothermal lines, Humboldt developed a theory that different countries and regions experienced different climates based on variables such as temperature, elevation, annual rainfall, and position relative to the equator. He was working on the fifth volume of his final publication, *Kosmos,* when he died at the age of 89 in 1859. He had pioneered the field of biogeography and greatly enhanced the Enlightenment understanding of man and his relationship to the environment by bridging science, politics, history, religion, and aesthetics to develop a holistic view of the human condition.[75]

Humboldt was among the first writers of his day to warn against humankind's environmental footprint, although that was not the language he used. A first-hand witness to colonial exploitation of indigenous peoples, he recognized that human beings were remaking the landscape with little or no understanding of the long-term effects. It was hardly the clarion call of a later age, and historians debate his role in advancing or retarding the growth of an environmental ethic, but Humboldt helped to publicize the human imprint on geography, no mean feat in any era.[76]

The Enlightenment: Friend or Foe of Environmentalism?

For all its advances in promoting learning and scientific discoveries, the Enlightenment paradigm also quickened the pace of "progress" as defined by Cartesian/Newtonian adherents and neoclassical economists. It was no accident that the Industrial Revolution ignited not long after the Enlightenment sparked the European imagination. Along the way, democratic ideals, capitalist aspirations, and humanitarian ethics became inextricably linked so that any negative critique was denounced as an assault on all, an indictment of western civilization and its undeniable accomplishments. It was ironic that in displacing the well-worn traditions of the past and in rooting out the myths and superstitions that had plagued civilized man for generations, the Enlightenment created its own proverbs and totems as replacements. When Enlightenment figures assumed positions of authority, they erected ramparts against would-be challengers to the new status quo. To challenge progress was to appear as a backward cretin with no appreciation of how much a new era had bettered human life following centuries of stagnation and decay.

Despite the efforts of a new ruling class to stave off a counter-Enlightenment, the denunciations came, first in a trickle and eventually in a torrent, especially as the frenetic undulations of Enlightenment defenders became apparent. In the nineteenth century, Charles Dickens voiced opposition to the dark side of urban life with its callous treatment of human beings who could not rise to the top of the capitalist heap. Where Dickens focused on the human drama of poverty and hopelessness among the English underclass, Karl Marx and Friedrich Engels constructed a collectivist critique of the classes in the capitalist system itself. Sigmund Freud added to the criticism of civilization and its discontents by exploring the neuroses of modern man as he was forced to survive in a stifling, competitive economic and political system of unrelenting inhumanity. By the late nineteenth century, the shine had worn off the Enlightenment and its implications for the human creature sometimes were viewed negatively—more on that point in subsequent chapters.[77]

As if to presage a dawning epoch that would question the foundations of civilization, Jean-Jacques Rousseau (Figure 5.2) emerged as a leading eighteenth-century critic of the Enlightenment. Reared in the city-state of Geneva, the center of Calvinism and a symbol of democracy, Rousseau always lived the life of an outsider. His mother died days after his birth, and he was passed around from one relative

Figure 5.2 Jean-Jacques Rousseau was a leading eighteenth-century critic of the Enlightenment. (Courtesy of the Library of Congress)

to another for much of his youth. He took up an itinerant life at an early age and never found his way into established society, although he flitted about the edges. He made his way to Paris, Venice, and back to Paris. He seems to have existed as a libertine in a time when such an existence was tolerated in certain portions of European society.

During the 1740s, the young man fell under Diderot's influence, eventually drifting away from a meandering career as a middling writer to become an insightful critic of society and western civilization. Diderot, a materialist, cautioned that the Enlightenment's march toward progress was a chimera. Human behavior, in his view, is genetically programmed and follows a predictable course. Consequently, faith in progress and civilization is not justified by the facts of existence. Rousseau came to share the belief that civilization is not a panacea for the human condition.[78]

Rousseau's emergence as a philosopher in his own right came after he submitted a piece to an essay contest on the question, "Has the restoration of the sciences and the arts contributed to refining moral character?" Publication of his winning essay, "A Discourse on the Moral Effects of the Arts and Sciences," in 1750 won him considerable public attention and solidified his persona as an iconoclast. Contrary to traditional Enlightenment discourse, he argued that human beings are naturally benevolent but their moral degeneration is hastened by prolonged exposure to the arts and sciences of a decadent, supposedly "civilized" society. Here was a contrarian's answer to the illuminating promises of the Renaissance and the Enlightenment. Rather than focus on the benefits of artistic cultivation and political self-participation, Rousseau intimated that the kind, sensitive, humble human animal found in nature eventually assumes the trappings of civilization and turns his back on human nature. A civilized man becomes concerned about his economic situation, his social position, and his ambition to climb up a ladder of success. In pursuing the all-consuming tasks of improving his societal lot, he is debased and corrupted. He no longer cares for his fellow man or his community; he becomes avaricious and self-absorbed. Natural man dies and gives way to a soft, pudgy, weak, ill-tempered, quietly desperate animal that has lost touch with the essential nature of his humanity. Rousseau would reiterate this theme in later works.[79]

In "Discourse on the Origin and Basis of Inequality among Men," commonly referred to as the Second Discourse, he expanded on his theory of man and his relationship with nature. Two types of inequality exist, he explained. The first is natural. One man is stronger than another owing to an accident of birth. Such natural inequalities were not the subject of his inquiry. Rather, he discussed inequalities created by men living in civil society. Where Locke and the American Founders championed the primacy of private property rights as a hallmark of the well-ordered society, Rousseau found the seeds of an ongoing, never-ending struggle: "The first man who, having enclosed a piece of land, said 'This is mine,' and found people simple enough to believe him, that man was the true founder of civil society," he wrote in the second part of the work. "From how many crimes, wars, and murders, from how many horrors and misfortunes might not any one have saved the human species, by pulling up the stakes, or filling up the ditch, and crying to his fellows: Beware of listening to this impostor; you are lost if you once forget that the fruits of the earth belong to us all, and the earth itself to nobody."[80]

Hobbes famously had argued that rational man would seek to remove himself from a state of nature as soon as possible. Natural processes are brutal and irrational and therefore to be avoided. Rousseau believed that Hobbes's view of human nature was preposterous and factually inaccurate. The Hobbesian man fears violent death in a chaotic war of all against all. The Rousseaun man is not a brute propelled forward by his fears. He is a kind, gentle fellow who lives in harmony with nature. Yes, he desires self-preservation, but he is not somehow distinct from nature. He lives as a part of nature. "I see him satisfying his hunger under an oak tree, quenching his thirst at the first stream, finding his bed at the foot of the same tree that supplied his meal," Rousseau wrote, "and thus all his needs are satisfied." Man is distinguished from other animals in his capacity to choose, to exercise his free will. He need not be corrupted by society; he can recapture his primitive self if he realizes the state of his predicament and chooses to resist.[81]

Rousseau's faith in the perfectibility of man was consistent with the Enlightenment, but he differed with mainstream thinkers in the steps involved in the conversion. Most Enlightenment philosophers argued that man was perfected when he moved away from nature and embraced the civilizing aspects of manmade institutions. Rousseau believed it was precisely those institutions that corrupted the human animal. As he wrote in *Emile*, "Everything is good as it leaves the hands of the Author of things; everything degenerates in the hands of man."[82]

He continued this argument in *Of the Social Contract, Or Principles of Political Right*. In perhaps his most widely quoted statement, Rousseau wrote, "Man is born free, and everywhere he is in chains." This bold statement contradicted the generally understood concept of a social contract. The standard version was most famously advanced by Hobbes and Locke during the seventeenth century, and echoed by the American Founders in Rousseau's time. Human beings voluntarily surrender part of their natural freedom in exchange for a political collectivity that will shield them from the vagaries of a state of nature. Although Locke did not share Hobbes' antipathy for nature, he recognized that an entity called the state was necessary as a neutral arbiter when the interests of men clashed and a decision had to be reached. The social contract was a metaphorical construct by which a government's authority for taking away some measure of individual freedom could be legitimized. Surrendering a small bit of freedom in exchange for a government was a small price to pay. Security trumps freedom in some instances.[83]

Rousseau believed, however, that western man had too readily surrendered his liberties to civil society. Once empowered, the sovereign engaged in all manner of mischief and seldom relinquished power absent violent insurrection led by the disaffected masses. A bloody scenario was not preordained if government was properly organized and operated. Rousseau explained how government ought to work: Within the social contract, the sovereign was to exercise the general will; that is, he was to determine what the citizenry desired and seek to meet those objectives. The government, which was separate and apart from the sovereign, was to fulfill particular tasks in support of the sovereign. The problem was that too often an obdurate sovereign did not exercise the general will. He transformed himself into a despot who ruled in his own interests. The task, therefore, was for the governed to control their governors.[84]

The difficulty in assessing Rousseau's work is that he prefigured the Romantic Movement with its emphasis on feeling and passion. So much of the Rousseaun enterprise was imprecise and occasionally contradictory. The golden age of primitive man was when he lived apart from the corrupting influence of cities, Rousseau contended, and yet at other times the philosopher extolled the virtues of life in small communities. Rousseau heralded the perfectibility of man even as he discounted the possibility of human beings reaching a higher state through the arts and sciences. The nadir of human existence was denounced, but Rousseau could never quite pinpoint when and where it occurred.

Of course, he would not see such inconsistencies as a death knell for his philosophical musings, for to be a counter-Enlightenment man was to reject the Enlightenment's ruthless logic, its inexorable push for scientific and mathematical certainty. Human beings are messy and imprecise, often inconsistent in their meandering quest for a better life. So be it. Yet to reject intellectual rigor in favor of a romantic, idyllic vision is to undermine the validity of virtually any coherent standard. Without consciously embracing nihilism, Rousseau risked a descent into relativism in his desire to free humankind from the chains of Enlightenment-generated knowledge and its attendant formal institutions of society and government.[85]

Rousseau's objections to the Enlightenment can be translated into modern terms by considering the claims of biocentric environmentalists that the Enlightenment project sought to establish a political/economic order based on unlimited economic growth in lieu of responsible limits on expansion. Biocentrists have embraced the counter-Enlightenment trend that Rousseau so famously championed. Knowingly or unknowingly, they mimic his assertions: Enlightenment proponents, building on the works of earlier mechanistic philosophers such as Descartes, Hobbes, and Bacon, were hostile to nature and failed to understand the need for a harmonious relationship between man and his environment. By advancing the concepts of the Cartesian/Newtonian model as well as neoclassical economics, Enlightenment supporters destroyed any chance of achieving harmony between man and nature. For the radical environmentalist, there can only be one answer to the question of whether the Enlightenment was friend or foe—foe.[86]

Yet the environmentalist objection to Western views is only one reading of the Enlightenment historiography. Another valid interpretation is to view the rebirth of western man in the seventeenth century and beyond as a new method for contemplating the meaning of human life. Rather than blindly accepting the established ways of the church, the monarch, or society, the Enlightenment man searched for new answers, a different understanding of ontology, and a more precise and nuanced view of reality. The truly enlightened thinker understands that the intellectual tradition is never staid; it is in constant motion. The mechanisms of the Enlightenment project are self-correcting; that is, they change as humankind's knowledge increases. They probe; they question; they evolve. According to this view, even an antagonistic environmental ethic can be "corrected."[87]

The difficulty in rejecting the Enlightenment project in the name of environmentalism—or in service of whatever goal is deemed sufficient—is that it risks what all romanticism risks, namely sliding back into an imprecise, relative, immeasurable morass. The radical environmentalist of the counter-Enlightenment touts the advantages of deep ecology, but without a means of measuring the success of the endeavor, it all smacks of mysticism and wishful thinking. As discussed in Chapter 3, some radical environmentalists risk ridicule or, worse, irrelevancy in their embrace of extreme positions.

Whatever else it achieved, the Enlightenment changed the Western intellectual tradition by encouraging imaginative, daring souls to produce new works in the arts and sciences, install responsive governments based on consensual self-rule, and explore untamed lands. By and large, most students of the western tradition have deemed these developments to be a positive step away from the stultifying rigidity of the Dark Ages. Yet as the eighteenth century gave way to the nineteenth and the Industrial Revolution commenced, humankind hastened the day of reckoning when its relationship with nature would become problematic. The mainstream Enlightenment's anthropocentric, utilitarian perspective on the natural environment eventually would have to become more nuanced and take into account counter-Enlightenment objections if *Homo sapiens* hoped to reconcile an inexhaustible appetite with an exhaustible resource.

Notes

1. Jean-Jacques Rousseau, *His Educational Theories Selected from Emile, Julie, and Other Writings* (Great Neck, N.Y.: Barron's Educational Series, Inc., 1964), 173.
2. The literature on the Enlightenment is voluminous and far beyond the scope of this chapter. Moreover, some scholars argue that the term "Enlightenment" is so vague and all encompassing that it should no longer be used to describe seventeenth and eighteenth century intellectual achievements in the West. The point is well taken; however, because the term retains its currency beyond the walls of academe, the descriptor appears in this book. See, for example, Bettina Dietz, "Making Natural History: Doing the Enlightenment," *Central European History* 43, 1 (March 2010): 25–46; Graeme Garrard, "The Enlightenment and Its Enemies," *The American Behavioral Scientist* 49, 5 (January 2006): 664–680; Darrin M. McMahon, *Enemies of the Enlightenment* (Oxford and New York: Oxford University Press, 2001); James Schmidt, "The Legacy of the Enlightenment," *Philosophy and Literature* 26, 2 (October 2002): 432–442.
3. Much has been written about Newton and his contributions to Enlightenment science and culture. See, for example, Patricia Fara, *Newton: The Making of a Genius* (New York: Macmillan/Columbia University Press, 2002); Jeffrey R. Wigelsworth, *Deism in Enlightenment England: Theology, Politics, and Newtonian Public Service* (Manchester, England: Manchester University Press, 2009). The excerpt from Alexander Pope appears on page 42 of Fara, *The Making of a Genius.*
4. Robert Pool, "Was Newton Wrong?" *Science* 241, 3909 (August 12, 1988): 789–790; Darren Staloff, *Hamilton, Adams, Jefferson: the Politics of Enlightenment and the American Founding* (New York: Hill & Wang, 2005), 261–267.

5. See, for example, Robert A. Goldwin, "John Locke/1632–1704," in *History of Political Philosophy*, 2nd ed., Leo Strauss and Joseph S. Cropsey, editors (Chicago: University of Chicago Press, 1972), 451–486.
6. Charles D. Tarlton, "Reason and History in Locke's *Second Treatise*," *Philosophy* 79, 308 (April 2004): 250–257.
7. John Locke, *Two Treatises of Government*, Thomas I. Cook, editor (New York: Hafner, 1947), 122.
8. For a discussion of Locke's views on property, see, for example, Alex Tuckness, "The Coherence of a Mind: John Locke and the Law of Nature," *Journal of the History of Philosophy* 37, 1 (January 1999): 73–90; Thea Vinnicombe and Richard Staveley, "John Locke, Thomas Hobbes and the Development of Political Economy," *International Journal of Social Economics* 29, 9/10 (2002): 690–705.
9. Locke, *Two Treatises of Government,* 134.
10. Ibid., 134–137. See also Traci Lynne Timmons, "Earth Jurisprudence and Lockean Theory: Rethinking the American Perception of Private Property," *Earth Jurisprudence and Environmental Justice Journal* 1, 1 (Summer 2011): 108.
11. Locke, *Two Treatises of Government,* 140.
12. Ibid., 144.
13. Ibid., 136; Tarlton, "Reason and History in Locke's *Second Treatise*," 269.
14. Timmons, "Earth Jurisprudence and Lockean Theory," 106–109.
15. Goldwin, "John Locke/1632–1704," 480–486.
16. Mark F. Griffith, "John Locke's Influence on American Government and Public Administration," *Journal of Management History* 3, 3 (1997): 224–237; Timmons, "Earth Jurisprudence and Lockean Theory," 103–116.
17. The literature on the Jeffersonian-Hamiltonian debate is voluminous. See, for example, Andrew Burstein and Nancy Isenberg, *Madison and Jefferson* (New York: Random House, 2010), 237–242; Gordon S. Wood, *Empire of Liberty: A History of the Early Republic, 1789–1815* (Oxford and New York: Oxford University Press, 2009), 150–158. For a more detailed discussion of Thomas Jefferson's and Alexander Hamilton's contrasting view of nature, see, for example, J. Michael Martinez, "Jeffersonian and Hamiltonian Views of Nature in the Early American Republic," *Politics & Policy* 33, 3 (September 2005): 522–552.
18. Quoted in Saul K. Padover, editor, *Thomas Jefferson on Democracy* (New York: The New American Library, 1939), 169. See also Jon Meacham, *Thomas Jefferson: The Art of Power* (New York: Random House, 2012), xxiii–xxiv, 216–219, 383–392; Staloff, *Hamilton, Adams, Jefferson*, 117–120, 332–350. For a fascinating account of Jefferson's relationship with slavery in general and Sally Hemings especially, see Annette Gordon-Reed, *The Hemingses of Monticello* (New York: W. W. Norton, 2008).
19. Henry Adams, *The United States in 1800* (Ithaca, N.Y.: Cornell University Press, 1974), 128.
20. Joseph J. Ellis, *American Sphinx: The Character of Thomas Jefferson* (New York: Vintage, 1998).
21. Thomas Jefferson, *Notes on the State of Virginia*, William Peden, editor (Chapel Hill, N.C.: University of North Carolina Press, 1982), 165. See also Fawn Brodie, *Thomas Jefferson: An Intimate History* (New York: W. W. Norton, 1974); Ellis, *American Sphinx*, 8; Richard Hofstadter, *The American Political Tradition* (New York: Vintage Books, 1948); Merrill Peterson, *The Jefferson Image in the American Mind* (New York: Oxford University Press, 1960).

22. Quoted in Padover, editor, *Thomas Jefferson on Democracy*, 106–107.
23. Charles A. Miller, *Jefferson and Nature: An Interpretation* (Baltimore and London: Johns Hopkins University Press, 1993), 251.
24. Jefferson, *Notes on the State of Virginia*, 165. See also Richard Hofstadter, *Anti-Intellectualism in American Life* (New York: Vintage Books, 1962), 155–156; Meacham, *Thomas Jefferson*, 159; William D. Richardson, *Democracy, Bureaucracy, & Character: Founding Thought* (Lawrence: University Press of Kansas, 1997), 76–79; Mark Sturgis, "Enclosing the Commons: Thomas Jefferson, Agrarian Independence, and Early American Land Policy, 1774–1789," *Virginia Magazine of History and Biography* 119, 1 (2011): 42–74.
25. Quoted in Padover, editor, *Thomas Jefferson on Democracy*, 70.
26. Quoted in Richardson, *Democracy, Bureaucracy, & Character*, 76. Emphasis in the original.
27. Quoted in Padover, editor, *Thomas Jefferson on Democracy*, 71. For more information on Jefferson's correspondence with Madison on virtue in the American republic, see, for example, Burstein and Isenberg, *Madison and Jefferson*, 73–74, 511–12; Richard Vetterli and Gary C. Bryner, "Religion, Public Virtue, and the Founding of the American Republic," in *Toward a More Perfect Union: Six Essays on the Constitution*, Neil L. York, editor (Provo, Utah: Brigham Young University, 1988), 91–117.
28. Lance Banning, "1787 and 1776: Patrick Henry, James Madison, the Constitution, and the Revolution," in *Toward a More Perfect Union: Six Essays on the Constitution*, Neil L. York, editor (Provo, Utah: Brigham Young University, 1988), 59–89; John Patrick Diggins, "Recovering 'First Principles': Critical Perspectives on the Constitution and the Fate of Classical Republicanism," in *Toward a More Perfect Union: Six Essays on the Constitution*, Neil L. York, editor (Provo, Utah: Brigham Young University, 1988), 119–143; Will Morrisey, "The Moral Foundations of the American Republic: An Introduction," in *The Moral Foundations of the American Republic*, 3rd ed., Robert H. Horwitz, editor (Charlottesville, Va.: University Press of Virginia, 1986), 1–23.
29. Richardson, *Democracy, Bureaucracy, & Character*, 76–79.
30. Staloff, *Hamilton, Adams, Jefferson*, 280–285.
31. Thomas Fleming, *Duel: Alexander Hamilton, Aaron Burr and the Future of America* (New York: Basic Books, 1999), 37–39.
32. Jefferson, *Notes on the State of Virginia*, 162.
33. Robert C. Jeffrey, "Southern Political Thought and the Southern Political Tradition," in *Confederate Symbols in the Contemporary South*, J. Michael Martinez, William D. Richardson, and Ron McNinch-Su, editors (Gainesville: University Press of Florida, 2000), 27.
34. Quoted in Fleming, *Duel*, 109. See also Jeffrey, "Southern Political Thought and the Southern Political Tradition," 27–29; Jay Winik, *April 1865: The Month That Saved America* (New York: HarperCollins, 2001), 6–11.
35. Staloff, *Hamilton, Adams, Jefferson*, 280–295. For an excellent discussion of Jefferson's "non-scientific" views of science, see especially I. Bernard Cohen, *Science and the Founding Fathers: Science in the Political Thought of Thomas Jefferson, Benjamin Franklin, John Adams & James Madison* (New York: W. W. Norton, 1995), 22–23.
36. Jefferson, *Notes on the State of Virginia*, 56.
37. Locke, *Two Treatises of Government*, 134. See also Andrew Heywood, *Political Ideas and Concepts: An Introduction* (New York: St. Martin's Press, 1994), 141–142.

38. See, for example, Paul Gates, *The Jeffersonian Dream: Studies in the History of American Land Policy and Development* (Albuquerque: University of New Mexico Press, 1996); Paschal Larkin, *Property in the Eighteenth Century: With Special Reference to England and Locke* (London: Cork University Press, 1930); Dumas Malone, *Jefferson and His Time: Jefferson and the Rights of Man*, Vol. II (New York: Little, Brown & Company, 1972).
39. Quoted in Fleming, *Duel*, 108–109. See also Gates, *The Jeffersonian Dream*, 97–120.
40. Martinez, "Jeffersonian and Hamiltonian Views of Nature in the Early American Republic," 537–538.
41. Staloff, *Hamilton, Adams, Jefferson*, 78–86.
42. Alexander Hamilton, James Madison, and John Jay, *The Federalist Papers*, Clinton Rossiter, editor (New York: New American Library, 1961). See also Ron Chernow, *Alexander Hamilton* (New York: The Penguin Press, 2004), 374–379; Forrest McDonald, "The Constitution and Hamiltonian Capitalism," in *How Capitalistic Is the Constitution?* Robert A. Goldwin and William A. Schambra, editors (Washington, D.C.: American Enterprise Institute for Public Policy Research, 1982), 49–74; Gordon Wood, "The Political Ideology of the Founders," in *Toward a More Perfect Union: Six Essays on the Constitution*, Neil L. York, editor (Provo, Utah: Brigham Young University, 1988), 7–27.
43. Diggins, "Recovering 'First Principles,'" 119–143; McDonald, "The Constitution and Hamiltonian Capitalism," 49–74; Vincent Ostrom, *The Political Theory of a Compound Republic: Designing the American Experiment* (Lincoln: University of Nebraska Press, 1987); Clinton Rossiter, *Conservatism in America* (New York: Random House, 1955), 108; Jean Yarbrough, "Rethinking 'The Federalist View of Federalism,'" *Publius* 15, 1 (Winter 1985): 31–53.
44. Quoted in Fleming, *Duel*, 68. See also Kenneth M. Dolbeare, *American Political Thought* (Chatham, N.J.: Chatham House, 1984), 154–157; Kenneth R. Hoover, *Ideology and Political Life*, 2nd ed. (Belmont, Calif.: Wadsworth, 1994), 71–74.
45. Fleming, *Duel*, 38–39. See also Hoover, *Ideology and Political Life*, 88–105; McDonald, "The Constitution and Hamiltonian Capitalism," 49–74.
46. Alexander Hamilton, "Report on Manufactures (1791)," in *American Political Thought*, Kenneth M. Dolbeare, editor (Chatham, N.J.: Chatham House Publishers, 1984), 166–170. See also Staloff, *Hamilton, Adams, Jefferson*, 98–102.
47. Hamilton, "Report on Manufactures (1791)," 168. Emphasis in the original.
48. Ibid., 168–169.
49. Quoted in Richard Brookhiser, *Alexander Hamilton: American* (New York: The Free Press, 1999), 94.
50. Hamilton, Madison, and Jay, *The Federalist Papers*, 91–92.
51. Quoted in Richardson, *Democracy, Bureaucracy, & Character*, 51.
52. Brookhiser, *Alexander Hamilton*, 93–95; McDonald, "The Constitution and Hamiltonian Capitalism," 49–74.
53. Martinez, "Jeffersonian and Hamiltonian Views of Nature in the Early American Republic," 537–538.
54. Cohen, *Science and the Founding Fathers*, 23–24.
55. See, for example, James Campbell, "The Pragmatism of Benjamin Franklin," *Transactions of the Charles S. Peirce Society* 31, 4 (Fall 1995): 745–792; Walter Isaacson, *Benjamin Franklin: An American Life* (New York: Simon & Schuster, 2003), 129.
56. Isaacson, *Benjamin Franklin*, 130–132.

57. Ibid., 129–145.
58. Campbell, "The Pragmatism of Benjamin Franklin," 745–748.
59. Herman A. Karl, Lawrence E. Susskind, and Katherine H. Wallace, "A Dialogue, Not a Diatribe: Effective Integration of Science and Policy through Joint Fact Finding," *Environment* 49, 1 (January/February 2007): 22; A. Michael McMahon, "'Small Matters': Benjamin Franklin, Philadelphia, and the 'Progress of Cities,'" *The Pennsylvania Magazine of History and Biography* 116, 2 (April 1992): 157–182.
60. The biographical material comes from Gay Wilson Allen and Roger Asselineau, *An American Farmer: The Life of St. John de Crèvecœur* (New York: Penguin, 1987), 2–96; Thomas Patchell, "J. Hector St. John de Crèvecœur," in *Early American Nature Writers: A Biographical Encyclopedia*, Daniel Patterson, Roger Thompson, and Scott Bryson, editors (Westport, Conn.: Greenwood Press, 2008), 96–104.
61. John Hector St. John de Crèvecœur, *Letters from an American Farmer* (Carlisle, Mass.: Applewood Books, 2006 [1782]), 39–40.
62. Ibid., 49, 50.
63. Ibid., 54. 55, 56.
64. Ibid., 56.
65. Ibid., 69–70.
66. Ibid., 74.
67. Ibid., 75.
68. Allen and Asselineau, *An American Farmer*, 102; Patchell, "J. Hector St. John de Crèvecœur," 96–104.
69. Geoffrey C. Bowker and Susan Leigh Star, *Sorting Things Out: Classification and Its Consequences* (Cambridge, Mass.: MIT Press, 1999), 10–16.
70. See, for example, Isabelle Charmantier, "Carl Linnaeus and the Visual Representation of Nature," *Historical Studies in the Natural Sciences* 41, 4 (October 2011): 365–404; H. C. J. Godfray, "Linnaeus in the Information Age," *Nature* 446, 7133 (March 15, 2007): 259–260.
71. See, for example, Alan Bewell, "Romanticism and Colonial Natural History," *Studies in Romanticism* 43, 1 (Spring 2004): 30; Lee Alan Dugatkin, *Mr. Jefferson and the Giant Moose: Natural History in Early America* (Chicago: The University of Chicago Press, 2009); Jacques Roger, *Buffon: A Life in Natural History* (Ithaca, N.Y.: Cornell University Press, 1997); Joanna Stalnaker, "Buffon on Death and Fossils," *Representations* 115, 1 (Summer 2011): 20–41; Charles W. J. Withers, "Geography, Natural History and the Eighteenth Century Enlightenment: Putting the World in Place," *History Workshop Journal* 39, 1 (Spring 1995): 140–142.
72. Bewell, "Romanticism and Colonial Natural History," 28–29; J. P. Lethbridge, "Religion and Science in the Eighteenth Century," *Historian* 77, 1 (Spring 2003): 23–26.
73. Quoted in Sandra Nichols, "Why Was Humboldt Forgotten in the United States?" *Geographical Review* 96, 3 (July 2006): 399. See also Nicolaas A. Rupke, *Alexander von Humboldt: A Metabiography* (Chicago and London: University of Chicago Press, 2008); Aaron Sachs, *The Humboldt Current: Nineteenth-Century Exploration and the Roots of American Environmentalism* (New York: Viking, 2006); John Silber, "Alexander von Humboldt's Natural History Legacy and Its Relevance for Today—Opening Remarks," *Northeastern Naturalist* 8, Supplement (January 2001): 5–8.

74. Quoted in Michael Dettelbach, "Alexander von Humboldt between Enlightenment and Romanticism," *Northeastern Naturalist* 8, Supplement (January 2001): 13, 17. See also Laura Dassow Walls, *The Passage to Cosmos: Alexander von Humboldt and the Shaping of America* (Chicago and London: University of Chicago Press, 2009); Laura Dassow Walls, "Rediscovering Humboldt's Environmental Revolution," *Environmental History* 10, 4 (October 2005): 758–760.
75. Sandra Rebok, "Enlightened Correspondents: The Transatlantic Dialogue of Thomas Jefferson and Alexander von Humboldt," *Virginia Magazine of History and Biography* 116, 4 (2008): 329–332; Ingo Schwarz, "Alexander von Humboldt's Visit to Washington and Philadelphia, His Friendship with Jefferson, and His Fascination with the United States," *Northeastern Naturalist* 8, Special Issue 1 (2001): 43–56; Walls, *The Passage to Cosmos*, 9–11; Karl S. Zimmerer, "Humboldt and the History of Environmental Thought," *Geographical Review* 96, (July 3, 2006): 456–458.
76. See for example, Aaron Sachs, "The Ultimate 'Other': Post-Colonialism and Alexander von Humboldt's Ecological Relationship with Nature," *History and Theory* 42, 4 (December 2003): 111–135; Walls, "Rediscovering Humboldt's Environmental Revolution," 759–760.
77. Garrard, "The Enlightenment and Its Enemies," 664–680; Heywood, *Political Ideas and Concepts*, 51–53; Lethbridge, "Religion and Science in the Eighteenth Century," 22–28; Staloff, *Hamilton, Adams, Jefferson*, 169–177.
78. Pat Duffy Hutcheon, *Leaving the Cave: Evolutionary Naturalism in Social-Scientific Thought* (Waterloo, Ontario, Canada: Wilfrid Laurier Press, 1996), 59.
79. For biographical information on Rousseau, see, for example, Henry Thomas, *Biographical History of Philosophy* (New York: Doubleday, 1969), 213–215.
80. Jean-Jacques Rousseau, *On the Origin of Inequality*, G. D. H. Cole, translator (New York: Cosimo, 2005), 62. See also, for example, Nathaniel Wolloch, "Animals in Enlightenment Historiography," *The Huntington Library Quarterly* 75, 1 (March 2012): 53–68; Nathaniel Wolloch, "Rousseau and the Love of Animals," *Philosophy and Literature* 32, 2 (October 2008): 293–302.
81. Rousseau, *On the Origin of Inequality*, 26.
82. Quoted in Laurence D. Cooper, *Rousseau, Nature, and the Problem of the Good Life* (University Park, Penn.: Pennsylvania State University Press, 1999), x.
83. Jean-Jacques Rousseau, *Of the Social Contract, Or Principles of Political Right*, G. D. H. Cole, translator (New York: Cosimo, 2008), 14.
84. Cooper, *Rousseau, Nature, and the Problem of the Good Life*, 95.
85. Garrard, "The Enlightenment and Its Enemies," 664–680; Heywood, *Political Ideas and Concepts*, 51–53; Lethbridge, "Religion and Science in the Eighteenth Century," 22–28; Jonathan Marks, "Rousseau's Discriminating Defense of Compassion," *American Political Science Review* 101, 4 (November 2007): 727–739.
86. The literature on this point is voluminous. See, for example, Daniel A. Coleman, *Ecopolitics: Building a Green Society* (New Brunswick, N.J.: Rutgers University Press, 1993); David Ehrenfeld, *The Arrogance of Humanism* (Oxford and New York: Oxford University Press, 1978); William Leiss, *The Domination of Nature* (New York: Braziller, 1972); Martin Lewis, *Green Delusions: An Environmentalist Critique of Radical Environmentalism* (Durham, N.C.; Duke University Press, 1992); Timothy Luke, *Eco-critique: Contesting the Politics of Nature, Ecology and Culture* (Minneapolis: University of Minnesota Press, 1997); Christopher Manes, *Green Rage: Radical Environmentalism and the Unmaking of Civilization* (Boston: Little, Brown, 1990); Carolyn Merchant,

The Death of Nature: Women, Ecology, and the Scientific Revolution (San Francisco: Harper & Row, 1980); William Orphuls, *Requiem for Modern Politics: The Tragedy of the Enlightenment and the Challenge of the New Millennium* (Boulder, Colo.: Westview Press, 1997); Dorothy Slusser and Gerald Slusser, *Technology: The God That Failed* (Philadelphia: Westminister Press, 1971).

87. Lewis P. Hinchman and Sandra K. Hinchman, "Should Environmentalists Reject the Enlightenment?" *The Review of Politics* 63, 4 (Fall 2001): 663–692.

Chapter 6

The Nineteenth Century

> Out-of-doors, n. That part of one's environment upon which no government has been able to collect taxes. Chiefly useful to inspire poets.
>
> **—Ambrose Bierce, *The Devil's Dictionary*[1]**

The history of American environmentalism leading up to the nineteenth century is a tale of sputtering, hesitant, staccato fits and starts. An environmental movement as it is thought of in the twenty-first century did not yet exist. Native Americans and the occasional lone dissenter decried the despoliation of nature, but their goals were not altogether coherent; as a result, alarms frequently fell on deaf ears. The natives touted the advantages of a simple, rustic lifestyle—man living in harmony with nature—yet how many individuals escaped from a primitive existence at their first opportunity? Rousseau praised the virtues of natural man, but the witty, urbane, educated, cosmopolitan philosopher could not be found living in a mud hut. Crèvecœur expounded on the democratic values of farming, but late in life he abandoned his adopted land to return to his ancestral home in Europe. The chasm between talk and action, between philosophy and practice, often appeared deep and wide.[2]

The march of progress often commences with unbridled enthusiasm before detractors call attention to the dark side of a human enterprise. Proponents and opponents of the new *Weltanschauung* must reach an accommodation whereby the benefits of the system are accentuated and its imperfections lessened or avoided. So it was with environmentalism. Eighteenth-century man—at least the elites living in European society—marched headlong into the warm embrace of a new culture of refinement, learning, and industrial advancement. With a few notable exceptions, he did not stop to ruminate on civilization and its discontents. That clear-eyed, devastatingly rational assessment was left for another day. Ironically, when the sobering assessment came, it frequently arrived not as the result of a systematic scientific investigation of cause and effect—although such critiques would follow in due course—but often as romantic observations on the wonders of nature and the destructive tendencies of civilization. Trafficking in metaphor and hyperbole, romantics employed overheated rhetoric as a means of exciting the heart through dramatic soliloquy in lieu of engaging the head through inductive inquiry. The emphasis on feeling and emotion as superior to reason and thought has carried forward to many modern-day nature-lovers, which at least partially explains why some environmental advocates remain committed to public policy initiatives unsupported by, even antithetical to, scientific evidence. If science fails to explain environmental dangers, modern supporters of the precautionary principle contend, why be limited to the constraints of studies and data? Intense feeling and emotion can bridge the gap between uncertainty and verifiable knowledge. Wistful, magical thinking displaces turgid, rationalistic discourse in some, although not all, instances. One of the great "what if" questions is whether the environmental message of an earlier epoch would have been received with greater affection had it been carried by mainstream anthropocentric scientists pointing to credible investigatory data.[3]

Who could blame eighteenth-century man for seeking to move into an industrialized society equipped to provide creature comforts? Life outside of a city was harsh and left little time for much higher-order activity. Apart from eking out a subsistence living, the average man had no space in his life for acculturation. An artist, musician, or philosopher does not create enduring works if his belly is empty. And nature was a stern taskmaster, a brutal, often unforgiving mistress who had yet been brought to heel. Blistering heat, torrential rain, violent windstorms, and unpredictable natural disasters wreaked havoc on human beings and their rural communities. Life for most generations living in the eighteenth century had changed little in a thousand years. On the frontier in virtually any country around the globe, men and women fashioned their own clothing from homespun cloth. They used farm implements that had been designed in an unknown, unimagined, distant past. Their homes, often built with their own hands, were poorly constructed and unable to keep out the wind and rain. Denizens seldom enjoyed luxuries such as a glass window or flooring comprised of anything other than dirt. Moreover, the coming century suggested that little would change. No wonder Hobbes, the seventeenth-century philosopher of nature-as-apocalypse, described the primitive life as nasty, brutish, and short. A rational being would pay almost any price to escape from the whims of Mother Nature. Better yet, from the safe distance of civilization, humankind could devise clever ways to harness nature for his own ends. If he could not quite control her moods and idiosyncrasies, he could make her dance and sing upon occasion.[4]

Yet if the coming century seemed as dark as its predecessors did at the outset, history has a way of surprising humanity. The nineteenth century witnessed industrial advancement far beyond what anyone could have imagined at the advent of the Enlightenment. The changes, when they came, came quickly; one technological advance snowballed into dozens more almost *ad infinitum*. Knowledge is a wheel that spins faster with each succeeding revolution. The grand theoretical suppositions of Newton and his progeny rolled into a new era where craftsmen and engineers fashioned machinery to automate production and gather momentum with each technological advance. With faster mechanisms, mighty corporations arose to churn out consumer goods at such a rapid rate that they flooded the marketplace, thereby lowering prices. As prices dropped, business owners compensated for what they had lost in unit cost with an ever-increasing volume. Competitors on the sidelines witnessing the promise of entrepreneurial riches flocked to the marketplace in droves. As competition grew, business owners necessarily found themselves compelled to innovate. A consumer class arose. The laboring classes that could never afford the work product of artisans suddenly found that mass produced, low-priced consumer goods were not beyond their means after all. Thus did the slow, steady, creaky engine of eighteenth-century mercantilism yield to the swift, sure machine of nineteenth-century capitalism. Momentum generated ever more momentum. So rapid was the development of industrial might in the 1800s compared to the previous industrial innovations of human beings in all previous centuries combined, a person who came of age in the early 1800s, assuming he lived long enough, would scarcely recognize his world by the time another century dawned.[5]

Discordant themes were drowned out of the marketplace of ideas. Now and again, a Rousseau would raise his voice above the din, but his message somehow seemed vague and amorphous, almost shrill and enervating, when considered in the context of industrial progress, especially as the nineteenth century advanced through the decades. Who had patience for primitivism when he could travel on steamboats and locomotives, capture a person's likeness on a daguerreotype, light his home with a new gas lamp, or purchase reliable, affordable products from a department store and still enjoy a modicum of leisure time? To look up from one's labors, to wipe the sweat from one's brow, to step away from the plow, and to enjoy convivial times with neighbors and relations while still acquiring goods without risking debtor's prison was a marvelous development in the life of man. For the masses, almost any price was worth paying for an improved quality of life.[6]

The swift, relentless march toward an unimaginable future was a risky business, for no one knew where civilization might lead. New technologies threatened to overwhelm their creators, in effect reversing the roles of master and servant. Literary tales in the nineteenth and twentieth centuries envisioned a world where human beings became slaves to the machine-masters they had invented. These insights crystallized especially after the Enlightenment crescendo. We have seen Dr. Frankenstein's monster, and he is us.[7]

This is the risk inherent in the Promethean enterprise: As human beings separate themselves from nature, which met their needs for millennia, they risk alienating themselves from the familial and community ties that sustained life for as long as human memory can recall. Technological change can insulate human beings from the cruel vicissitudes of life in a quasi-state of nature, but it can also isolate persons from the familiar landmarks of their lives. Rousseau presciently warned of the dehumanization that accompanies the loss of traditional authority and kinship relations. Legal-rational authority supplants traditional patriarchal authority, as Max Weber later explained. As this subtle shift occurs, human beings no longer feel connected with each other. Twentieth-century Rousseauns point to the murder of Kitty Genovese in Queens, New York, in 1964 as evidence that human beings stuffed into crowded metropolises surrender their feeling for one another in exchange for minuscule modern conveniences and creature comforts. While the young woman was brutalized near her apartment building, 38 or so witnesses (the number varies) refused to get involved by lending a hand or even calling the police, believing the rape and murder of a person outside their immediate family was not their problem. They had become desensitized to the sufferings of their neighbors.[8]

Despite contrarians' warnings, most Enlightenment figures did not deem the risk of alienation a high price to pay for advances in human knowledge. Nineteenth-century man was even less inclined to call a halt to human progress than his counterparts had been a century earlier. The list of achievements was simply too spectacular to discount: Authoritarian governments became marginally more responsive to the citizenry on fear of succumbing to popular revolutions. Giant new buildings arose, towering over the landscape with a magnificent luminescence, spurring bold new designs in architectural aesthetics. Long-accepted religious dogma crumbled in the face of great awakenings in the Protestant faith. Insights into art, music, and literature led to innovations in creative expression of human longings and desires. Crop yields increased as farmers applied scientific practices to their fields. Human life was vastly improved owing to inventions that spread across the Eurocentric world of the eighteenth and nineteenth centuries. If a downside existed, such questions were best left for another time.[9]

The nineteenth century commenced with a promise of limitless progress as a continuation of the new discoveries of an earlier epoch. In the beginning, almost no one could foresee the destructive effects of the Industrial Revolution with its displacement of the lower classes, its increasingly mechanistic, dehumanizing machineries of commerce, and the enormity of the pollution that would soon be evident as smokestacks belched out ever-larger dark plumes of soot. Yet cracks in the façade existed if anyone cared to delve closely. Revolutions that rippled across Europe in 1848 were both a symptom and a cause of rapid change, so much so that beginning in that year, a new class of intelligentsia identified and explored the alienation that lower classes had long endured. Old empires, founded as they were on outmoded claims of administrative efficacy and discarded principles of autocratic rule, collapsed in the face of sustained assaults on the status quo. The passing of the old guard did not automatically lead to democratic self-rule—often one form of despotism replaced another as when Napoleon emerged from the ashes of King Louis XVI's displaced regime in eighteenth-century France—but it suggested that hereditary rule, whatever its dubious virtues in an earlier time, would no longer suffice as a foundational principle unless coupled with superior arms or military alliances forged through Faustian bargains with like-minded tyrants.

Nineteenth-century industrialization enjoyed the support of self-styled democratic regimes, none more vehement in singing its praises than the Anglo-Americans. Even today, persons of a certain political stripe refuse to acknowledge the dangers of unbridled capitalism, citing unconstrained growth as well as faith in the almost miraculous advances of technology as magic elixirs. Such enthusiasm had to be met by equally strident critics if their message was to be heard above the constant din of boosterism. Yet when the nineteenth century marched past the halfway mark, moderation lost ground, especially after the American Civil War demonstrated the folly of an underdeveloped, largely agricultural section of the country challenging the industrial might of its well-developed, richly endowed kin. American Southerners had argued that Northern factories created a type of slave labor that robbed the average citizen of economic choices that well-to-do white Southerners vigorously defended as their distinctive province, at least in their public rhetoric. The argument never prevailed north of the Mason-Dixon Line. Factory labor was still free labor no matter how assiduously apologists for slavery sought to expound on the poor menu of choices offered to the lower classes. In any event, calling attention to the dire circumstances of one subset of impoverished Americans was hardly a logically defensible justification for continuing to force another group into poverty. It is ironic, however, that the central message employed by the Southern planter elite to bolster its argument in favor of limited government—increasing the options for the working class to avoid the dehumanization of factory labor means greater human freedom (at least for persons of a lighter hue)—was the argument that Communism's greatest defenders would eventually employ against elites.[10]

Marx and Engels were by no means the only intellectuals to pen an incendiary manifesto against the evils of capitalist expansion, but they were among the more influential writers to spell out the societal ills that accompanied unfettered growth of the industrial machine. They also resided across the pond from the Americans, and that made all the difference. If anyone hoped to put the brakes on industrialization, it would be left to Europeans. With the exception of Thoreau, the Transcendentalists, and other occasional oddballs outside the mainstream, Americans of the early to mid-nineteenth century were unwilling to criticize industrial progress. And so the leviathan grew ever larger while the citizenry worshipped its form and function, seemingly oblivious to its serpentine flaws and dangers. Yet with each improvement in industrialization, something was gained and something lost. As Marx and Engels pointed out, the leisure classes always enjoyed the benefits while the laboring classes invariably bore the burden. What they did not spell out in their zeal for faulting the bourgeoisie, but others would soon add to the dialogue, was that eventually even elites who had been protected by their wealth from the vagaries of poverty, disease, and social problems could not build walls high enough to prevent environmental degradation from invading the sanctuary. Time would erode the seemingly impenetrable wall erected to protect the haves from the have-nots, for environmental degradation, as Americans would learn to their dismay, respects no manmade borders.[11]

Despite its inauspicious beginnings, the nineteenth century represented a turning point in American environmental history. As events unfolded and natural resource destruction spread, the spotty environmentalism of Benjamin Franklin, John Hector St. John de Crèvecœur, and Jean-Jacques Rousseau gave way to the constancy of a new breed of thinker. With the works of George Caitlin, John James Audubon, and Henry David Thoreau, among others, the new century would find the dedicated environmental activists lacking in the Age of Enlightenment. The English naturalist Charles Darwin's landmark 1859 book, *On the Origin of Species by Means of Natural Selection, or the Preservation of Favoured Races in the Struggle for Life*, augmented the writings and drawings of this new breed of proto-environmentalists, radically altering western man's sensibilities by suggesting that *Homo sapiens*, rather than existing apart from the environment, is fundamentally tied to the earth and its animal life. The American environmental movement, still no more than a distant idea existing on a far-off horizon, was at least possible. From these humble beginnings, American environmentalism would rise even as the Gilded Age, with its ostentatious consumption, defined the postwar American landscape.

Lewis and Clark and the Advent of Manifest Destiny

Before the dark side of industrialization could be appreciated, a new paradigm had to develop. It began with two men who set out on a journey proposed by an American president anxious to unite a developing nation. They were products of the late eighteenth-century American frontier, these two explorers of America's western lands. Native Virginians and veterans of the Indian wars, Meriwether Lewis, along with his partner William Clark, accepted a call from President Thomas Jefferson to explore the area inside the Louisiana Purchase and determine both its commercial and scientific possibilities. Jefferson had long been interested in a potential water route across the continent, but he also possessed an abiding curiosity in the flora and fauna encountered in a western trek. The assignment was a blend of the practical and the esoteric, a desire to buttress commercial possibilities as well as accumulate scientific knowledge for the sake of knowledge. Other expedition goals included asserting sovereign authority over Native Americans and discouraging foreign powers from intervening inside an American sphere of influence.[12]

White explorers had ventured into the western United States in earlier centuries, but their travels had been haphazard affairs that yielded few tangible rewards. Cabeza de Vaca had tramped through parts of what became central Texas in the sixteenth century, not long before Hernando De Soto explored the southern territory and Francisco Coronado traipsed through the Southwest. A century later, French adventurers in search of furs and promising minerals followed in the footsteps of the Spanish conquistadores. Englishmen had undertaken expeditions before Jefferson commissioned the Corps of Discovery to embark on the celebrated journey in 1804, but much of the area remained unexplored.[13]

Lewis and Clark would improve on their predecessors' efforts by organizing an expedition bearing the imprimatur of the U.S. government, not vague assurances from a far-off colonial empire. Their goals were clearly defined and eminently achievable. They were not in search of elusive elixirs or cities of gold. They knew exactly what they sought: In exploring a contiguous land mass that stretched from ocean to ocean, they were mapping out the terrain necessary for national expansion. For almost two years, the band of more than 30 tracked the Missouri River along a meandering route across the Continental Divide. The men, accompanied part of the way by a 15-year-old Shoshone Indian interpreter and guide, Sacagawea, reached the Pacific Ocean near present-day Oregon in 1805. The expedition was not the headline-grabbing enterprise it would become in the judgment of history, but Jefferson and members of his government recognized the necessity of exploring interior lands as a prelude to annexation.[14]

It was unfortunate that in exploring the untamed wilderness, Lewis and Clark, to some extent, hastened its demise. They entered the unknown forests and emerged to tell the tale. Nature, so wild and unpredictable, so inscrutable to eighteenth-century white Europeans, was not so mysterious after all. If familiarity does not quite breed contempt, it does divest the heretofore-unknown lands of their rich allure and, in some quarters, engenders apathy for the aesthetics of untamed nature. The coming century would see a new appetite for exploration, industrialization, and utilitarian manipulation of nature. By 1893, when Frederick Jackson Turner famously lamented the closing of the American frontier, he could point to the preceding nine decades as the concluding chapters in the story of a pristine America filled with untapped resources and unspoiled lands. The denouement was ugly and ignominious, at least in the judgment of environmentally conscious historians. If Americans of the late eighteenth and early nineteenth centuries struggled to delineate the national boundaries, citizens of a later age would be left to grapple with the limitations of those natural resources and pay the price for profligate practices.[15]

The Lewis and Clark expedition seemed to awaken a gigantic hunger in white Americans of the 1830s and 1840s. Having fought the War of 1812 and somehow survived, having limped through the Panic of 1819, and having witnessed the transition of political authority from the founding generation to its successors, Americans of the time expressed giddy confidence in their success—a confidence that hardly seemed justified by external events. Their heads were bloody but unbowed. Daunting issues remained: Infrastructure improvements were sorely needed; the federal government remained small and institutionally weak; stable political parties had not yet been established; foreign enemies still menaced the young republic along disputed boundary lines; and the peculiar institution of slavery threatened to tear the nation apart. Despite these lurking issues—or perhaps because of them since political leaders, then as now, were anxious to shift the focus away from apparently intractable problems toward more promising fare—a new spirit was alive in the land. Belief in American exceptionalism was the order of the day.[16]

That spirit has come to be known as manifest destiny. As a convenient label coined during the 1840s, the concept represented a broad, multi-layered ideal. Confident, forward-looking Americans, mostly freed from European entanglements because a vast ocean and limited transportation options provided a 3,000-mile buffer zone, could turn their attention to the west. A huge continent stretched before them. It had yet to be fully explored, but Lewis and Clark demonstrated that it could be traversed, and what could be traversed could be conquered. What could be conquered could be governed if it was annexed in a way that foiled Mexican and European competitors from staking out defensible claims. The race was on to spread a special brand of American imperialism to the Pacific Ocean, come what may. Any impediments, natural or artificial, might pose temporary obstacles, but a way must be found—a way *would* be found—to surmount those obstacles through whatever means were required. In the first volume of *History of the United States of America*, published in 1834, the great historian George Bancroft included the motto "Westward the star of empire takes its way." It was the credo of the age.[17]

The signs of man's destructive tendencies in the western trek were visible to anyone who cared to take note. The American painter George Caitlin, who traveled with William Clark along the Mississippi River during the great explorer's later years, was one such careful observer. A faithful student of Native American tribes, Caitlin prided himself on the authenticity of his portraits, the result of numerous trips undertaken during the 1830s. He also chronicled his experiences in lectures and in a two-volume work, *Letters and Notes on the Manners, Customs, and Condition of the North American Indians*. In Volume I, he commented on the negative impact that human beings visited on the natural environment even before the West had been won. "Many are the rudeness and wilds in Nature's work, which are destined to fall before the deadly axe and desolating hands of cultivating man," he presciently wrote. "And so amongst her ranks of *living*, of beast and human, we often find noble stamps, or beautiful colours, to which our admiration clings; and even in the overwhelming march of civilizing improvements and refinements do we love to cherish their existence, and lend our efforts to preserve them in their primitive rudeness." Subsequent students of nature would seize on Caitlin's recommendations in establishing the country's first national parks as a means of preserving areas of profound and remarkable beauty.[18]

A love of nature inspired some men to seek solace in the storied western lands. Henry David Thoreau, the much-beloved naturalist, eloquently expressed the longing that many Americans felt for a trip across the North American continent:

> Eastward I go only by force; but westward I go free. Thither no business leads me. It is hard for me to believe that I shall find fair landscapes or sufficient wildness and freedom behind the eastern horizon. I am not excited by the prospect of a walk thither; but I believe that the forest

> which I see in the western horizon stretches uninterruptedly toward the setting sun, and there are no towns nor cities in it of enough consequence to disturb me. Let me live where I will, on this side is the city, on that the wilderness, and ever I am leaving the city more and more, and withdrawing into the wilderness. I should not lay so much stress on this fact, if I did not believe that something like this is the prevailing tendency of my countrymen. I must walk toward Oregon, and not toward Europe. And that way the nation is moving, and I may say that mankind progress from east to west.[19]

As Thoreau and nature lovers came to understand, this longing for westward expansion, while shared by numerous fellow citizens, did not necessarily translate into an abiding respect for nature. The race across the continent would not evince a reverence for wilderness lands or the non-human inhabitants there. Innumerable depredations would be the order of the day as fortune-seekers pursued riches at all costs.

Thoreau failed to appreciate the righteousness and enthusiasm with which zealots of manifest destiny would pursue their objectives, their single-minded purposefulness in the face of hardships. They were nothing if not tenacious, these men of gargantuan dreams and impossible schemes. The obstacles they faced would have dissuaded less desperate souls from undertaking foolhardy enterprises, but their personalities drove them to take risks. Western explorers frequently were the dregs of society or thrill-seekers in search of the next big adventure, the adrenaline rush of trudging through unexplored wilderness. They had nothing to lose but their sorry, sordid pasts back east. As Native Americans and anyone else who stood in the way of "progress" quickly discovered, the ravenous appetite for new and better lands, for greater space to expand, for a fulfillment of God's plan for the American people would not be satiated for decades to come. The amorphous, ambiguous goal of persons who followed Horace Greeley's advice to "go west, young man" justified all manner of heavy-handedness. If forests had to be knocked down, so be it. If Native Americans had to be swept to one side regardless of treaty obligations and the rights supposedly afforded to sovereign nations, the price would be paid. If European nations that still fielded vast, well-equipped armies had to be cajoled or threatened—thereby bringing the United States to the brink of war on more than one occasion—the burden would be borne in service of a higher cause. For some Americans, this quest for space was an unquenchable thirst for excitement in new, unexplored environs. Others sought to establish homesteads with vast acreage far removed from the stifling effects of civilization. Still others wished to transport the peculiar institution into the territories to ensure its perpetual existence without interference from eastern do-gooders who would unconscionably interfere with a man's liberty and property rights. Whatever their motives, many denizens, even those initially disinclined to alight toward a distant horizon, perceived advantages in westward expansion.[20]

James Knox Polk became the most visible disciple of manifest destiny after he ascended into the presidency of the United States in 1845 following a bitter campaign and a close election. A young man—the youngest occupant of the executive chair up until that time—he was a colorless personality known to be interested in one thing only, namely realizing his own ambition for high office. Despite his relative youth, he had enjoyed a lengthy political career before he emerged as a dark-horse presidential aspirant. He was an admirer of the antebellum South's leading soldier-politician, Andrew Jackson, the legendary frontiersman and exemplar of the energetic executive. Polk had served as governor of Tennessee and speaker of the U.S. House of Representatives; as a result, aside from emulating Old Hickory, he knew a thing or two about political machinations from his own hard-won experience. His practical background and willingness to engage in *realpolitik* would serve him well.[21]

Polk, sometimes called "Young Hickory" in homage to his political mentor, revealed his imperialist intentions in his inaugural address. After expressing his gratitude to "that Almighty Ruler of the Universe in whose hands are the destinies of nations and of men to guard this Heaven-favored land," the new president explained that he would be governed by the principle of limited, clearly delineated power. "It will be my first care to administer the Government in the true spirit of that instrument, and to assume no powers not expressly granted or clearly implied in its terms," he wrote. "The Government of the United States is one of delegated and limited powers, and it is by a strict adherence to the clearly granted powers and by abstaining from the exercise of doubtful or unauthorized implied powers that we have the only sure guaranty against the recurrence of those unfortunate collisions between the Federal and State authorities which have occasionally so much disturbed the harmony of our system and even threatened the perpetuity of our glorious Union." It was a masterful opening to an address that would articulate a strong case for expansive American power far beyond anything expressed in a strict constitutional interpretation. Thus in this opening salvo did President Polk serve notice of his administration's intent to gobble up lands at the slightest pretext. No one reasonably could claim surprise at the administration's expansionist policies even if the particulars were not known beforehand.[22]

In considering whether the Republic of Texas should become a part of the United States and whether the disputed Oregon Territory belonged to Great Britain, Polk unequivocally expressed his opinion that no territory was too large to be brought into the fold. "In the earlier stages of our national existence the opinion prevailed with some that our system of confederated States could not operate successfully over an extended territory, and serious objections have at different times been made to the enlargement of our boundaries," he wryly observed. "These objections were earnestly urged when we acquired Louisiana. Experience has shown that they were not well founded." He then arrived at the crucial issue:

> Nor will it become in a less degree my duty to assert and maintain by all constitutional means the right of the United States to that portion of our territory which lies beyond the Rocky Mountains. Our title to the country of the Oregon is "clear and unquestionable," and already are our people preparing to perfect that title by occupying it with their wives and children. But eighty years ago our population was confined on the west by the ridge of the Alleghenies. Within that period—within the lifetime, I might say, of some of my hearers—our people, increasing to many millions, have filled the eastern valley of the Mississippi, adventurously ascended the Missouri to its headsprings, and are already engaged in establishing the blessings of self-government in valleys of which the rivers flow to the Pacific. The world beholds the peaceful triumphs of the industry of our emigrants. To us belongs the duty of protecting them adequately wherever they may be upon our soil. The jurisdiction of our laws and the benefits of our republican institutions should be extended over them in the distant regions which they have selected for their homes.[23]

Here was a blueprint for Americans to spread the gospel of a republican form of government from sea to shining sea. Polk was serious in fulfilling his territorial ambitions; he oversaw the slave state Texas joining the Union at the end of 1845. After initiating and winning a war with Mexico, the president could point to the acquisition of a vast swath of land in the Southwest, including California and parts of what would later become nine additional states, as a realization of the dreams of even the most enthusiastic supporters of manifest destiny. His successors occasionally cited Polk as a model of efficient public administration and prudent executive leadership. He limited himself to a single term in office, set forth a manageable set of concrete goals, and spent his tenure in office checking off the list of promises until he could boast of an enviable record of accomplishments.[24]

Yet for all of his influence, Polk was not the only force propelling white settlers toward the Pacific coast. When gold was discovered in 1848, adventurers rushed to California in search of overnight fortunes. It did not matter that few opportunists struck it rich. The promise of wealth without the slow, steady acquisition of capital, however improbable, was too enticing to ignore. Newly transplanted West Coast residents built San Francisco into a major city as a steady stream of new inhabitants flowed into what became the Golden State in 1850. That same year, Utah became a territory. With the Gadsden Purchase in 1854, some 50,000 square miles of land acquired from Mexico were available for the construction of a transcontinental railroad. The addition of Alaska in 1867 continued the wave of territorial expansion unleashed by the Louisiana Purchase 64 years earlier.[25]

It had happened in a whirlwind. In the space of a single human lifetime, the United States had burst from its original confines to consume everything in its path. The floodgates had not been wrenched open without risk or consequence. A nation dedicated to consensual self-rule and the higher-law background undergirding the U.S. Constitution could not expand rapidly without occasionally allowing pragmatic considerations to trump more esoteric matters such as cherished philosophical ideals and legal niceties. That white Americans entertained notions of empire was never in serious dispute until later historians seeking a suitable rejoinder to progressive critics reinterpreted the meaning of the antebellum era to fashion a benign explanation for the turbulence of those years. Yet Jefferson himself had spoken of an "empire of liberty," and lesser mortals had championed policies to suck Canadian, Cuban, and Mexican lands into the vortex.

The notion of an American empire was not without its critics even when manifest destiny remained a vital concept. Even as the Democratic Party, comprised mostly of slaveholders desperate to open the territories to human bondage, promoted manifest destiny as God's divine plan for a favored people, the competing national party, the Whigs, was dismayed by the rapidity of change. With eastern states still stymied by the lack of well-constructed roads, bridges, and canals, it made little sense to move westward until the nation could provide for the citizenry struggling to live in the East. The Whigs wanted the nation's time, attention, and treasure to be devoted to improving life in cities and towns desperate for unadulterated food and water, adequate housing, and affordable transportation options. Moreover, the anti-slavery wing of the party, the "Conscience Whigs," argued that manifest destiny was a convenient pretext to preach about the gospel of slavery across a larger canvas, thereby dooming the United States to wrestle with this national disgrace for the near future. The most divisive issue of nineteenth-century politics was reflected in the issue of westward expansion, albeit somewhat obliquely.[26]

Aside from the debate over whether the peculiar institution should be expanded or restricted, Americans in the antebellum years faced a crisis of numbers. Census data revealed the scale of the problem. How could the country accommodate a population explosion without sacrificing the national values that attracted people to the New World in the first place? In the first U.S. census, conducted in 1790, the U.S. population was slightly under four million people, of whom almost 18 percent were slaves. Forty years later, almost 12.9 million people (excluding Native Americans), of whom 15.6 percent were slaves, populated the country. By 1850, 23,191,876 people lived in the United States—close to 36 percent more than had been counted a decade earlier. Of that figure, 3,204,313, or 13.8 percent, were slaves, a smaller percentage than in 1830 owing to the large influx of non-slave-owning immigrants arriving on the nation's shores. On the eve of the Civil War in 1860, 30 million people had crowded into the country. Growing from 4 million to 30 million people in seven decades taxed the national infrastructure to such an extent that simply keeping up with societal needs consumed the time and energy of political and economic leaders. The United States was figuratively bursting at the seams.[27]

Lost in the expansive quest across newly available western lands in the wake of the population boom was a concern for the environment. For a country desperate to meet the needs of its people and not split apart over the slavery issue, abstract questions about nature received scant attention. As Abraham Maslow would observe in the twentieth century, lower-order needs such as food, clothing, and shelter must be met before the human animal can focus on external needs, in this case a clean environment. As discussed in later chapters, environmental degradation can grow so pervasive that it interferes with the satisfaction of human beings' plans and desires. When that occurs, protecting the natural environment assumes a new urgency. In the pre-Civil War era, however, the bifurcation was complete: Human needs and environmental protection were logically divisible and necessarily separate.[28]

John James Audubon—The Ornithologist as Environmental Steward

Against the backdrop of a rapidly developing country, if political leaders were not situated or predisposed to call attention to natural-resource preservation, leadership emanated from other quarters—and from unlikely sources. The new America pioneered by Lewis and Clark was a land freed from the vestiges of aging European empires with their emphasis on the primacy of aristocracy. It was true that Europeans had sent the first explorers and settlers to the New World, but the intrepid adventurers were risk-takers who were not vested in the stifling ways of Old Europe. In the aftermath of the first white Europeans' explorations, a relatively classless society developed, especially on the frontier that spread westward throughout the nineteenth century. Persons who would never have been afforded an opportunity to reach their full potential under the aristocratic strictures of hierarchical societies suddenly carved out niches for themselves on the American frontier.[29]

John James Audubon was one such figure. The illegitimate son of a French naval officer, he was born in 1785 on a sugar plantation operating on an island later known as Haiti. Raised in France during the era of the French Revolution, the boy turned his back on the turbulent events of his nation and immersed himself in nature. His father desired a naval career for the young man, but the son did not find a life at sea to his liking, a propitious turn of events for his subsequent career as a naturalist and illustrator. John James developed an affinity for birds early in life. His artistic skill served him well as he sketched numerous scenes of birds in flight, nursing their young, and defending their nests from aggressive predators.

He first came to America in 1803 to escape the Napoleonic misadventures. At 18, he was ready for anything. Early on, he labored in his father's employ in Pennsylvania, but eventually young Audubon moved to the frontier and opened a general store with a business partner. His frequent travels provided access to numerous species of birds, which encouraged his ornithological career to grow. At first, life was good. He married, started a family, and renounced his French citizenship in favor of his adopted homeland.

The Panic of 1819, the first major financial recession in the United States, forced Audubon into bankruptcy and debtor's prison. It was the low point of his life. Later, after his release, he embraced an itinerant existence, financing his travels by sketching likenesses of frontier men and women in an age before photography was available. Despite his growing proficiency in producing life-like images of the human form, his first love remained birds. In fact, at age 35 he resolved to paint every species of bird in North America.[30]

Audubon knew he would need major financing if he hoped to produce the lavish-colored drawings he envisioned. After failing to arouse sufficient enthusiasm in eastern U.S. cities, he sailed for England in 1826. The English were captivated by Audubon's skill and ambition. For three years, he traveled around the island exhibiting his paintings and raising money to produce a book. *Birds in America*, published serially between 1827 and 1838, was a seminal work containing 435 life-sized, hand-colored prints of the 497 species of birds that Audubon had identified. A later work, *Ornithological Biography, or, An Account of the Habits of the Birds of the United States of America*, featured a history of the lives of many birds he had encountered. He co-authored the book with the acclaimed Scottish naturalist William MacGillivray. Owing to his travels as well as these works, he is remembered as an influential figure in American environmentalism.

John James Audubon (Figure 6.1) is revered today as an inspirational figure, a committed nature lover who was not afraid to get his hands dirty as he explored ornithology up close and personal. He was not an environmentalist in the modern sense of the term. He did not lobby legislators to enact measures favorable to natural-resource protection nor did he devote his life to philosophizing on nature. He left that work to those who followed him. His contribution was to raise awareness of the wondrous species of birds that populated the New World and, by implication, he highlighted the virtues of nature.[31]

His devotees were legion. One influential admirer was George Bird Grinnell, a Yale-educated naturalist and historian who hobnobbed with elites such as Theodore Roosevelt and Gifford Pinchot in the late nineteenth and early twentieth centuries. Concerned about the wholesale slaughter of millions of birds each year to satisfy women's fashions—feathered hats were the style—he co-founded the Audubon Society in 1886, 35 years after John James's death. The society was an early prototype of the National Audubon Society, organized in 1905. Although Grinnell had not personally known the great man, John James's widow, Lucy, had tutored him. Familiar with tales of the naturalist's zeal for birds and wildlife, Grinnell chose the name "Audubon" to evoke a love and appreciation of nature.[32]

In 2012, the National Audubon Society included this statement on its website about the naturalist's influence:

> Audubon's story is one of triumph over adversity; his accomplishment is destined for the ages. He encapsulates the spirit of young America,

Figure 6.1 John James Audubon was an ornithologist and environmental steward. (Courtesy of the Library of Congress)

> when the wilderness was limitless and beguiling. He was a person of legendary strength and endurance as well as a keen observer of birds and nature. Like his peers, he was an avid hunter, and he also had a deep appreciation and concern for conservation; in his later writings he sounded the alarm about destruction of birds and habitats. It is fitting that today we carry his name and legacy into the future.[33]

Henry David Thoreau on Government and Society

Even more than John James Audubon, Henry David Thoreau's legend looms large across the canvas of nineteenth-century American history (Figure 6.2). His willingness to embrace civil disobedience, his fiercely abolitionist sentiments, and his well-known antipathy for the dehumanizing effects of modernity transformed him into the most widely recognized figure in American environmentalism from the antebellum era. For anti-government partisans, tax resisters, and economic libertarians of the twenty-first century, he is a shining example of the good citizen who is able and willing to speak the truth to power. For detractors, he has come to represent a quaint sort of New England intellectual, a symbol of the contrarian who revels in dissent for the sake of dissent but who is woefully ignorant of the challenges inherent in governing a heterogeneous, highly industrialized nation dependent on twenty-first-century technology. In the popular imagination, he is portrayed as a man of peace, a practitioner of civil disobedience, a writer and thinker of elegant simplicity who foreshadowed Mohandas Gandhi and Martin Luther King, Jr., as well as an out-of-touch idealist who tilted at windmills and produced a body of highly individualistic, oddly idiosyncratic polemics that appeals to disciples and infuriates non-believers. He was all of these things and more.[34]

Born in Concord, Massachusetts, in 1817, Thoreau hailed from a humble family of no particular station. Despite his modest means, it was clear the child was intelligent and of an intellectual bent. He graduated from Harvard University after taking a brief leave of absence to teach school. Unlike most other men of his generation who finished the august college, Thoreau was not interested in pursuing a career in a traditional field such as law, medicine, or business. He eventually stumbled into a writing career under the tutelage of Ralph Waldo Emerson, another idiosyncratic New Englander who, like his protégé, refused to fall into line with societal expectations. Thoreau's early works reflected a nascent form of Transcendentalism that Emerson and his circle advocated. Today it might be dismissed as a form of New Age mysticism, but at the time, Transcendentalism appealed to a certain class of New Englander as a distinctly American, cutting-edge philosophical rumination on spiritualism. While not altogether consistent or systematic in developing their precepts, Transcendentalists adhered to the notion that life consists of more than the world of physicality. Self-reflection is crucial to self-mastery and an understanding of the external world. Unlike religion, which developed dogmas and sacred texts

Figure 6.2 Henry David Thoreau was the leading American environmentalist of the antebellum era. (Courtesy of the Library of Congress)

that must be mastered to comprehend spiritual matters, Transcendentalism emphasized the primacy of personal intuition. Sober, clear-eyed self-assessment was an antidote to the soul-sickness of increasingly stressed, harried humankind. It was a fiercely independent creed for a fiercely independent group of people disgusted with the degeneracy found in parts of the country, especially the slaveholding South.

Whatever the merits of this idealistic school of thought, a man must still earn a living. In addition to serving as a tutor, editor, and sometime gardener for the Emerson clan, Thoreau occasionally labored in a pencil factory owned by his family. Aside from pioneering new techniques in pencil production, Thoreau in his early years was satisfied to float through his life with no clear direction apart from innocuous pursuits that appeared to be little more than hobbies. By the mid-1840s, however, he had grown restless with his life. He came to see the passing of the years as a kind of slow, steady demise. Concerned that he was living a life of "quiet desperation," he chose to withdraw from his comfortable but numbing routine and march into the woods. There he could reflect on the nature of his life, and life in general, in the best Transcendentalist fashion.[35]

Unlike the popular image of the alienated, mentally disturbed, misanthropic hermit nestled in deep hinterlands far removed from civilization, Thoreau was not physically isolated nor did he consciously turn his back on organized society. The little cabin where he chose to live was situated on the outskirts of Concord on land that Emerson owned. Thoreau's purpose, as he carefully explained, was not to trudge into an unspoiled acreage that might imperil his existence. He desired something more spiritually uplifting and less physically torturous than fighting each day for mere survival; he sought a chance for uninterrupted reflection in keeping with the tenets of Transcendentalism, such as they were. As he wrote in a famous passage in *Walden*:

> I went to the woods because I wished to live deliberately, to front only the essential facts of life, and see if I could not learn what it had to teach, and not, when I came to die, discover that I had not lived. I did not wish to live what was not life, living is so dear; nor did I wish to practice resignation, unless it was quite necessary. I wanted to live deep and suck out all the marrow of life, to live so sturdily and Spartan-like as to put to rout all that was not life, to cut a broad swath and shave close, to drive life into a corner, and reduce it to its lowest terms, and, if it proved to be mean, why then to get the whole and genuine meanness of it, and publish its meanness to the world; or if it were sublime, to know it by experience, and be able to give a true account of it in my next excursion.[36]

The problem with most human beings, Thoreau decided, is they spend their time focused on the wrong things, the minutiae in their lives, the petty day-to-time entanglements that serve no genuine good, no long-term purpose. "Our life is frittered away by detail," he observed. The cure for the disease of modern man, the brake on the frenetic pace at which he lives, is "simplicity, simplicity, simplicity!" Discarding superfluous possessions, eating enough to live but not succumbing to gluttony, focusing on the important questions of life, and undertaking activities that are crucial to survival in lieu of engaging in silly amusements that serve no real purpose and seldom provide genuine entertainment will result in a rich, satisfying, ennobling human life.[37]

Anyone who desires a rigorously scientific explanation for the vagaries of natural processes will be disappointed with Thoreau's work. He is a romantic—more poet and artist than scientist, more Rousseau than Newton. He does not seek life's answers with measured precision. A man who deigns to "suck out all the marrow of life" and "drive life into a corner" is not interested in empirical explanations of observable data. He is an artist of perspectivism. *Walden* is filled with florid prose on the beauties of nature. Indeed, much of the book is a celebration of the primitive life. Sucking the marrow from life requires a constant appreciation of bountiful natural resources. Nature, properly understood, can meet virtually all human needs. Even more importantly, nature must be appreciated in the here and now, not studied and dissected in an overly intellectual manner. "The earth is not a mere fragment of dead history, stratum upon stratum like the leaves of a book, to be studied by geologists and antiquaries chiefly," he asserted, "but living poetry like the leaves of a tree, which precede flowers and fruit,—not a fossil earth, but a living earth; compared with whose great central life all animal and vegetable life is merely parasitic. Its throes will heave our exuviae from their graves."[38]

Thoreau's time living deeply and deliberately taught him about the wretchedness of modern life, the inexorable horrors of humankind's "progress" up from clay toward industrialized civilization. He did not like what he learned. In the hustle and bustle of eking out a living and impressing the neighbors with one's achievements, a human being loses his connection with the earth. He fails to see the importance of communing with nature. "We can never have enough of Nature," he explained. "We must be refreshed by the sight of inexhaustible vigor, vast and titanic features, the sea-coast with its wrecks, the wilderness with its living and its decaying trees, the thunder-cloud, and the rain which last three weeks and produces freshets." With his penchant for intemperate language and apocalyptic, biocentric-sounding pronouncements of unusual eloquence, it is little wonder that modern environmentalists find much to admire in Thoreau's writing.[39]

Thoreau's critics, then and now, contend that withdrawing from social intercourse—even to the limited degree that he chose to remove himself to Walden Pond—is to engage in a thinly disguised form of narcissistic navel-gazing that is far too self-indulgent. Instead of producing good citizens, such adolescent silliness encourages self-absorption and dissent for the sake of dissent. To resist societal mores and governmental action as principled protest in pursuit of creating a more perfect union may be the highest form of patriotism—at great personal sacrifice, a person turns away from friends and family and willingly risks becoming a social pariah in service of a higher good—but Thoreau's philosophy can easily reverse means and ends. Resisting government and revering nature, rather than the means of calling attention to the ills of modern society, may become ends in themselves. Self-abnegation—a noble kind of asceticism—can enrich the human psyche if the denial is tied to a cause worth pursuing; however, it also can degenerate into one of two extremes—the worship of oneself or its opposite, self-loathing. In any case, isolating the individual from society does not appear to be a realistic course of action for the mass of those men leading lives of quiet desperation.[40]

What is modern man to make of Henry David? Is he a radical quack spewing forth dangerous anti-social ideas that threaten the life and health of the polity? Is he a lovable, quirky iconoclast with a gift for turning a phrase—a wonderfully entertaining historical figure that can be quoted at parties by enthusiastic college undergraduates anxious to impress members of the opposite sex with their intellectual profundity or by loony environmentalists desperate to shore up their bona fides to their equally loony peers? Is he a thoughtful philosopher of nature that called attention to the threats posed by an economic and political system spiraling out of control with little regard for the dehumanizing aspects of unfettered growth? Is he a combination of these things? How one answers these questions reveals much about the respondent's values with respect to the natural environment, for Thoreau has been transformed into an icon, a Rorschach test for judging one's perspective on the continuum of American environmentalism.

Gilded Age and American Urbanization

The Civil War remade nineteenth-century America on multiple levels. Aside from the obvious changes with the death of a peculiar institution and the ratification of the Thirteenth, Fourteenth, and Fifteenth Amendments, the conflict marked a turning point between a nation comprised of rural households and small villages where a more-or-less homogenous population depended on agriculture to scratch out a living in the antebellum era and a country characterized by burgeoning cities filled with heterogeneous groups of ethnically diverse people devoted to manufacturing and industrial concerns in the postbellum age. Before the war, government existed largely on the local level, and many citizens spent their entire lives with little or no contact with elected officials. After the war, the federal government began a long, steady rise to prominence, culminating in the explosion of the administrative state during Franklin Roosevelt's New Deal in the 1930s. Among other things, postwar industrialization hastened the urbanization of cities and the closing of the American frontier. In the East, great factories arose, their smokestacks reaching for the skies as ever more people moved away from working on farms and in family enterprises and accepted employment in industries. A construction boom financed with borrowed capital led to the creation of the Transcontinental Railroad, completed in 1869, which united east and west. A revolution in mining allowed for minerals to be extracted from the ground more efficiently, which encouraged rapid expansion of energy-intensive industries and further convinced mainstream political and economic leaders that nature's role was to serve as a convenient repository for man's needs.[41]

Mark Twain and Charles Dudley Warner coined the deliberately sardonic term "the Gilded Age" to describe the period between the end of Reconstruction in the 1870s and the mid-1890s. It was an era when many denizens were concerned with making money, although only the genuinely affluent upper classes achieved sustained prosperity. In the heartland, citizens were anxious to put the hard times of the war and Reconstruction behind them and build stable lives. Stability required a predictable economy, a safe, sound currency, and a consolidation of government power to introduce uniformity into the political and economic system. Citizens longed for a stress-free society where they could come of age, raise a family, build a career, retire, and pass from the scene absent the social upheaval that had characterized their lives during earlier decades of the century. Yet despite initial promises of a return to normality—whatever that might mean after the cataclysm of civil war—the Gilded Age produced a mixed record. The increasing number of railroad lines and the demand for more miners and factory workers brought new opportunities, but the Panic of 1873, followed by the Panic of 1893, led to debilitating recessions as well. Industries rose and fell rapidly as innovations occurred. The kerosene market, for example, grew steadily until Thomas Edison's electric light bulb propelled the industry into a downturn at the end of the 1870s. Kerosene producers eventually turned to oil and gas production as gasoline-powered engines became commonplace after 1886. Before Karl Benz and Gottlieb Wilhelm Daimler created new engines for automobiles in Europe during the 1880s, gasoline had been an inexpensive solvent produced as a byproduct of kerosene production. By the end of the nineteenth century, as kerosene production declined precipitously, a new industry had emerged to take its place and further fuel industrial capacity.[42]

As America's industrial might developed during the late nineteenth century, environmental preservation emerged as a national issue when adventurers fanned out across western lands offering colorful reports on the pristine beauty they encountered. One frequent correspondent, George Perkins Marsh, was so eloquent that he became known as one of America's first environmentalists. A career diplomat and experienced overseas traveler, Marsh understood that European forestland had been decimated by an unrestrained zeal for industrial development. He was worried that the United States, despite its seemingly infinite treasure trove of natural resources, was headed for exactly the same sad fate as had befallen western European nations. As early as 1864, Marsh warned in his book *Man and Nature; or Physical Geography as Modified by Human Action* that human actions were destructive and inevitably harmful to nature. In a famous passage, he observed, "Man is everywhere a disturbing agent. Wherever he plants his foot, the harmonies of nature are turned to discords." Although all creatures affect their environment, Marsh understood that man was different because he saw himself as different. "The fact that, of all organic beings, man alone is to be regarded as essentially a destructive power, and that he wields energies to resist which, nature—that nature whom all material life and all inorganic substance obey—is wholly impotent, tends to prove that, though living in physical nature, he is not of her, that he is of more exalted parentage, and belongs to a higher order of existences than those born of her wombs and submissive to her dictates."[43]

The concept of man existing separate and apart from nature, so often touted as a benefit, may be a detriment in the context of American environmentalism. In the twenty-first century, Marsh's thesis that human actions are destructive to the natural environment is so widely accepted as to be beyond serious argument, but in his day, it was a revolutionary insight. Most Americans who thought about technology during the Gilded Age tended to see it as a measure of progress, a friend who led the way out of darkness and into the light. Marsh explained that technological progress, for all of its well-known benefits, carried a high price tag, especially when employed to raze forests, blast through mountain terrain while building railroad trestles and roads, change the course of rivers, and reclaim land without considering the attendant consequences.

Although nature generally is "wholly impotent" to resist man's activities, Marsh explained, "she avenges herself upon the intruder, by letting loose upon her defaced provinces destructive energies hitherto kept in check by organic forces destined to be his best auxiliaries, but which he has unwisely dispersed and driven from the field of action." When trees are felled at a prodigious rate, the denuded acres are subject to soil erosion and potential desertification. When rivers are rerouted, they can overflow their banks and destroy valuable farmland, frustrating poorly designed human plans. In the final analysis, it is in human beings' best interests to preserve nature because "the whole earth, unless rescued by human art from the physical degradation to which it tends, becomes an assemblage of bald mountains, of barren, turfless hills, and of swampy and malarious plains." Human beings create environmental problems and they can solve them. They lack the will, not the ability.[44]

Another late nineteenth-century naturalist, the one-armed Civil War veteran John Wesley Powell, joined Marsh in calling attention to a serious problem lurking on the horizon. Powell was among the more prominent explorers who touted the wonders of the stunning vistas he encountered in his western travels. Despite having lost most of his right arm during ferocious fighting at the Battle of Shiloh in Tennessee in 1862, he remained in uniform until war's end. Later, Powell became a professor of geology at Illinois Wesleyan University. A co-founder of the Illinois Museum of Natural History, he refused a full-time appointment so he could head to the western United States. Powell was far too restless to remain rooted in a university setting. Longing for adventure, he felt compelled to explore the area around the Rocky Mountains. In his 1875 book *The Exploration of the Colorado River and Its Canyons,* Powell described his 1869 travels along the Colorado River through the Grand Canyon. He spoke of a strikingly beautiful but desolate landscape "so destitute of animal and vegetal life" that it "would not support large numbers." Anyone who settled there would barely manage "but a precarious and scant subsistence." Yet nature need not be hospitable to human habitation to hold value. The primitive beauty, colorful sunsets, and breathtaking topography should be preserved regardless of their value to short-term human needs. Accordingly, humankind must be careful not to destroy natural vistas through reckless actions that befoul the environment.[45]

Marsh and Powell were lonely voices figuratively crying out from the wilderness, but not for long. As a new century loomed and industrialization altered the American landscape, a growing number of critics emerged to publicize the dangers of environmental degradation. Their warnings served as a foundation for the twentieth-century environmental movement. Yet calling attention to nature's plight was not equivalent to resolute action. Addressing specific environmental problems and implementing workable programmatic solutions would be a task slated for denizens of a later epoch. For nineteenth-century environmentalists, identifying the issues and offering recommendations for a course of action were the only reasonably achievable goals. With the accelerating pace of industrial development, however, it was increasingly clear that unfettered consumption could not continue indefinitely. Wise environmental management practices would be required if human progress were to continue.

Notes

1. Ambrose Bierce, *The Devil's Dictionary of Ambrose Bierce: Complete and Unabridged*, Special Edition (El Paso, Tex.: El Paso Norte Press, 2010), 143.
2. Benjamin Kline, Ph.D., *First Along the River: A Brief History of the U.S. Environmental Movement,* 3rd ed. (Lanham, Md.: Rowman & Littlefield, 2007), 23–24; Philip Shabecoff, *Earth Rising: American Environmentalism in the 21st Century* (Washington, D.C. and Covelo, Calif.: Island Press, 2001), 2–3.
3. For more on Romanticism, the eighteenth century, and nature, see, for example, Alan Bewell, "Romanticism and Colonial Natural History," *Studies in Romanticism* 43, 1 (Spring 2004): 5–10; Christopher Hitt, "Ecocriticism and the Long Eighteenth Century," *College Literature* 31, 3 (Summer 2004): 123–147; Onno Oerlemans, *Romanticism and the Materiality of Nature* (Toronto: University of Toronto Press, 2002).
4. Henry Adams, *The United States in 1800* (Ithaca, N.Y.: Cornell University Press, 1974), 12.
5. The literature on this point is voluminous. See, for example, Stuart M. Blumin, "Driven to the City: Urbanization and Industrialization in the Nineteenth Century," *Magazine of History* 20, 3 (May 2006): 47–53; John H. Hamer, "Money and the Moral Order in Late Nineteenth Century and Early Twentieth-Century American Capitalism," *Anthropological Quarterly* 71, 3 (July 1998): 139–149; Naomi R. Lamoreaux, "Rethinking the Transition to Capitalism in the Early American Northeast," *The Journal of American History* 90, 2 (September 2003): 437–461.
6. See, for example, M. Susan Barger and William B. White, *The Daguerreotype: Nineteenth-Century Technology and Modern Science* (Washington, D.C.: Smithsonian Institution Press, 1991); Kendall Beaton, "Dr. Gesner's Kerosene: The Start of American Oil Refining," *Business History Review* 29, 1 (March 1955): 28–53; James W. Ely, Jr.,

Railroads and American Law (Lawrence: University Press of Kansas, 2001), 28–53; Daniel Walker Howe, *What Hath God Wrought: The Transformation of America* (Oxford and New York: Oxford University Press, 2007), 214–216, 532–538, 562–569.

7. See, for example, Kim Hammond, "Monsters of Modernity: Frankenstein and Modern Environmentalism," *Cultural Geographies* 11, 2 (April 2004): 181–198.
8. Much has been written about the Genovese case. See, for example, A. M. Rosenthal, *Thirty-Eight Witnesses: The Kitty Genovese Case* (Berkeley and Los Angeles: University of California Press, 1999 [1964]); Percy Smith, "Silent Witness," *The Safety & Health Practitioner* 30, 5 (May 2012): 43–46.
9. See, for example, James Schmidt, "The Legacy of the Enlightenment," *Philosophy and Literature* 26, 2 (October 2002): 432–442.
10. See, for example, Helge Berger and Mark Spoerer, "Economic Crises and the European Revolutions of 1848," *The Journal of Economic History* 61, 2 (June 2001): 293–306; Owen Connelly, *The Wars of the French Revolution and Napoleon, 1792–1815* (New York: Routledge, 2006); Pierre Desrochers, "How Did the Invisible Hand Handle Industrial Waste? By-Product Development before the Modern Environmental Era," *Enterprise & Society* 8, 2 (June 2007): 348–374; Howe, *What Hath God Wrought*, 544–546.
11. The literature on this issue is voluminous. See, for example, Paul Burkett, "Nature in Marx Reconsidered," *Organization & Environment* 10, 2 (June 1997): 164–183; Brett Clark and John Bellamy Foster, "The Environmental Conditions of the Working Class: An Introduction to Selections from Frederick Engels's *The Condition of the Working Class in England in 1844*," *Organization & Environment* 19, 3 (September 2006): 375–388; John Bellamy Foster, "Ecology and the Transition from Capitalism to Socialism," *Monthly Review* 60, 6 (November 2008): 1–12; Michael Lowy, "From Marx to Ecosocialism," *Capitalism, Nature, Socialism* 13, 1 (March 2002): 121–133; Jason W. Moore, "Capitalism as World-Ecology," *Organization & Environment* 16, 4 (December 2003): 431–458.
12. See, for example, William A. Koelsch, "Thomas Jefferson, Geographers, and the Uses of Geography," *Geographical Review* 98, 2 (April 2008): 260, 264–269; Ingo Schwarz, "Alexander von Humboldt's Visit to Washington and Philadelphia, His Friendship with Jefferson, and His Fascination with the United States," *Northeastern Naturalist* 8, Special Issue 1 (2001): 43–56.
13. See, for example, Ida Altman, "The Revolt of Enriquillo and the Historiography of Early Spanish America," *The Americas* 63, 4 (April 2007): 587–614; Richard Flint, *No Settlement, No Conquest: A History of the Coronado Entrada* (Albuquerque: University of New Mexico Press, 2008); Andrés Reséndez, *A Land So Strange: The Epic Journey of Cabeza de Vaca: The Extraordinary Tale of a Shipwrecked Spaniard Who Walked across America in the Sixteenth Century* (New York: Basic Books, 2007); Richard White, "Discovering Nature in North America," *The Journal of American History* 79, 3 (December 1992): 877–885.
14. Many fine works exist on the Lewis and Clark expedition. Perhaps the most accessible is Stephen E. Ambrose, *Undaunted Courage: Meriwether Lewis, Thomas Jefferson, and the Opening of the American West* (New York: Touchstone Books, 1996).
15. On Turner's thesis about the closing of the American frontier, see Frederick Jackson Turner, "The Significance of the Frontier in American History," *Annual Report of the American Historical Association for 1893* (Washington, D.C.: U.S. Government Printing Office, 1894): 199–227. See also Robert H. Block, "Frederick Jackson Turner and

American Geography," *Annals of the Association of American Geographers* 70, 1 (March 1980): 31–42; Ronald H. Carpenter, "Frederick Jackson Turner and the Rhetorical Impact of the Frontier Thesis," *Quarterly Journal of Speech* 63, 1 (April 1977): 117–129; T. R. C. Hutton, "Beating a Dead Horse: The Continuing Presence of Frederick Jackson Turner in Environmental and Western History," *International Social Science Review* 77, 1/2 (2002): 47–57; Alex Wagner Lough, "Henry George, Frederick Jackson Turner, and the 'Closing' of the American Frontier," *California History* 89, 2 (2012): 4–23; Wilbur Zelinsky, *The Cultural Geography of the United States* (Englewood Cliffs, N.J.: Prentice Hall, 1973), 34–35.

16. Mike Dunning, "Manifest Destiny and the Trans-Mississippi South: Natural Laws and the Extension of Slavery into Mexico," *Journal of Popular Culture* 35, 2 (Fall 2001): 111–127; Sally Frahm, "The Cross and the Compass: Manifest Destiny, Religious Aspects of the Mexican-American War," *Journal of Popular Culture* 35, 2 (Fall 2001): 83–99; Howe, *What Hath God Wrought*, 762–764; Michael A. Morrison, *Slavery and the American West: The Eclipse of Manifest Destiny and the Coming of the Civil War* (Chapel Hill: University of North Carolina Press, 1997).
17. Quoted in Howe, *What Hath God Wrought*, 704.
18. Quoted in George Caitlin, *Letters and Notes on the Manners, Customs, and Condition of the North American Indians*, vol. I (London: Tosswell and Myers, 1841), 260. Italics included in the original. See also Kline, *First Along the River*, 34–35. For an in-depth analysis of Caitlin's life and work, see especially John Hausdoerffer, *Caitlin's Lament: Indians, Manifest Destiny, and the Ethics of Nature* (Lawrence: University Press of Kansas, 2009).
19. Henry David Thoreau, "Walking," in *The Writings of Henry David Thoreau, Vol. V: Excursions and Poems*, Bradford Torrey, editor (Boston and New York: Houghton Mifflin and Company, the Riverside Press, Cambridge, 1906 [1865]), 217–218.
20. Howe, *What Hath God Wrought*, 702–708; Joseph R. Stromberg, "Imperialism, Noninterventionism, and Revolution: Opponents of the Modern American Empire," *The Independent Review* 11, 1 (Summer 2006): 81–82.
21. Fred I. Greenstein, "The Policy-Driven Leadership of James K. Polk: Making the Most of a Weak Presidency," *Presidential Studies Quarterly* 40, 4 (December 2010): 726–727; Howe, *What Hath God Wrought*, 701–702.
22. Quoted in "James K. Polk," *U.S. Presidential Inaugural Addresses* (Whitefish, Mont.: Kessinger Publishing, 2004), 79. See also Terri Bimes and Quinn Mulroy, "The Rise and Decline of Presidential Populism," *Studies in American Political Development* 18, 2 (Fall 2004): 143. Polk also included an oblique defense of slavery as well as a denunciation of the rabble rousers who would threaten the integrity of the Union with their abolitionist sentiments: "It is a source of deep regret that in some sections of our country misguided persons have occasionally indulged in schemes and agitations whose object is the destruction of domestic institutions existing in other sections—institutions which existed at the adoption of the Constitution and were recognized and protected by it. All must see that if it were possible for them to be successful in attaining their object the dissolution of the Union and the consequent destruction of our happy form of government must speedily follow." "James K. Polk," 82.
23. Quoted in "James K. Polk," 86–87.
24. Bimes and Mulroy, "The Rise and Decline of Presidential Populism," 138; Howe, *What Hath God Wrought*, 708–710.

25. See, for example, Karen Clay and Randall Jones, “Migrating to Riches? Evidence from the California Gold Rush,” *The Journal of Economic History* 68, 4 (December 2008): 997–1027; Jacques G. Richardson, “The Sweeping Changes of 1867—What Became of Them, and How,” *Foresight: The Journal of Future Studies, Strategic Thinking and Policy* 5, 3 (2003): 52–53; Stacey L. Smith, “Remaking Slavery in a Free State: Masters and Slaves in Gold Rush California,” *Pacific Historical Review* 80, 1 (February 2011): 28–63.
26. See, for example, Marc Egnal, “The Beards Were Right: Parties in the North,” *Civil War History* 47, 1 (March 2001): 30–56; Daniel M. Mulcare, “Restricted Authority: Slavery Politics, Internal Improvements, and the Limitation of National Administrative Capacity,” *Political Research Quarterly* 61, 4 (December 2008): 671–685.
27. See, for example, United States Census Bureau Website: “History: 1790 Fast Facts,” http://www.census.gov/history/www/through_the_decades/fast_facts/1790_fast_facts.html (Accessed August 16, 2012); United States Census Bureau Website: “History: 1830 Fast Facts,” http://www.census.gov/history/www/through_the_decades/fast_facts/1830_fast_facts.html (Accessed August 16, 2012); United States Census Bureau Website: “History: 1860 Fast Facts,” http://www.census.gov/history/www/through_the_decades/fast_facts/1860_fast_facts.html (Accessed August 16, 2012); United States Census Office, *Eighth Census: Population* (Washington, D.C.: U.S. Government Printing Office, 1860); Thomas Weiss, “Estimates of White and Nonwhite Gainful Workers in the United States by Age Group, Race, and Sex: Decennial Census Years, 1800–1900,” *Historical Methods* 32, 1 (Winter 1999): 21–35; Thomas Weiss, “Demographic Aspects of the Urban Population, 1800 to 1840,” in *Quantity and Quiddity: Essays in U.S. Economic History*, Peter Kilby, editor (Middletown, Conn.: Wesleyan University Press, 1987), 171–213.
28. Kline, *First Along the River*, 35, 37–39. See also Patrick A. Gambrel and Rebecca Cianci, “Maslow's Hierarchy of Needs: Does It Apply in a Collectivist Culture?” *Journal of Applied Management and Entrepreneurship* 8, 2 (April 2003): 143–161.
29. See, for example, David M. Wrobel, “Global West, American Frontier,” *Pacific Historical Review* 78, 1 (February 2009): 1–26.
30. Many sources provide information on Audubon's life. See, for example, Michael K. Steinberg, “Audubon Landscapes in the South,” *The Mississippi Quarterly* 63, 1/2 (Winter 2010): 314–315.
31. Albert Boime, “John James Audubon: A Birdwatcher's Fanciful Flights,” *Art History* 22, 5 (December 1999): 728–755; Steinberg, “Audubon Landscapes in the South,” 313–314.
32. Douglas Brinkley, *The Wilderness Warrior: Theodore Roosevelt and the Crusade for America* (New York: HarperCollins, 2009), 10–11, 184–189.
33. National Audubon Society Website: “Home Page,” http://www.audubon.org/john-james-audubon (Accessed August 7, 2012).
34. As with many figures discussed in this book, the literature on Thoreau's life is voluminous. See, for example, Daniel Botkin, *No Man's Garden: Thoreau and a New Vision of Civilization and Nature* (Washington, D.C. and Covelo, Calif.: Island Press/Shearwater Books, 2001).
35. Botkin, *No Man's Garden*, 88–90; Curtis Kent, “The Virtue of Thoreau: Biography, Geography, and History in Walden Woods,” *Environmental History* 15, 1 (January 2010): 40–41.

36. Henry David Thoreau, *Walden; or, Life in the Woods* (New York: Thomas Y. Crowell & Company, 1910 [1854]), 118.
37. Ibid., 119.
38. Ibid., 408.
39. Ibid., 419.
40. Botkin, *No Man's Garden*, xvi; Kent, "The Virtue of Thoreau," 31–33; Kline, *First Along the River*, 32–34.
41. See, for example, Charles W. Calhoun, *From Bloody Shirt to Full Dinner Pail: The Transformation of Politics and Governance during the Gilded Age* (New York: Hill & Wang, 2010), 4–9; Ballard C. Campbell, "Understanding Economic Change during the Gilded Age," *Magazine of History* 13, 4 (Summer 1999): 16–20.
42. See, for example, Bryant Morey French, *Mark Twain and "The Gilded Age": The Book That Named an Era* (Dallas: Southern Methodist University Press, 1965); Christopher D. Morris, "Hermeneutic Delusion in Mark Twain and Charles Dudley Warner's *The Gilded Age*," *Nineteenth-Century Literature* 66, 2 (September 2011): 219–243; Michael Robbins, "The Early Years of Electric Traction: Invention, Development, Exploitation," *The Journal of Transport History* 21, 1 (March 2000): 94–95.
43. George Perkins Marsh, *Man and Nature; or Physical Geography as Modified by Human Action* (New York: Charles Scribner, 1864), 36.
44. Ibid., 42. See also Kline, *First Along the River*, 46–47; Shabecoff, *Earth Rising*, 2.
45. John Wesley Powell, *The Exploration of the Colorado River and Its Canyons* (New York: Penguin Books, 2003 [1875]), 83. See also Kline, *First Along the River*, 47.

Chapter 7

The Twentieth Century

What have they done to the earth?
What have they done to our fair sister?
Ravaged and plundered and ripped her and bit her
Stuck her with knives in the side of the dawn
And tied her with fences and dragged her down.

—The Doors, "When the Music's Over"[1]

Comes now twentieth-century man. Is he poised on the edge of a calamitous precipice of environmental degradation or ascending to a higher plane of human existence owing to his mastery of technology and its promise of one day mitigating his destructive behavior? With progress at his back, propelling him forward into a brave new world of human achievement, must he stop to take account of the woes his actions have visited on the natural world or are such questions prudently left to as-yet undreamt of devices and processes? For a creature that can overcome the boundaries and impediments of his world, the future appears limitless. In the face of such change, is man a marvel of enlightened innovation or a disastrous blight on a polluted planet? Can his existence be both a blessing and a curse? These dichotomous questions lurked behind the accelerating pace of industrialization as the twentieth century dawned.

Progressivism and Twentieth-Century Social Movements

The question of twentieth-century man's relationship with, and responsibility for, the natural environment arose concomitantly with the birth of the Progressive movement late in the nineteenth century. The roots of Progressivism and its emphasis on curing all manner of social ills stretched back for decades. In the aftermath of the Civil War, Americans of modest means expressed frustration at their limited economic prospects. They joined fraternal organizations such as the National Grange of the Order of Patrons of Husbandry, a post-Civil War group dedicated to promoting the economic and political interests of rural farming families. During the 1870s and 1880s, the Greenback Party called for a currency system not backed by gold as a means of ensuring that more money would circulate, thereby providing relief to less-affluent citizens who desperately needed specie to survive in a depressed economic climate following the Panic of 1873. In the 1890s, the Populist Party arose to champion "the people" over the moneyed elite that seemed to control political power and economic policy, often to the detriment of the little man. Although the party faded during the twentieth century, a concern for the needs of the lower and middle classes remained a staple of American politics for generations to come.[2]

From these nineteenth-century roots, Progressivism prospered. Progressives believed that the growth of corporations and systematic corruption of the country's political parties and governmental institutions were stifling the middle class. Unlike classical liberals who believed that government is best which governs least, Progressives argued that government is either part of the problem or part of the solution. To be part of the solution, government must intervene to ensure that citizens enjoy a level playing field with elites. No longer could elected leaders manipulate the electorate with impunity. No longer could big business expand with few limitations on growth or evince a lack of concern for the citizenry. Government would intervene to hold the power of the private sector in check. If elected officials stuffed ballot boxes and fixed elections, government officials would police them. If products were defective, the companies that produced them would be forced to pay rather than force social costs onto society. If corporate interests polluted the countryside, they would be penalized and the costs would be borne by the entity that created them.[3]

Social movements were dedicated to combating the perceived inequities of the time, which meant that Progressives involved themselves in numerous activities. Concerned about the unfettered power of corporations, they called for "trust-busting" laws to curb the rise of monopolies. To counteract corporate heft, Progressives promoted labor unions as a means of empowering labor to deal effectively with management. Part of their concern about poor workers stemmed from tales of children laboring in dangerous factories for long hours under appalling conditions. As the nature and extent of the problems were publicized, the public was horrified, especially following calamities such as the Triangle Shirtwaist Factory fire of March 1911, in which 146 people, mostly girls and young women, died after a fire spread throughout the factory and workers could not escape because the exit doors were locked. Poor children did not have an opportunity to attend public school—they had to work so their impoverished families could earn enough money to survive—thus, the cycle of poverty would never end unless child labor laws restricted the age and conditions under which they worked. Progressives lobbied to enact new laws that would correct these deficiencies.[4]

The settlement house movement, one of the most celebrated of Progressive endeavors, originated in England in the 1880s as Victorians, increasingly worried about the effects of poverty, assisted the poor in receiving food, shelter, and clothing. The movement spread to the United States, where Progressives established a series of settlement houses. Over time, settlement house activists created a professional class of social worker to replace well-meaning amateurs. Hull House, a settlement house established in Chicago in 1889 by Jane Addams and Ellen Gates Starr, was perhaps the most famous settlement house in America and became emblematic of the new social activism of the Progressive era.[5]

Progressives also turned their attention to resolving pressing environmental issues. With a surging population—some 9.5 million immigrants came to the United States between 1900 and 1915—and a burgeoning economy, consumerism blossomed. As consumers used more resources, environmental degradation increased. Reformers recognized that government could serve as an effective mechanism for reducing or eliminating environmental damage, even if members of the environmental community—preservationists and conservationists—could not always agree on the ends.[6]

John Muir, Gifford Pinchot, Theodore Roosevelt, and the Preservation–Conservation Debate

For a growing chorus of Progressive activists, modern man could not slough off the problems of environmental degradation for future generations to tackle. Nature must be protected, and the time to act was at hand. Perhaps no one epitomized the new thinking as well as John Muir, a larger-than-life character who championed environmental preservation toward the end of the nineteenth century. Explorer, author, founder of the Sierra Club, environmental activist before anyone thought to coin the phrase "environmental lobby," he was among the most powerful and influential naturalists of his, or any, era. He came to see that the accelerating rate of industrialization and the scope of environmental destruction presented a clear, credible threat to America's abundant, but finite, natural resources. Someone must do something soon—and he was that someone. He would act immediately. In his approach to environmental preservation, he did much to promote America's national park system and usher in twentieth-century environmentalism.[7]

Born in Dunbar, Scotland, in 1838, Muir emigrated from Glasgow to Marquette County, Wisconsin, with his family when he was 11 years old. He was always an odd, intensely driven fellow who marched to the beat of his own drummer. As a boy growing up on a farm, he displayed a love of the out-of-doors and a wide-eyed wonder at nature's charms. He also demonstrated an aptitude for the rough-and-tumble outdoors life, a willingness to "rough it" no matter how stark the accommodations or how bleak the circumstances. On most days, young John could be found tinkering with farm machinery, exploring nature, and reflecting on man's relationship with his world. An inquisitive, hands-on explorer, Muir constantly sought explanations for natural and manufactured phenomena, pulling apart and reassembling machinery and gadgets to find answers to all manner of questions. It was not enough to accept the world as it was; he sought an explanation for the *how* and *why*. He wanted to understand what others seemed to accept as given.[8]

For all his excursions into the outdoors, he was no unschooled mountain man. Muir was a voracious reader and a self-taught inventor. While he was a student at the University of Wisconsin-Madison, he developed a desk that retrieved and reshelved books. His lessons showed him that historical figures also shared his concern for the environment. While studying American literature, he encountered Thoreau's reflections on nature as well as Emerson's uniquely American Transcendentalism. Muir discovered kindred spirits lurking among these quirky New Englanders. Thoreau's distinctive voice—especially his ruminations on man's relationship with nature—spoke to the young immigrant as nothing else had, and the pragmatic, intuitive nature of Transcendentalist philosophy appealed to Muir's strong sense of self as well as his rich interior life. Owing to his studies, he developed a passion for botany during those years, an interest that would push him to explore untamed lands in the United States, Canada, and parts of South America after he left university. He was satisfied to live frugally and pursue odd jobs to finance his wilderness adventures in his twenties, for he knew his future required him to be conversant with nature in all her beauty and glory.[9]

Throughout his travels, Muir felt a special bond for the area around what later became Yosemite National Park. He first explored the region in 1868 and was enthralled by the beauty and diversity of the plant and animal life he encountered. Muir's extensive studies in "the University of Wilderness," as he lovingly called it, convinced him that he must devote his life to preserving and protecting the nation's scenic wonders. The challenge was to convince citizens to support public policies for protecting pristine nature when many Americans did not enjoy the leisure time to encounter the majesty of sites such as the Sierra Nevada Mountains. Preserving nature absent a desire to harvest her bounty made for a tough sell; many Americans saw the environment as instrumentally valuable and little more. Describing the indescribable beauty of forest or meadow or valley was virtually impossible despite the celebrated eloquence of Muir's pen. As he wrote in his influential 1894 book *The Mountains of California*, "But, write as I may, I cannot give anything like an adequate idea of the exquisite beauty of the mountain carpets as they lie smoothly outspread in the savage wilderness. What words are fine enough to picture them?"[10]

Although he believed that words could never capture the innate beauty and aura of nature, John Muir never stopped trying. Following *The Mountains of California*, he produced numerous writings about his experiences in the wild. In *Our National Parks*, a 1901 polemic urging the U.S. government to carve out large swaths of land for protection, he extolled the virtues of the country's natural resources, especially western lands. Yet Muir was worried that too few people appreciated the dangers posed by manmade environmental destruction. "In the noblest forests of the world, the ground, once divinely beautiful, is desolate and repulsive, like a face ravaged by disease. This is true also of many other Pacific Coast and Rocky Mountain valleys and forests," he observed. "The same fate is awaiting them all, unless awakening public opinion comes forward to stop it."[11]

Later in that same work, he lamented the destruction of old-growth forests by loggers who cared nothing for the consequences of their shortsighted, misguided actions. "Any fool can destroy trees," Muir wrote. "They cannot run away; and if they could, they would still be destroyed—chased and hunted down as long as fun or a dollar could be got out of their bark hides, branching horns, or magnificent bole backbones." He explained that few loggers ever plant new trees after they cut the old ones. Even if they did, it would take generations to repair the damage. "It took more than a thousand years to make some of the trees in these Western woods—trees that are still standing in perfect strength and beauty, waving and singing in the mighty forests of the Sierra." Yet felling such majestic forests can be accomplished in a matter of days or weeks unless steps are taken to prevent indiscriminate cutting. "Through all the wonderful, eventful centuries since Christ's time—and long before that—God has cared for these trees, saved them from drought, disease, avalanches, and a thousand straining, leveling tempests and floods," Muir wrote in the conclusion to his book, "but He cannot save them from fools." With man's arrival, God alone cannot stop the destruction: "Only Uncle Sam can do that."[12]

Fortunately for John Muir, he had friends working in high places inside Uncle Sam's government. One man, Gifford Pinchot, was especially influential in the American environmental movement during the late nineteenth and early twentieth centuries. A Connecticut native and a graduate of Yale University, Pinchot was interested in wise forest management practices throughout his life. After he finished Yale, he studied at the French National School of Forestry. His travels through France and Germany while still in his twenties convinced him that he could improve American forestry by instituting fundamental changes in land management practices. Pinchot was a rare character in any age—charismatic, handsome, strong, intelligent, and indefatigable, a man with the ability and will to carry out his plans. Early in his career, he experimented on the family estate, Grey Towers, near Milford, Pennsylvania, and the grounds of the Biltmore Estate outside Asheville, North Carolina, demonstrating how to plant and preserve forestlands and avoid deforestation.[13]

During the 1890s, while still a young man, Pinchot began to make a name for himself as a self-styled forester. At the time, no U.S. college or university offered a program in forestry, but Pinchot believed it was time to correct that oversight. A disciple of George Perkins Marsh, the author of *Man and Nature* discussed in Chapter 6, Pinchot came to believe that the wise use of wooded acreage did not preclude cutting trees; it required a policy to ensure that enough forests were left for later generations. This anthropocentric perspective put him at odds with environmentalists who argued that a Utilitarian view of nature was misguided and ultimately unsustainable. In time, it put him at odds with John Muir.[14]

Pinchot rose through the ranks to become a member of the National Forest Commission created by the National Academy of Sciences. Later, he served as chief of the Division of Forestry at the U.S. Department of Agriculture. After the U.S. Forest Service was established, he served as head from 1905 until 1910. Late in life, he was elected governor of Pennsylvania. A lasting friendship with Theodore Roosevelt aided his career during these years.[15]

Roosevelt was a pivotal figure in environmentalism during the Progressive era—one of the few nature lovers powerful enough to act on his love of the outdoors. Recognizing the importance of cultivating an influential benefactor, John Muir enjoyed a cordial relationship with Roosevelt although the two men never became close friends. In the same month that Muir penned the preface to *Our National Parks*—September 1901—Theodore Roosevelt became president of the United States after President McKinley was assassinated. When the president visited Yosemite National Park during a three-day excursion in 1903, he and Muir spent hours sitting around the campfire discussing their respective views on nature. Waving away other companions, Roosevelt spent long hours on horseback with the famed naturalist. An iconic photograph of the two men posing at Glacier Point overlooking Yosemite Valley has become one of the most famous advertisements for the country's national parks and has been reproduced countless times in books, magazines, postcards, and on websites. It can be found here as well (Figure 7.1). Roosevelt and Muir maintained their mutual admiration for each other even in instances when they disagreed on policy.[16]

The same could not be said for Muir and Pinchot. The former believed that humankind's presence was invariably destructive and could seldom, if ever, be ameliorated, while the latter believed that human beings could use natural resources effectively even if natural lands were not left in a pristine condition. Their rift can be thought of as the tension between biocentrism and anthropocentrism: Muir wanted to preserve nature and Pinchot wanted to conserve it.[17]

Their falling-out involved numerous issues. The two men argued over sheep grazing in national parks. Pinchot initially assured Muir that the practice was anathema, but later Pinchot changed his mind. Fearful that the Wool Growers Association had lobbied the forester, Muir confronted Pinchot. Railing against the hypocrisy of allowing sheep to disturb nature in a protected area, Muir assured his former friend, "I don't want anything more to do with you."

Figure 7.1 This famous photograph of President Theodore Roosevelt and naturalist John Muir was taken on Glacier Point overlooking Yosemite Valley, California, in 1903. (Courtesy of the Library of Congress)

Matters only grew worse from there. As San Francisco's population exploded early in the twentieth century and the need for water became critical, politicians developed a plan to dam the Tuolumne River that flows through Yosemite National Park to serve as a freshwater reservoir. Muir was incensed that man was so arrogant that he would presume to improve on nature. Using his political connections with

President Roosevelt, the naturalist raised a hue and cry. Normally a friend to Muir and other environmentalists, Roosevelt refused to intercede because he believed that the need for water to serve human interests outweighed the need to preserve the Hetch Hetchy Valley in Yosemite. Muir had been disturbed by Roosevelt's friendship with the anthropocentric Gifford Pinchot, and this rift seemed to confirm Muir's worst fears.[18]

Despite differences of opinion with President Roosevelt, Muir managed to delay the project for years by initiating a public relations campaign on multiple fronts. In an article he wrote for *The Century Magazine* in 1909, he sought to drum up support for preserving the pristine area. "Excepting only Yosemite," he wrote, "Hetch Hetchy is the most attractive and wonderful valley within the bounds of the great Yosemite National Park and the best of all camp-grounds." To emphasize the instrumental value of preserving the valley, he observed, "People are now flocking to it in ever-increasing numbers for health and recreation of body and mind." Aside from its recreational value, Hetch Hetchy offered a priceless aesthetic value. "Though the walls are less sublime in height than those of Yosemite, its groves, gardens, and broad, spacious meadows are more beautiful and picturesque.... It is one of God's best gifts and ought to be faithfully guarded."[19]

After Roosevelt left office, his successor, William Howard Taft, suspended the Department of the Interior's right of way, a necessary component in the dam construction process. Yet Muir's victory did not last. On December 19, 1913, President Woodrow Wilson signed a bill authorizing construction of the dam. It was Muir's last major battle on behalf of the natural environment. He died a little over a year after Wilson signed the bill, and so he did not live to see completion of the dam in 1923.[20]

The debate between John Muir and Gifford Pinchot on the virtues of preservation versus conservation persists in the environmental community to this day. Preservationists believe that nature must be left alone in its primitive state without interference from humankind. Diverting or damming a river causes consequences that human beings do not understand and cannot foresee. Conservationists believe that human needs supersede the needs of other species. The fruits of nature can be harvested to meet those needs, but care must be taken not to exhaust a resource. Clear-cutting forests, polluting rivers and streams, or overfishing rivers destroys the resource so that future needs cannot be satisfied. Despite their differences, preservationists and conservationists agree that humankind is a change agent, and seldom an agent of change for the best.[21]

From Progressivism to the New Deal: The Evolving Role of Government in Natural-Resource Protection

The Gilded Age gradually gave way to a new era. Following the end of World War I in 1918, the United States emerged as one of the leading nations of the world. Yet Americans were weary of global involvement. They had paid a high price in blood and treasure to intervene in European affairs during the Great War. U.S. citizens were anxious to put the horrors of trench warfare behind them and live lives of tranquility insulated from political and social strife. In that time and place, few Americans had the stomach for delving into complex issues such as protecting the natural environment, which involved seemingly intractable problems.[22]

They got their wish for a time. With the triumph of presidential candidate Warren G. Harding in the 1920 election, Americans achieved their earnest desire for a "return to normalcy." This longing for a simpler, less-worrisome life translated into the Roaring Twenties, a frivolous, carefree, exuberant decade characterized by a laissez-faire attitude toward the private sector and an emphasis on popular culture—Vaudeville; motion pictures; "flappers"; public figures such as the suave, debonair movie idol Rudolph Valentino; the fascinating magician and escape artist Harry Houdini; and "Lucky" Lindy, the aviation pioneer Charles Lindbergh who completed a hazardous transatlantic airplane crossing despite long odds and formidable obstacles. Illicit resistance to Prohibition's laws against imbibing alcoholic spirits and a less inhibited view of sexual mores lent an edge of danger and salaciousness to the era.[23]

President Harding epitomized the age: With his copious, silver-haired locks and easy-going manner, he appeared to be a character sent over from central casting in Hollywood to portray the part of the wise, empathetic, "regular-guy" president. He played his part well. Like all good actors, the man knew how to move gracefully through the landscape and delight the crowd with his beguiling performance. Yet beneath the façade of a competent chief executive, he was utterly vacuous, a placeholder ideally suited to an age that did not wish to challenge the citizenry to tackle big problems in a big way. Harding understood his role. He was not interested in pursuing a Progressive agenda. Aside from the issues that time and circumstances required him to face, he limited his administration's initiatives to a call for tax cuts to stimulate private investment and a naïve, wishful belief that all would be well in a highly industrialized society if government was as weak and ineffectual as he was.

Owing to the lack of executive leadership, the Harding administration was rife with abuses of power. Indeed, the most infamous example of government corruption before the Watergate episode of the 1970s occurred during Harding's tenure. The president's Secretary of the Interior, Albert Fall, had leased oil deposits at Elk Hills, California, and Teapot Dome, Wyoming, to petroleum companies in exchange for kickbacks. The details of the corrupt scheme were unknown by most Americans until after Harding died suddenly in office in 1923. When Congress investigated the affair during Calvin Coolidge's administration, the scope of the malfeasance became clear and the late president's image was forever tarnished in the history books. It was but one example, albeit an extreme one, of the vast indifference the Harding administration demonstrated toward the environment and the nation's natural resources.[24]

The dead president's successor, Calvin Coolidge, while not as indifferent to environmental problems as Harding, did not make natural resources a national priority of his government. For all of his sojourns into the wooded acreage of South Dakota, Coolidge was interested in letting the good times roll while pursuing the same hands-off policy that seemed to place his predecessor in good stead. What neither he nor his contemporaries could foresee was that the good times would not last indefinitely. The Great Depression arrived at decade's end to top off the decadence of the Twenties. A grim reality soon set in, namely that the American way of life and standard of living would change radically.[25]

Herbert Hoover, Commerce Secretary under Harding and Coolidge, ascended into the White House in 1929, less than a year before the stock market crash that triggered the worldwide economic downturn. Historians often have castigated Hoover—and, in most cases, rightly so—but he was not without his virtues, one of which was his desire to preserve and protect natural resources to a greater extent than had the previous two administrations. The 1920s had witnessed enormous increases in demand for automobiles, electricity, and agricultural commodities. This voracious appetite for resources stimulated private-sector companies to churn out record numbers of cars as well as build additional electric-generating power plants and cut down forests so wheat fields could flourish in the American heartland. Such developments might have been a godsend for the national economy during the 1920s, but without a sound environmental management plan, no measures were in place to regulate, much less ameliorate, environmental despoliation. By the time Hoover took office, it was clear that the country's natural resources were under attack as Americans' burgeoning love affair with fossil fuels led to unmitigated air and water pollution. Hoover had fought for the development of salmon fisheries in Alaska, the construction of the St. Lawrence waterway, and the Boulder Canyon project (which was renamed Hoover Dam) before he came to the presidency. He knew that leadership and direction were needed to ensure natural-resource protection.[26]

Alas, Herbert Hoover was not the man to provide executive leadership on environmental issues. As the Great Depression dawned and consumed his administration's time and attention, he did not devote resources to promoting environmentalism. Even if his administration had been less consumed by economic disaster, Hoover might not have spent time and energy on natural-resource management. In keeping with his political philosophy and the era in which he lived, President Hoover was not a champion of a robust, activist federal government. He understood that government had a role in setting the national agenda, but it is doubtful that Hoover would have offered far-reaching policies aimed at making the federal executive branch a leader in policing private-sector behavior and possibly curtailing industrial development in the name of environmental protection.[27]

Franklin D. Roosevelt, Hoover's successor, would lead the charge from a bully pulpit that was vastly different from the one Hoover used during his White House years. Roosevelt understood that the economic calamity facing the nation during the 1930s was unprecedented in its scope and destructive effect. Business cycles always rise and fall, but the Great Depression was something altogether different. If the suffering of millions of Americans was to be lessened, government must respond in new and creative ways. Although the economic policies of Roosevelt's first term, the so-called "New Deal," did not resolve every problem for which they were designed, the new administration demonstrated a willingness to use the engine of government in flexible, innovative ways. The federal executive branch would not sit on the sidelines and allow the private sector to take the lead on crucial policy issues. Government would respond energetically. No issue was too remote to escape the scrutiny of the newly revitalized, undeniably robust federal authority.

Franklin Roosevelt's early years in the executive chair were devoted to finding a fix for economic and social concerns—rampant unemployment, hunger, the lack of a suitable infrastructure, and social security for the elderly and disabled, among other issues—but the president was at heart a nature lover, and he never lost sight of the issue despite other priorities. Indeed, he believed that government programs could address multiple problems—the proverbial killing of two birds with one stone. A prime example was the Civilian Conservation Corps (CCC), one of the alphabet-soup agencies created to put Americans back to work. The primary purpose was to increase employment, but if the CCC could improve the natural environment along the way, so much the better. The CCC was modeled on a hierarchical military organization, but its peacetime mission was to marshal the resources and expertise of the unemployed to build new parks, beautify existing hiking trails, plant trees, and repair dilapidated roads and bridges.[28]

President Roosevelt was pleased with the CCC's successes, but his administration's environmental efforts did not stop there. He turned to Harold Ickes, his Secretary of the Interior, to assist in effective conservation management. Ickes also directed an alphabet-soup agency known as the Public Works Administration (PWA). Among other things, the PWA sited large public works projects in depressed areas to help shore up the local economy. During Ickes' 13-year tenure at Interior, he successfully pushed the U.S. government to establish some of the nation's best-loved national parks: Kings Canyon in California, Shenandoah in Virginia, and Olympic in Washington state. He also helped to acquire land that eventually became the Grand Tetons National Park. In 1940, the federal government established the U.S. Fish and Wildlife Service for "conserving the nature of America" by establishing wildlife refuges and sanctuaries.[29]

The Dust Bowl of the American Southwest, where severe dust storms in the 1930s decimated millions of acres of farmland following record droughts, was a calamitous and heartbreaking crisis during the era. If farmers were to improve their lot, effective agricultural practices would have to be implemented, for the crisis was at least partially manmade. Soil erosion caused by a sustained drought, overplanting crops that robbed the soil of precious nutrients, and overgrazing by cattle could be combated, but the might of the federal government was required. Thus, Congress passed and the president signed the Taylor Grazing Act of 1934 to limit cattle grazing to designated areas. The Soil Conservation Act established the Soil Conservation Service, which developed plans for constructing dams, providing much-needed irrigation in specified fields, and advising farmers on proper soil management techniques. The Tennessee Valley Authority (TVA) created a hydroelectric dam that provided construction jobs, low-cost electricity to the poor and needy, and helped to improve environmental quality in a distressed region (although this last point was a debatable point for environmentalists). Well into the next century, the TVA acronym would be synonymous with activist public policy.[30]

Not all of the Roosevelt administration's plans would prove to be popular. The Agricultural Adjustment Administration (AAA) devised a plan to improve market prices for certain commodities. A fundamental principle of economics dictates that when a surplus of goods floods the marketplace, prices drop. Wheat, corn, dairy products, and pigs could be found in plentiful supply in some places throughout the country in the 1930s. Consequently, the price for these products continued to drop at a time when farmers could barely eke out a living. The AAA plan would pay farmers not to produce these products. The surplus would disappear and, presumably, prices would increase. At a time when some Americans were hungry and desperate for economic relief, the idea that their government was paying farmers *not* to produce goods was met with vehement protest.[31]

Environmental Activism Comes of Age

Outside of the formal institutional structure of government, the New Deal era saw the birth of a new class of environmental activist. Aldo Leopold arguably was the most famous and influential of this breed. Born into a middle-class Iowa family in 1887, Leopold came of age at the turn of the century, not long after the closing of the American frontier. His family members were outdoor enthusiasts who taught the young man the benefits of living, working, and playing in the open air. Just as a young John James Audubon demonstrated an aptitude for observing birds existing in their natural habitats, Aldo Leopold found joy in bird watching from an early age. When Leopold learned that Gifford Pinchot, the renowned ecologist who headed the Division of Forestry in the Department of Agriculture, had donated funds to Yale University to establish a much-needed school of forestry, the young fellow knew what he must do. Anxious to secure admission into the prestigious university, he persuaded his parents to enroll him in the Lawrenceville School, a preparatory academy in New Jersey. He later entered Yale. Although the forestry program only granted graduate degrees, Leopold completed several related courses during his undergraduate years.[32]

When he was 22 years old, Aldo Leopold joined the U.S. Forest Service. Initially, he drew an assignment in District 3, which covered Arizona and New Mexico. He held various positions in that region of the country until 1924. The wide-open spaces and plentiful federal lands afforded the budding ecologist with numerous opportunities to observe nature and man's relationship to the indigenous plants and animals.[33]

One of his early assignments required him to kill mountain lions, wolves, and bears because the animals preyed on livestock and caused problems for ranchers and farmers. In his classic work *A Sand County Almanac*, published in 1949, a year after Leopold's death, he described a moment of conversion when he realized he had no business shooting wild animals. In a passage titled "Thinking Like a Mountain," he remembered eating lunch with his colleagues when they spotted a family of wolves frolicking near a turbulent river:

> In those days we had never heard of passing up a chance to kill a wolf. In a second we were pumping lead into the pack, but with more excitement than accuracy: how to aim a steep downhill shot is always confusing. When our rifles were empty, the old wolf was down, and a pup was dragging a leg into impassable slide-rocks. We reached the old wolf in time to watch a fierce green fire dying in her eyes. I realized then, and have known ever since, that there was something new to me in those eyes—something known only to her and to the mountain. I was young then, and full of trigger-itch; I thought that because fewer wolves meant more deer, that no wolves would mean a hunters' paradise. But after seeing the green fire die, I sensed that neither the wolf nor the mountain agreed with such a view.[34]

It was a revelatory moment. Leopold understood that nature had a balance and harmony that human beings often failed to appreciate. Shooting an animal because humanity judges that animal to be harmful does not consider the natural order of things.

In a later passage, titled "The Land Ethic," Leopold discussed the interconnectedness of all things and "that the individual is a member of a community of interdependent parts." This "land ethic," in his view, "simply enlarges the boundaries of the community to include soils, waters, plants, and animals, or collectively: the land." Despite subsequent attempts to portray him as a starry-eyed mystic, Leopold was anything but a naïve, New-Age guru spouting out clichés about Mother Earth's inviolable rights. Instead, he appreciated values other than human needs and interests. "A land ethic of course cannot prevent the alteration, management, and use of these 'resources,'" he conceded, "but it does affirm their right to continued existence, and, at least in spots, their continued existence in a natural state." Stated another way, "Conservation is a state of harmony between men and land."[35]

A Sand County Almanac was to the twentieth-century environmental movement what *Walden* was to nineteenth-century ecology: It was a defining work, a watershed, a call to incorporate an environmental ethic into decision-making. Edited by Aldo's son, Luna, the work appeared not long after the elder Leopold died from a heart attack he suffered while fighting a wildfire on a neighbor's property. Written in eloquent, simple prose and easily accessible to the masses, the book has sold more than two million copies and has been translated into a dozen languages. When pressed to identify the most influential works of popular nature writing from the last hundred years, most American environmentalists identify *A Sand County Almanac* and Rachel Carson's *Silent Spring*.[36]

Although he was not as well known as Aldo Leopold, Robert "Bob" Marshall was another influential twentieth-century environmental activist. Marshall hailed from a privileged Jewish family—his father, Louis, was a partner in the powerful Washington, D.C. law firm of Guggenheimer, Untermeyer, and Marshall—and grew up with politically liberal values of protecting minority rights, combating racial and ethnic discrimination, and advocating on behalf of civil liberties. Even as a child, he participated in strenuous outdoor activities, emulating his father's example. The elder Marshall was active in efforts to protect the Adirondack Forest Preserve in New York. Young Marshall developed a deep love of nature and a belief that he must do what he could to persuade the federal government to protect American forestland from unscrupulous timber interests.[37]

Bob Marshall and Aldo Leopold co-founded an environmental group, the Wilderness Society, in 1935 to "protect wilderness and inspire Americans to care for our wild places." Later, Marshall served as head of the Division of Recreation and Lands for the U.S. Forest Service and worked briefly in the Forestry Division of the Bureau of Indian Affairs. During his time in both the public and private sectors, Marshall, echoing Henry David Thoreau, came to see a forest as a place to retreat "from the encompassing clutch of a mechanistic civilization." He feared that the encroaching forces of a culture largely predicated on consuming natural resources would trigger a "ruination of the forest beauty" if steps were not taken to curb environmental abuses. Sounding a clarion call in the grand tradition of Jean-Jacques Rousseau, he argued that "people cannot live generation after generation in the city without serious retrogression—physical, moral, and mental."[38]

Despite his close association with, and tenure at, the U.S. Forest Service, Marshall criticized the organization's policy of allowing development of forestlands. Siding with the preservationist school of John Muir against the conservationist ethic of Gifford Pinchot and his progeny, Marshall contended that any level of industrial development threatened forests. Recognizing that the U.S. attitude toward nature tended to shift over time in favor of development, he worried that prohibitions against logging in certain areas gradually would relax, thereby allowing incremental encroachment of timber-cutters as time progressed. He also argued that the National Park Service's efforts to promote recreation in the parks only hastened the pace of environmental destruction. No matter how light the touch or how well meaning the intentions, human beings invariably spoil the natural environment.[39]

Marshall articulated a radical ethic sometimes labeled a "green utopia." In his 1930 pamphlet *The Social Management of American Forests* as well as his 1933 book *The People's Forests*, Marshall argued on behalf of public ownership of all U.S. timberlands. A self-identified socialist, he believed that public stewardship was the most effective method of preserving wilderness acreage in perpetuity. To combat the criticism that only upper-income Americans possess the necessary resources to visit protected federal lands, he suggested that if recreational use of parks and wildlife areas continued, a provision could be made to subsidize visits by persons of modest means. This latter point proved to be controversial and subjected him to charges of inconsistency. If he sought to protect forests from unwarranted intrusion from human beings, empowering less-affluent Americans to visit the parks would increase the level of destruction.[40]

Marshall was a radical, enigmatic figure; he paid little attention to developing logically consistent positions, instead devoting his time and energy to publicizing the plight of nature and her resources. "Rocks and mountains may be ageless, but man and society are emphatically of the present, and they cannot wait for the slow process of nature to retrieve the catastrophe caused by their unthinking destructiveness," he wrote in *The People's Forests*. As for his plan promoting public ownership of forestlands, he explained, "The fundamental advantage of public ownership of forests over private ownership is that in the former social welfare is substituted for private gain as the major objective of management."[41]

In 1938, this radical environmentalist attracted the attention of an investigatory committee from the U.S. House of Representatives delving into "un-American activities." Committee chairman Martin Dies, a reactionary Democratic congressman from Texas known for his intense hatred of Franklin Roosevelt's New Deal, told the *New York Times* that the committee was investigating eight federal officials with ties to organizations sympathetic to Communist causes. Bob Marshall was among the officials the Dies Committee was investigating. Frequently away on wilderness travels, Marshall avoided unpleasant publicity surrounding the congressional witch-hunt, but it was clear his socialist leanings might harm the future of federal wilderness management. Wilderness Society executive director Robert Sterling Yard expressed his concern that Marshall's "democratic wilderness" proposal, aside from being logistically unworkable and economically impracticable, would politicize the issue and play into the struggles between conservatives seeking to discredit the New Deal and liberals battling their reactionary critics.[42]

The issue became moot. While on an overnight train trip from Washington, D.C. to New York City the following year, the 38-year-old activist unexpectedly died in his sleep. Independently wealthy and with no family to provide for, Marshall left his $1.5 million estate (equivalent to $25 million in 2012 dollars) to several causes near and dear to his heart, including a trust established for "preservation of wilderness conditions in outdoor America, including, but not limited to, the preservation of areas embracing primitive conditions of transportation, vegetation, and fauna."[43]

Marshall's abrupt death forestalled further scrutiny that might have tarnished the Wilderness Society with a charge of subversion. He left behind a rich legacy that culminated in one high-profile environmental achievement in particular. In 1964, when President Lyndon B. Johnson signed the Wilderness Act into law to protect nine million acres of wilderness lands, Marshall's Wilderness Society had been instrumental in lobbying efforts that helped to enact the law. That same year, the federal government established the Bob Marshall Wilderness in the Flathead and Lewis and Clark National Forests in Montana.[44]

Alice Hamilton was another twentieth-century activist representing a new breed of environmentalist. Unlike Aldo Leopold and Bob Marshall, she was not a forester. Trained in medicine, she became the first woman to serve on the faculty of Harvard University. Many books on environmentalism ignore her contributions, but she was an important bridge between naturalists of the early and late parts of the twentieth century. Dr. Hamilton became vitally interested in occupational health and safety. A supporter of the settlement house movement and a former resident of Hull House, she had spent time with poor residents and therefore understood that some of their health problems were caused, at least in part, by prolonged exposure to toxic chemicals. Today, Dr. Hamilton is remembered for her contention that workers in manufacturing plants were unnecessarily exposed to numerous chemicals about which little information was known. Some toxic substances, she argued, have no acceptable threshold of safety. She became a prominent advocate for reducing or eliminating commonly encountered hazardous materials such as tetraethyl lead in gasoline. "It makes me hope that the day is not far off when we shall take the next step and investigate a new danger in industry before it [i.e., tetraethyl lead] is put into use, before any fatal harm has been done to workmen," she wrote in a 1925 article. She did not want industrial dangers to be "demonstrated on the bodies of workmen."[45]

In her classic book *Industrial Poisons in the United States*, Hamilton explained, "The sources of our knowledge of industrial poisoning in the United States are neither full nor, for the most part, accurate. We lack the sickness insurance system which obtains in all industrial countries in Europe and which brings to light the incidence of illness of all kinds in all groups of workers." *Industrial Poisons in the United States* as well as her increasingly public pronouncements did much to publicize the need for federal regulations on occupational health and safety. By the time she retired, Dr. Hamilton had broadened American environmentalism beyond simply preserving forestland or combating pollution, as important as those efforts are in protecting the natural environment. She had made possible the work of pioneers such as Rachel Carson in the area of public health and safety. So great was her influence that in 1987, the National Institute for Occupational Safety and Health (NIOSH) formally dedicated its research facility as the Alice Hamilton

Laboratory for Occupational Safety and Health. NIOSH also began presenting an annual Alice Hamilton Award to researchers who study occupational safety issues. In 1995, 25 years after her death at the age of 101, the U.S. Post Office issued a postage stamp in her honor as part of the "Great Americans" series.[46]

If Hamilton was the figure that united environmentalists' efforts to preserve natural resources with medical practitioners' concerns for the prevalence of toxic chemicals in the environment, Rachel Carson was the logical heir apparent. This popular nature writer and sometime environmental scientist took up the mantle after Hamilton had retired from the scene. Carson's 1962 book, *Silent Spring*, became one of the most important, famous, and influential environmental tracts ever written. It transformed her into a household name and a pioneering environmentalist to be emulated. Almost every successive work touting the dangers of pollution and toxic substances would be advertised as "the new *Silent Spring*." Dying from breast cancer as she penned the work, Carson's dire warnings appeared all the more poignant because the doomed heroine managed to scrawl a warning before she succumbed to a dread disease. As one commentator later remarked, "It is said that ladies often have the last word. Rachel Carson will have hers from beyond the grave."[47]

She was born on May 27, 1907, on a family farm in Springdale, a small town on the Allegheny River near Pittsburgh, Pennsylvania. An avid reader with a strong, inquisitive mind, young Rachel divided her time between exploring the nooks and crannies of the 65-acre farm and plowing through as many books on nature and wildlife as she could find. Early in life, she settled on a career in science. Owing to her life-long passion for investigating natural phenomena, it was little wonder that she studied biology in college and went on to Johns Hopkins University to complete graduate studies in zoology and genetics. She also displayed an unusual gift for producing clear, crisp prose coupled with an ability to reduce complex subjects to easily understood ideas and concepts. A precocious child, when she was eight years old Carson began writing stories about animals she had encountered on the farm. She published her first story at the age of 11.[48]

Carson had intended to earn a doctorate after she completed a master's degree in 1932, but she was forced to leave the university before she wrote her dissertation so she could help care for her family. Eventually, she landed a temporary position working at the U.S. Bureau of Fisheries. One of her jobs at the Bureau was to write copy for a series of short radio programs on educational topics. Carson's supervisor was impressed with the young woman's ability to synopsize complicated issues in a popular, easily accessible format. She later earned a permanent position as an aquatic biologist with the Bureau of Fisheries.[49]

Throughout the 1930s, Carson wrote about nature and wildlife for a variety of publications, including popular magazines. The editors at Simon & Schuster were impressed with an essay she wrote for *The Atlantic Monthly* and asked her to consider producing a full-length work on marine life. As a result, her first book, *Under the Sea Wind*, appeared in 1941. Reviewers were kind to the work, but readers were slow to respond. Nonetheless, Carson kept at it. A subsequent book on ocean life,

The Sea Around Us, proved to be far more popular. It sold more than two million copies and won the National Book Award, a stunning achievement for an academic work. Once again, Carson's ability to convey difficult, complex ideas in compelling prose proved to be a successful formula.

The Sea Around Us made Rachel Carson famous, especially among a small group of budding environmentalists. In the years that followed, she explored a variety of environmental topics for diverse audiences. Having left the Bureau of Fisheries (which was renamed the U.S. Fish and Wildlife Service), she could devote herself to writing full time. Her divergent interests eventually led her to the topic for which she would remain famous long after her death: pesticides. She first became interested in the subject around 1945. Post-World War II agricultural advances included widespread use of dichlorodiphenyltrichloroethane (DDT), an organochlorine insecticide and a seemingly miraculous substance that had helped to combat malaria and typhus during the war. It was tasteless and virtually odorless; consequently, persons exposed to the chemical often were unaware of its presence. Later, Carson learned of a 1957 lawsuit concerning indiscriminate DDT spraying in Long Island, New York. She became convinced that human beings were blithely using DDT, as well as other toxic chemicals, without adequately protecting themselves or even understanding the negative side effects.[50]

Through four long years of arduous research, Carson compiled and analyzed data on pesticide use and the resultant physiological and environmental effects. She also reached out to the broader community of scientists for advice and assistance. The Audubon Society had long opposed pesticide spraying absent data from further testing. Pleased with Carson's efforts, the group worked to publicize her concerns and her work. Even before *Silent Spring* appeared in print, however, her published essays on the issue generated immense controversy, exposing a rift between scientists who believed that pesticide use was dangerous and those who argued that the benefits outweighed the potential risks.[51]

Poor health delayed the project for years. As Carson seemed to be on the mend and was completing her work, she discovered cysts in her left breast. She was devastated. After undergoing a mastectomy, she hoped her health would improve. Unfortunately, doctors informed her that at least one cyst was a malignant tumor and it had metastasized. As she finished revising *Silent Spring*, Rachel Carson knew she was dying.[52]

Struggling to dream up a sufficiently evocative title that would appeal to a popular readership, Carson listened when her literary agent, Marie Rodell, suggested *Silent Spring*. Rodell realized it would capture the book's main argument memorably. As Carson had written early in Chapter 1, "There was a strange stillness. The birds, for example—where had they gone? Many people spoke of them, puzzled and disturbed. The feeding stations in the backyards were deserted. The few birds seen anywhere were moribund; they trembled violently and could not fly. It was a spring without voices."[53]

Silent Spring was filled with equally eloquent, engaging passages to communicate a serious, alarmist message. Throughout the book, Carson explained that pesticides ideally affect only a target population, namely the insects and rodents that destroy agricultural crops. Yet DDT and similar substances are seldom limited to the targets. Instead, owing to the process of bioaccumulation, small amounts of synthetic chemicals enter the bloodstream of all types of animals, including human beings. Over time, as quantities increase, the chemicals affect animals in unpredictable ways, often interfering with biological processes and severely damaging or even killing the animals. In her words, "One of the most sinister features of DDT and related chemicals is the way they are passed from one organism to another through all the links of the food chains."[54]

According to Carson, when pressed to defend the safety of pesticides, the chemical companies that manufacture and market those products either refused to acknowledge the risks or provided fabricated data. In the meantime, government officials and members of the general public uncritically accepted industry claims in exchange for promises of miraculous crop yields and a higher standard of living for everyone. The long latency period between exposure and the presence of symptoms ensured that the damage was done before anyone could correct the catastrophic effects. Aside from these dangers, invasive species that were not affected as much as the target population displaced the original pests. Eventually, target populations developed a resistance to the pesticides and thereby required stronger chemicals to control their presence, thus initiating another, perhaps more damaging cycle of chemical use. Instead of engaging in this ever-increasing arms race, Carson recommended the use of alternative, organic materials. Relying on manufactured chemicals to control unwanted insects and rodents was a recipe for disaster.[55]

Publication of *Silent Spring* in September 1962 represented a milestone in the modern environmental movement, igniting a firestorm of strong support and bitter criticism that persists to this day. Sales skyrocketed, propelling the book to the top of the bestseller lists when more than 600,000 copies sold during the fall. Environmentalists promoted the book as one of the most important tracts ever printed and a wakeup call for all Americans worried about the unknown and possibly dangerous effects of modernization. The *New York Times* published a positive review, observing, "Miss Carson will be accused of alarmism, or lack of objectivity, of showing only the bad side of pesticides while ignoring their benefits." Yet such criticism would not blunt the impact of her message. "If her series helps arouse enough public concern to immunize government agencies against the blandishments of hucksters and enforce adequate controls, the author will be as deserving of the Nobel Prize as was the inventor of DDT." *The Washington Post* praised the work by applauding her courage in the face of what surely would be a massive wave of criticism. No one appreciates hearing bad news, but "Carson's negative case is virtually as powerful as the poisons she deplores."[56] Recognizing the galvanizing public effect of the book, President John F. Kennedy promised that his administration would launch a thorough investigation into the charges and take appropriate action. The publicity was so massive that many environmental laws and regulations enacted in the decade after *Silent Spring*'s publication were aimed at addressing problems discussed in the book.[57]

For captains of industry, *Silent Spring* was a cleverly disguised, beautifully written polemic based on insufficient evidence designed to scare the public with a crisis *du jour* that conveniently ignored contrary data. Some of the harshest critics argued that Rachel Carson was not a genuine scientist in the sense that she published cutting-edge research in peer-reviewed scientific journals. She wrote eloquent nature pieces for popular magazines. She did not even hold a Ph.D. in her field! In an unsuccessful effort to prevent publication before *Silent Spring* hit the bookstores, Velsicol Chemical Corporation, manufacturer of the insecticide chlordane, threatened publisher Houghton Mifflin with a libel lawsuit. The editors demonstrated enough intestinal fortitude to withstand such attacks. When the threats did not achieve the desired effect, the National Agricultural Chemicals Association earmarked $25,000 for a public relations campaign questioning Rachel Carson's methodology and conclusions.[58]

Industry representatives were worried about the loss of market share while government scientists feared the book made them appear foolish and negligent because they had not raised appropriate concerns about pesticides. Detractors responded with a level of vitriol that surprised even Rachel Carson, who had known her conclusions would stir deep emotions. Several critics questioned her patriotism, slyly asking whether she was an agent of Communist forces. John Maddox, a theoretical physicist who savaged the book in his review, typified the industry response. "The most seriously misleading part of the narrative is the use of horror stories about the misuse of DDT to create an impression that there are no safe uses worth consideration," he wrote in denouncing her "literary trick." "Miss Carson's sin was the use of 'calculated overdramatization.'"[59] William Darby of the Vanderbilt Medical School submitted a review to the industry journal *Chemical and Engineering News*. Titled "Silence, Miss Carson!" the piece lambasted the book and Carson in openly misogynistic language that belittled the female non-scientist who dared to assail the chemical industry with her overly emotional appeals to readers such as "the organic gardeners, the anti-fluoride leaguers, the worshippers of 'natural foods,' and those who cling to the philosophy of a vital principle, and pseudo-scientists and faddists." Another physician, Frederick J. Stare, castigated Carson in *Nutrition Reviews* by intimating that she had waded into subject matter way over her head. "In Miss Carson's case," he wrote dismissively, "research limited to selective reading, plus the urging of 'friends' with special interests, is certainly no diploma of equivalency for the academic training and experience required for authority." On and on it went in a litany of angry denunciations of the woefully misguided, poorly researched, unnecessarily hysterical diatribe.[60]

Despite the fire-and-brimstone reaction from certain sectors, the book was a success, and Rachel Carson enjoyed the fame and fortune that accompanied *Silent Spring*. She lived to see the accolades pour in and enjoy her success in the marketplace. The work became a Book-of-the-Month Club featured selection and was serialized in many publications. A cultural landmark, it became fodder for discussion in households across the country and around the world. The story of birth defects caused by the drug thalidomide broke around the same time that *Silent Spring* hit the bookshelves, underscoring the possibility that the use of chemicals without proper safeguards posed a clear-and-present danger to human beings. Rachel Carson's name, it seemed, would live forever (Figure 7.2).[61]

The same could not be said for Rachel Carson herself. Even as *Silent Spring* went to press, she was suffering through rounds of cancer treatments that, in the end, did little to stave off the inevitable. She died on April 14, 1964, at the age of 56, equally loved and reviled for her pioneering work in American environmentalism.[62]

Rise of the Modern American Environmental Movement

It would be an exaggeration to claim that Rachel Carson created the modern American environmental movement, but it would not be much of an exaggeration. *Silent Spring* galvanized public opinion and lavished attention on environmental issues unlike anything else written until that time. In the decades that followed publication, events designed to preserve and protect the natural environment unfolded in many areas—in Congress, state legislatures, the courts, and the public arena.[63]

Carson's work supported a well-known thesis that could be traced back at least to Rousseau: Civilization has a dark side. Perhaps the most prominent American work on the subject before *Silent Spring* was *Technics and Civilization* by the esteemed historian and philosopher of science Lewis Mumford. In the 1930s, Mumford argued that the negative effects of technological advances originated much earlier than the Industrial Revolution, which is the traditional view of when environmental degradation became a critical concern. In his view, technology can be traced back to the medieval period and its development can be shown to evolve through overlapping phases. Although Mumford recognized the benefits of technology, he also warned that humanity must be careful to make decisions based on all relevant factors in the long and short terms, not simply based on a desire for rapid economic gain. By ignoring the negative effects of industrialization on natural resources, human beings risk permanently destroying the environment, to say nothing of the dehumanizing impact on workers who must toil in dirty, dangerous industries. "The cost of this indifference to the environment as a human resource—who can measure it?" he asked rhetorically.[64]

Mumford's argument, amplified by Rachel Carson's popular work, gained additional converts after 1962. A multitude of publications—articles, stories, pamphlets, books, and technical reports—warned of the dark side of human development following the initial appearance of *Silent Spring*. Even John F. Kennedy's

Figure 7.2 Author Rachel Carson holds a copy of *Silent Spring* in this photograph taken around the time the book was first published. (Courtesy of the Granger Collection, New York)

Secretary of the Interior, Stewart L. Udall, got into the act, penning a popular book, *The Quiet Crisis*, arguing for the wise use of natural resources and the development of, in his words, "a land ethic for tomorrow."[65]

René Dubos, a French-born microbiologist, won the Pulitzer Prize for general non-fiction for his 1968 book *So Human an Animal*, which used Carson's thesis as the starting point for a broader discussion. Echoing Thoreau's famous comment that the "mass of men lead lives of quiet desperation," Dubos observed that most people in modern society are trapped "in the midst of noise, dirt, ugliness, and absurdity." They have become dehumanized to a variety of ills, especially environmental destruction. *So Human an Animal* opened with these fiery, uncompromising words:

> This book should have been written in anger. I should be expressing in the strongest possible terms my anguish at seeing so many human and natural values spoiled or destroyed in affluent societies, as well as my indignation at the failure of the scientific community to organize a systematic effort against the desecration of life and nature.[66]

Barry Commoner, an influential biologist who once ran for president of the United States on the Citizens Party ticket, a pro-environmental organization he founded in 1979, was another twentieth-century iconoclast of the environmental movement. A prominent anti-nuclear activist, Commoner wrote a series of books espousing an ecological philosophy that "the age of innocent faith in science and technology may be over." In his 1966 work *Science and Survival*, he referred to DDT as an example of a technological innovation that had been touted as beneficial to humanity but ultimately caused more harm than good. He also expressed fears about nuclear technology as well as the proliferation of nuclear weapons around the world. In his view, human beings place too much faith in technology before they fully understand the possible repercussions of supposed advances, thus leading to an impending global environmental crisis. "Science can reveal the depth of this crisis, but only social action can resolve it," he argued.[67]

Commoner's most popular and often-cited book, *The Closing Circle*, first published in 1971, set forth the "four laws of ecology." As Commoner explained, the first law posits that everything is connected to everything else. The world cannot be compartmentalized, and ecological disaster areas cannot be walled off from the rest of the natural environment. One ecosphere exists for all living organisms. What affects one living thing affects all things. The second law is that everything must go somewhere. "Waste" is a term that human beings use to justify discarding unwanted items, but nature does not recognize this concept. No otherworldly place—"away" from here—exists to house the waste. A third law holds that nature knows best. The long-held belief that human beings are somehow separate from, and superior to, their environment is nonsensical. Technology does not improve on nature; it removes part of nature from our view, but whatever problems existed prior to the technological change still exist. Commoner argued that technological change within the ecological system is "likely to be detrimental to that system." His final observation was that there is no such thing as a free lunch. Just as matter can be neither created nor destroyed, only transferred from one form to another, environmental problems do not disappear. They are exchanged—one problem can be converted into another problem. Exploiting nature for whatever purposes appear expedient in the short run invariably creates problems that will be manifested somewhere else.[68]

In a later work, *Making Peace with the Planet*, Commoner lamented the state of American environmentalism circa 1990. Despite all the time, money, and attention devoted to nature in the wake of Rachel Carson's work, the environmental crisis has not been solved. If natural resources are to be preserved and protected, he explained, the method for producing goods in the United States—and ideally around the globe—fundamentally will have to be redesigned. In practice, the redesign will require greater reliance on solar energy, the development of ethanol fuel for cars, the commercialization of electric-powered automobiles, the institution of a massive recycling campaign for a variety of goods, and the wholesale replacement of most, if not all, petrochemical products with substitutes made from wood, cotton, paper, and other renewable resources. Despite the conventional view that either technology advances or the environment is protected, Commoner suggested that a choice is not required because it is "possible to achieve economic growth *and* improve the environment—by sensibly applying the principles of ecology."[69]

Alvin Toffler, arguably the most popular social critic to rise to prominence in the decade after *Silent Spring*, raised a public outcry over the negative features of modern industrialization. Toffler billed himself as a "futurist," someone who predicts how and why society will evolve in coming years. In his 1970 international bestseller *Future Shock*, he echoed Rousseau by arguing that too much change in too few years disorients and confuses individuals. The accelerating rate of change in modern society threatens to overwhelm the individual and leave him powerless to control his environment. As Toffler explained in the opening pages, "Future shock is no longer a distantly potential danger, but a real sickness from which increasingly large numbers already suffer. This psycho-biological condition can be described in medical and psychiatric terms. It is the disease of change."[70]

Aside from academics and authors who scrambled to write the "next *Silent Spring*," environmental organizations used Carson's findings as the impetus to lobby for stricter legal controls over polluters and to preserve and protect America's natural resources. The Sierra Club, a venerable group co-founded by John Muir at the end of the nineteenth century, is a prime example. The club found dynamic new leadership in revered nature photographer Ansel Adams and his colleague David Brower. A member of the group's board for 37 years, Adams frequently published his early photographs in the *Sierra Club Bulletin*. Originally a mountaineer, Brower later founded Friends of the Earth (FOE), but during his years as the Sierra Club's first executive director, beginning in 1952 and extending through

1969, he published books on nature and narrated films on endangered wilderness lands. He was instrumental in pressuring Congress to pass the Wilderness Act in 1964. Brower eventually had a falling out with Sierra Club members and left to form FOE. Later, he left FOE to form another group, the Earth Island Institute, which published books on ecology. He reconciled with the Sierra Club leadership and served on the group's board of directors until he resigned shortly before his death in 2000. In 2006, Kennesaw State University in Kennesaw, Georgia, erected a 350,000-pound monument sculpted by the celebrated Finnish-American artist Eino on the campus in Brower's memory.[71]

In addition to influencing the nascent environmental movement, *Silent Spring* affected the legislative process. The federal government held hearings on potential air pollution problems beginning in 1963, the year after the book first appeared in print. As a result of the hearings, Congress enacted the Clean Air Act of 1963, part of an ongoing series of laws aimed at curtailing air pollution from mobile and stationary sources. In 1964, Congress passed, and President Johnson signed, the Wilderness Act permanently barring certain tracts of land from industrial development. In 1966, Congress enacted the Endangered Species Preservation Act (a forerunner of the Endangered Species Act of 1973) as a means of protecting "endangered and threatened species of life."[72]

The culmination of these government activities came when President Richard M. Nixon signed an executive order creating the U.S. Environmental Protection Agency (EPA) in 1970. The purpose was to create a single coordinating organization that would cut across existing agency lines to provide a coherent national policy for the environment. In an unusual organizational arrangement, EPA was not housed inside a federal executive department but existed as a standalone independent agency with regulatory authority over air pollution, water pollution, drinking water contamination, hazardous waste disposal, pesticides, and toxic substances. The agency head, called an "administrator," is appointed by the president of the United States with the advice and consent of the U.S. Senate. The administrator oversees a staff of thousands as well as 10 regional offices.[73]

Since its halcyon days in the early 1970s, EPA has struggled to fulfill its mandate. The agency long ago developed a reputation of saying "no" to industries that pollute the natural environment, and the negative backlash harmed the organization's image. For some businesses, EPA has become a pesky, hypercritical traffic cop: The new manufacturing plant cannot release effluents into the waterways; the waste disposal facility cannot accept additional hazardous materials shipments in a particular community; the recycling facility cannot expand faster than allowed by its stringent air-permit requirements. When officers at companies that provide employment and grow the economy believe that a public agency has interfered with their economic goals, they cry foul. The EPA, they complain, is a bureaucratic agency awash in red tape, an out-of-control organ of government staffed by unelected officials that promulgates rules hampering the private sector without necessarily improving environmental quality. Despite the general public's professed desire to clean up the environment, when it comes to making difficult decisions about curbing pollution, which sometimes means putting the brakes on increased jobs and a higher short-term standard of living, Americans do not always live up to their ideals. In short, the EPA is the agency where concern for the environment bumps up against the standard neoclassical economic precepts discussed in Chapter 2.[74]

EPA administrators often see the agency as besieged by enemies simply because it attempts to fulfill its organizational mission. The agency lacks a cabinet-level advocate, which means that the administrator does not always enjoy full access to the president, especially in Republican administrations. President Reagan, for example, was extremely critical of the organization, installing a series of administrators who were hostile to the agency they headed. Funding has been cut repeatedly throughout the years, at least since the Reagan administration, to such an extent that sometimes the agency cannot effectively police the regulations that require enforcement. The lack of appropriate oversight and enforcement leads to charges that the agency is inefficient and takes too much time to sign off on new projects. Such criticism triggers legislative distrust, which can lead to a new round of budgetary cuts, thereby perpetuating the cycle of inefficiency and cynicism.[75]

For all its challenges and controversies, EPA has successfully provided a check on damage to the natural environment that would have continued unimpeded but for the agency's intervention. If it is not the ideal agency that environmentalists would hope for, it is preferable to no agency at all. As discussed in Chapter 9, the organization became an integral factor in implementing the litany of environmental laws and regulations created in the 1970s.[76]

Despite the achievements of the regulatory regime constructed late in the twentieth century, much of the success of the American environmental movement occurred outside of the public sector. Even as government agencies stepped in to regulate environmental quality during the 1960s and 1970s, activists who took up the mantle from Aldo Leopold, Bob Marshall, Alice Hamilton, and Rachel Carson sought innovative ways to publicize the need for "greening" the planet. Perhaps the most well known of these efforts is the event called "Earth Day."

The first Earth Day occurred on April 22, 1970, and it has become an annual event since that time. The brainchild of U.S. Senator Gaylord Nelson of Wisconsin, the concept was born amid the activism that swept across college campuses during the late 1960s. Nelson had grown up in rural Wisconsin—an area he referred to as a "magically mysterious" landscape—with a deep aversion to clear-cutting of forests by opportunistic timber companies. While serving as governor of Wisconsin from 1959 until 1963, he earned a pithy sobriquet—the "conservation governor"—for his reorganization of the state bureaucracy to create a single agency, the Department of Resource Development, designed to administer state parks more efficiently. He also created the Youth Conservation Corps to employ 1,000 young people in "green jobs." At a time when the environmental movement was still in its infancy, Governor Nelson supported a program to devote $50 million for the Outdoor Recreation Action Program (ORAP), a state initiative to acquire land for conversion into public parks and wilderness areas.[77]

A popular Democrat in the tradition of Robert M. La Follette, Wisconsin's famed progressive governor and senator, Nelson followed in his hero's footsteps by moving from the governorship to the U.S. Senate. He was elected in 1962, less than two months after the publication of *Silent Spring*. Nelson was an environmentally conscious legislator from an environmentally conscious state who shared Rachel Carson's concern for the indiscriminate use of DDT and other chemicals. After reading *Silent Spring*, he sponsored national legislation to ban DDT use. Despite his best efforts, he could not persuade his congressional colleagues to support the measure. Until the public railed against a variety of national crises, enacting laws on environmental causes was politically unpopular. Consequently, during his first seven years in the Senate, Gaylord Nelson often appeared to be a lone voice crying in the wilderness, a sad, silly Cassandra surrounded by rational, indifferent neoclassical economists.[78]

As times changed, Nelson's approach changed as well. If he could not succeed in the legislative arena, he could carry his message through alternative outlets. Rather than pursue a fruitless top-down approach, perhaps a bottom-up grassroots campaign would succeed. In 1969, while the senator was sitting on an airplane flying back from a visit to California to witness the after-effects of an oil spill, he read an article about a "teach-in" organized by college students protesting American involvement in the Vietnam War. He experienced an epiphany. If students could call attention to anti-war messages through a teach-in, perhaps they would be willing to organize a teach-in for the environment as well. Congress might ignore a lone liberal senator decrying natural-resource destruction, but they would ignore a populist grassroots call for change at their peril.[79]

In a series of speeches he delivered during the autumn of 1969, Nelson mentioned his idea repeatedly, although he did not call it "Earth Day." As the press covered the speeches and reported on his recommendation, the senator was astonished to learn that major publications such as *Time* and the *New York Times* had published stories on the proposed teach-in. Recognizing that momentum was building, Nelson selected April 22, 1970, as the date for a "National Teach-In on the Crisis of the Environment." By tapping into the youthful energy of the anti-war movement and scheduling a day of activism sandwiched between college spring breaks and final exams, the senator sought to trigger an unprecedented national media event that would capture a wide audience similar to the way *Silent Spring* had appealed to the populace in the early 1960s.[80]

Nelson's office was inundated with letters and telephone calls offering support for the teach-in on college campuses from coast to coast. Public school students, churchgoers, and community organizers also pledged fidelity to the effort. Overwhelmed by the mounting press coverage, in November 1969 Senator Nelson convened a steering committee composed of academics, scientists, environmentalists, and students to serve as an information clearinghouse and a resource for coordinating the national efforts. Liberal Republican Congressman Paul McCloskey of California signed on to serve as the co-chair for the event. Treating the teach-in as a political event, Nelson and McCloskey tasked their staffs with developing a platform, recruiting student volunteers, holding fundraisers, and developing media materials, including posters, buttons, an official documentary, and even a popular song.[81]

As a new year dawned and teach-in plans intensified, Senator Nelson insisted that he was not running a synchronized political event. The purpose was to mobilize support through a loose affiliation of grassroots organizations that would highlight environmental concerns in their respective communities. Deliberately eschewing a "one-size-fits-all" model, Nelson encouraged Americans across the country to participate "in any way they want." To ensure that he would not become the *de facto* boss of the gathering, Nelson eventually established Environmental Teach-In, Inc., an independent non-profit organization located in Washington, D.C. Denis Hayes, a former Stanford University student-body president and a Harvard Law School student, became a leader in the effort after the senator asked for help. For their part, the Environmental Teach-In staff initiated a public relations campaign that included, among other things, a newsletter, resource packets, posters, and a campaign to encourage citizens to write their legislators requesting action to protect the natural environment. On January 18, 1970, they sponsored a full-page advertisement in the *New York Times*. For the first time, organizers used the phrase "Earth Day."[82]

The advertisement was a lesson in how to operate a provocative public relations campaign aimed at capturing maximum attention. "A disease has infected our country," the copy read. "It has brought smog to Yosemite, dumped garbage in the Hudson, sprayed DDT in our food, and left our cities in decay. Its carrier is man." After reciting a list of environmental problems, the group set forth its goals: "Earth Day is a commitment to make life better, not just bigger and faster; to provide real rather than rhetorical solutions. It is a day to re-examine the ethic of individual progress at mankind's expense. It is a day to challenge the corporate and governmental leaders who promise change, but who short-change the necessary programs." This challenge was all the more remarkable because it had been initiated by a governmental leader serving at one of the highest levels of federal service but who had voluntarily ceded control to the masses.[83]

On April 22, 1970, more than 20 million people congregated at thousands of colleges, universities, parks, and other public spaces throughout the United States. Although the timing and variety of events differed, most gatherings included speeches and songs by students and local activists denouncing oil spills, pesticides, toxic chemicals, the loss of wilderness lands, and air pollution. In New York City, Mayor John Lindsay closed off Fifth Avenue and made Central Park available for a massive rally of perhaps one million people. Movie stars Paul Newman and Ali MacGraw attended. Other big cities featured large crowds that participated in all manner of community activities. Protest singer Pete Seeger spoke in Washington,

D.C. before regaling the crowd with his music. In Philadelphia, students called for an entire week devoted to publicizing environmental degradation because one day was insufficient to accomplish their ambitious goals. As expected, the events generated a barrage of media coverage. CBS News ran a one-hour prime-time television special titled "Earth Day: A Question of Survival" narrated by Walter Cronkite, one of the most venerated reporters in America.[84]

Senator Nelson appeared at an event in Denver, Colorado. "Earth Day is dramatic evidence of a broad new national concern that cuts across generations and ideologies," he told his audience. "It may be symbolic of a new communication between young and old about our values and priorities." Environmental efforts must not end with one symbolic day of speeches, music, and parades, however. To succeed, Nelson explained, the event must lead to a new awareness and eventually to legislative action. Linking environmental issues to other problems of the era, Nelson remarked, "Earth Day can—and it must—lend a new urgency and a new support to solving the problems that still threaten to tear the fabric of this society... the problems of race, of war, of poverty, of modern-day institutions." As bad as those problems were, he explained, "Winning the environmental war is a whole lot tougher challenge by far than winning any other war in the history of man." Winning the war was difficult because the objectives were so broad. "Our goal is not just an environment of clear air and water and scenic beauty. The objective is an environment of decency, quality and mutual respect for all other human beings and all other living creatures."[85]

Earth Day met Gaylord Nelson's original objectives, although the event evolved from a teach-in to a campaign for changing national environmental policy. By organizing a peaceful mass protest aimed at raising awareness of environmental problems without undermining the grassroots character of the event, he and his fellow organizers allowed the citizenry to voice its hopes and fears regarding natural-resource preservation and protection. Many environmental laws and regulations promulgated after 1970 would have been difficult to enact were it not for the increased awareness and the bottom-up grassroots activism that resulted from the initial event. When asked why he did not pursue a top-down effort with assistance from political leaders, celebrities, and organized interest groups such as the Sierra Club, the Audubon Society, the Izaak Walton League, or the National Wildlife Federation, Senator Nelson commented, "No one could organize 20 million people, 10,000 grade schools and high schools, 2,500 colleges and 1,000 communities in three and a half months, even if he had $20 million. The key to the whole thing was the grass-roots response."[86]

Yet everyone involved knew a single event could not guarantee that environmental issues would remain on the national agenda and permanently enter the public consciousness. Awareness alone does not a policy make. Denis Hayes immediately moved to formalize the Earth Day Network, a new organization dedicated to promoting environmental awareness year-round and campaigning for policy changes at all levels. The group included non-governmental organizations (NGOs), quasi-governmental agencies, local governments, activists, and other citizens interested in promoting environmental education and improving public policy through public relations campaigns and grassroots celebrations in addition to Earth Day. By the beginning of the twenty-first century, the Network could boast of an international reach: Over 19,000 organizations in 192 countries worked with the group.[87]

Other organizations sprang up to take advantage of new interest in environmentalism. A group called Environmental Action began an ongoing campaign to coordinate future Earth Day activities and call attention to the "dirty dozen," manufacturing and chemical companies with abysmal records of polluting the environment. The organization's website employed tough, muscular prose to discuss its agenda:

> We turn ideas into impassioned, results-oriented activism that protects our environment from polluters and their allies in government. We never flinch from a fight. We always put the planet first. We hold the powerful accountable and speak for the trees.[88]

The League of Conservation Voters formed in 1970, drawing members and resources from several existing groups, "to turn environmental values into national priorities," according to the non-profit organization's website. "To secure the environmental future of our planet, LCV advocates for sound environmental policies, elects pro-environment candidates who will adopt and implement such policies, and provides state LCVs with the resources and tools to accomplish and sustain their mission." During the twenty-first century, the group's slate of issues included global warming, clean energy, clean water, toxics, wildlife, public lands, and oceans and coasts.[89]

Not everyone was smitten with the Earth Day concept. Aside from industry representatives who contended that the effort smacked of anti-business opportunism that was unnecessarily alarmist and histrionic, some environmentalists worried that Earth Day served, and continues to serve, as an annual celebration of vague, amorphous ideals and collective liberal guilt. In some instances, Americans drive their gas-guzzling sport utility vehicles to the local park, stopping off at the nearest organic health-food store to purchase disposable picnic products in theoretically "green" disposable packaging along the way, only to arrive at the festivities, bow their heads, and curse humanity's wicked, wicked ways. After chatting with their friends and neighbors about the ball game or the local school-board elections, joining hands to sing a rousing chorus of "Kumbaya," and listening to strident speeches delivered by idealistic young students with no life experience, they return home to bury their faces in electronic gadgets that suck fossil-fuel-generated electricity from their home outlets at record levels. If this is the ultimate legacy, critics contend, Earth Day was a failure. Environmentalists were co-opted by mainstream Americans who care about environmental issues as long as their standard of living is not imperiled and their lives are minimally inconvenienced.[90]

Adding to the criticism, other environmentalists argued that the time and energy devoted to tilting at windmills through meaningless, symbolic Earth Day celebrations could be devoted to genuine, substantive issues such as overpopulation. A Stanford University biology professor, Paul Ehrlich, published a groundbreaking book, *The Population Bomb*, in 1968, shortly before the first Earth Day. The book called attention to the crisis of overpopulation facing the world. An updated version of Thomas Malthus's eighteenth-century thesis that population growth eventually will outpace the world's ability to feed its people, thereby leading to mass starvation, *The Population Bomb* set forth a provocative, alarming proposition. Because "Americans are beginning to realize that the underdeveloped countries of the world face an inevitable population-food crisis," Ehrlich wrote in a section titled "Too Many People," the problem could no longer be ignored. "Each year food production in these countries falls a bit further behind burgeoning population growth, and people go to bed a little bit hungrier. While there are temporary or local reversals of this trend, it now seems inevitable that it will continue to its logical conclusion: mass starvation." Although the worst predictions of a Malthusian catastrophe failed to materialize, the fear that the "human footprint" was decimating the planet fit neatly with the alarmism of many environmental activists.[91]

Earth Day may not have been the substantive victory that every ecologist hoped for, but it occurred at the beginning of a decade that represented the high-water mark of American environmentalism. During the 1970s, environmental groups became more influential in the policymaking process while, for their part, policymakers enacted a multitude of environmental laws and regulations. These developments are discussed at length in Chapters 8 and 9.

Yet the success of the environmental movement also triggered a politically conservative backlash, especially during the 1980s. When Ronald Reagan became president in 1981, he expressed disgust with the excesses of modern liberalism and the oft-stated desire to use the instruments of government to resolve environmental problems. He installed cabinet members openly skeptical of environmentalists' claims that natural-resource degradation and depletion had reached critical levels. Reagan's first Secretary of the Interior, James G. Watt, led the Sagebrush Rebellion, a movement initiated by western ranchers, miners, and loggers to sell or lease public lands to private interests. Donald P. Hodel, Secretary of Energy before moving over to the Interior Department, once remarked that the dangers of ozone depletion could be ameliorated by applying copious amounts of suntan lotion. The president's decision to tap Anne Gorsuch Burford, an attorney and former Colorado state legislator, to serve as the EPA administrator was especially galling to the environmental community. Her marching orders were to oversee massive budget cuts and reduce regulatory enforcement of environmental laws. Congress eventually cited her for contempt for refusing to produce documents about possible mishandling of funds designated for the Comprehensive Environmental Response, Compensation, and

Liability Act (CERCLA), or "Superfund," program. Not all Reagan administration officials were hostile to environmental issues—William Ruckelshaus, the first EPA administrator from 1970 until 1973, returned to head the agency for a brief period beginning in 1983—but it was clear that a new era of counter-environmentalism had dawned.[92]

The environmental movement regained some of its lost luster during the 1990s, although it never completely recovered from the Reagan years. George H. W. Bush promised to be the "environmental president" when he campaigned for the White House in 1988. Once he was in office, he appointed a well-respected environmentalist, William K. Reilly, to administer EPA. Yet Bush faced a host of other issues, especially the end of the Cold War and a deteriorating economy, which turned his attention away from the natural environment as a pressing concern. Numerous environmental problems emerged during Bush's four-year tenure—alarm about oil spills in the wake of the 1989 *Exxon Valdez* accident in Prince William Sound, Alaska; concerns about ocean pollution and solid waste; the problem of disposing of radioactive waste from nuclear power operations; and fears over global warming, among other things—but he was unwilling or unable to develop a comprehensive environmental policy to address these matters.[93]

It was left to the Clinton administration to reverse 12 years of politically conservative environmental policies. Activists were heartened when candidate Clinton chose Tennessee Senator Al Gore, a self-professed environmentalist with a strong record of natural-resource protection, as his running mate. Gore became the administration's unofficial environment czar, persuading the president to appoint sympathetic officials to positions of authority. Gore's close associate, Carol Browner, for example, served as the EPA head for eight years. During that time, the administration sought to balance a growing economy with wise environmental stewardship. Officials promoted energy conservation, established guidelines for government purchases of fuel-efficient vehicles and "ozone-friendly" products, and committed to reducing national greenhouse gas emissions. Clinton did not reverse all of the policy initiatives of his immediate predecessors, but he set the stage for environmental policies in the years that followed.[94]

The twentieth century had witnessed tremendous changes from the days when John Muir, Gifford Pinchot, and Theodore Roosevelt debated preservationist versus conservationist approaches to environmentalism. In those early days, environmentalists were lonely iconoclasts denouncing American consumerism as the country raced toward increasing rates of industrialization and urbanization. By the end of the century, environmentalism had become a mainstream topic of debate among a large group of concerned citizens. No longer would calls for nuanced environmental sensitivity be laughed off or ignored. Environmentalism had become a crucial area of American public policy.

Notes

1. The Doors, "When the Music's Over," words and music by the Doors. Copyright © 1967 Doors Music Co. Copyright renewed. All rights reserved. Used by permission. Reprinted by permission of Hal Leonard Corporation.
2. The literature on Progressivism is voluminous. For a general introduction, see, for example, Daniel Gaido, "The Populist Interpretation of American History: A Materialist Revision," *Science & Society* 65, 3 (Fall 2001): 350–353; Gabriel Gherasim, "Introductory Sketch to the Analytics and Pragmatism of American Progressivism," *Studia Universitatis Babes-Bolyai. Studia Europaea* 56, 3 (September 2011): 119–130.
3. David Lee Amstutz, "Nebraska's Live Stock Sanitary Commission and the Rise of Progressivism," *Great Plains Quarterly* 28, 4 (Fall 2008): 259–260; Eldon J. Eisenach, "Progressivism as a National Narrative in Biblical-Hegelian Times," *Social Philosophy & Policy* 24, 1 (January 2007): 61–64.
4. Nancy J. Barrett, "The Struggles of Women Industrial Workers to Improve Work Conditions in the Progressive Era," *OAH Magazine of History* 13, 3 (Spring 1999): 43–48; Elizabeth V. Burt, "Working Women and the Triangle Fire: Press Coverage of a Tragedy," *Journalism History* 30, 4 (Winter 2005): 189–199; Allison D. Murdach, "Does American Social Work Have a Progressive Tradition?," *Social Work* 55, 1 (January 2010): 82–86.
5. Jean Bethke Elshtain, "A Return to Hull House: Reflections on Jane Addams," *Feminist Issues* 15, 1–2 (1997): 105–113.
6. Gherasim, "Introductory Sketch to the Analytics and Pragmatism of American Progressivism," 136; Benjamin Kline, Ph.D., *First Along the River: A Brief History of the U.S. Environmental Movement*, 3rd ed. (Lanham, Md.: Rowman & Littlefield, 2007), 51–54.
7. Steven J. Holmes, *The Young John Muir: An Environmental Biography* (Madison: The University of Wisconsin Press, 1999), 3–4.
8. Holmes, *The Young John Muir*, 17–18; James W. Shores, "A Win-Lose Situation: Historical Context, Ethos, and Rhetorical Choices in John Muir's 1908 'Hetch Hetchy Valley' Article," *The Journal of American Culture* 29, 2 (June 2006): 191; Donald Worster, "John Muir and the Modern Passion for Nature," *Environmental History* 10, 1 (January 2005): 9.
9. Douglas Brinkley, *The Wilderness Warrior: Theodore Roosevelt and the Crusade for America* (New York: HarperCollins, 2009), 540–541; Holmes, *The Young John Muir*, 138–139; Shores, "A Win-Lose Situation," 191.
10. John Muir, *The Mountains of California* (New York: Penguin Books, 1985 [1894]), 88.
11. John Muir, *Our National Parks* (Boston and New York: Houghton Mifflin and Company, the Riverside Press, Cambridge, 1901), 5–6. See also Kline, *First Along the River*, 48–49.
12. Muir, *Our National Parks*, 364–365.
13. Brian Balogh, "Scientific Forestry and the Roots of the Modern American State: Gifford Pinchot's Path to Progressive Reform," *Environmental History* 7, 2 (April 2002): 198–207; Brinkley, *The Wilderness Warrior*, 340; Char Miller and V. Alaric Sample, "Gifford Pinchot: A Life in Progress," *Journal of Forestry* 97, 1 (January 1999): 28.

14. Paul V. Ellefson, "The Safety of Our Forests and the Prosperity of Our People," *Journal of Forestry* 98, 5 (May 2000): 15; Kline, *First Along the River*, 55–56; Char Miller, "Thinking Like a Conservationist," *Journal of Forestry* 100, 8 (December 2002): 44–45; Gifford Pinchot, "The Economic Significance of Forestry," *The North American Review* 213, 783 (February 1921): 157–167.
15. Brinkley, *The Wilderness Warrior*, 340–347; Kline, *First Along the River*, 55–58; Stewart L. Udall, *The Quiet Crisis* (New York: Holt, Rinehart, and Winston, 1963), 126–145.
16. Brinkley, *The Wilderness Warrior*, 340–347, 541–545; Miller and Sample, "Gifford Pinchot," 29–30; Philip Shabecoff, *Earth Rising: American Environmentalism in the 21st Century* (Washington, D.C. and Covelo, Calif.: Island Press, 2001), 3.
17. Brinkley, *The Wilderness Warrior*, 544; Kline, *First Along the River*, 55–58.
18. Brinkley, *The Wilderness Warrior*, 789–790; Kline, *First Along the River* 58–60; Shores, "A Win-Lose Situation," 192–194.
19. John Muir, "On Saving Hetch Hetchy," in *American Environmental History*, Louis S. Warren, editor (Oxford, U.K.: B. H. Blackwell, 2004), 239.
20. Bill Devall, "The End of American Environmentalism?" *Nature and Culture* 1, 2 (Autumn 2006): 158; Kline, *First Along the River*, 58–60; Shores, "A Win-Lose Situation," 199.
21. Kline, *First Along the River*, 57–60, 84.
22. Bernard Fensterwald, Jr., "The Anatomy of American 'Isolationism' and Expansionism, Part I," *The Journal of Conflict Resolution* 2, 2 (June 1958): 111–112, 121–123.
23. See, for example, Heather Addison, "Capitalizing Their Charms: Cinema Stars and Physical Culture in the 1920s," *The Velvet Tray—A Critical Journal of Film and Television* 50, 1 (Fall 2002): 15–35; Susan Currell, *American Culture in the 1920s* (Edinburgh, Scotland: Edinburgh University Press, 2009).
24. Kline, *First Along the River*, 60–61; M. R. Werner and John Starr, *Teapot Dome* (New York: The Viking Press, 1959).
25. Hugh S. Gorman, "Efficiency, Environmental Quality, and Oil Field Brines: the Success and Failure of Pollution Control by Self-Regulation," *Business History Review* 73, 4 (Winter 1999): 602–604; Robert R. Keller, "Supply-Side Economic Policies during the Coolidge-Mellon Era," *Journal of Economic Issues* 16, 3 (September 1982): 775–777; Kline, *First Along the River*, 61, 64.
26. See, for example, Kendrick A. Clements, *Hoover, Conservation, and Consumerism: Engineering the Good Life* (Lawrence: University Press of Kansas, 2000).
27. Keller, "Supply-Side Economic Policies During the Coolidge-Mellon Era," 602–640; Kline, *First Along the River*, 61.
28. Neil M. Maher, *Nature's New Deal: The Civilian Conservation Corps and the Roots of the American Environmental Movement* (Oxford and New York: Oxford University Press, 2008); Neil M. Maher, "A New Deal Body Politic: Landscape, Labor, and the Civilian Conservation Corps," *Environmental History* 7, 3 (July 2002): 435–461; Edward J. Martin, "Economic Rights, Sustainable Development, and Environmental Management," *Public Administration & Management* 16, 2 (April 2011): 126–127, 137–138.
29. Kline, *First Along the River*, 64–67; Robert D. Leighninger, Jr., *Building Louisiana: The Legacy of the Public Works Administration* (Jackson: University Press of Mississippi, 2007).
30. Richard Hornbeck, "The Enduring Impact of the American Dust Bowl: Short- and Long-Run Adjustments to Environmental Catastrophe," *American Environmental Review* 102, 4 (June 2012): 1479–1483; Rhonda Skaggs, et al., "Vegetation Maps at the Passage of the Taylor Grazing Act (1934): A Baseline to Evaluate Rangeland Change

after Regime Shift," *Rangelands* 33, 1 (February 2011): 13–16; Milburn F. Wilson, "Will the Administration's Soil Conservation Act Prove to be Legal?" *Congressional Digest* 15, 3 (March 1936): 83–85; Edwin C. Hargrove, *Prisoners of Myth: The Leadership of the Tennessee Valley Authority, 1933–1990, Studies in American Politics: Historical, International, and Comparative Perspectives* (Princeton, N.J.: Princeton University Press, 1994); E. M. Hughes-Jones, "The Tennessee Valley Authority Fifty Years On," *Political Quarterly* 54, 3 (July-September 1983): 283–291.

31. Chester C. Davis, "Is the A.A.A. Experiment Proving a Success?" *Congressional Digest* 13, 12 (December 1934): 306, 308; Edwin C. Nourse, Joseph S. Davis, and John D. Black, *Three Years of the Agricultural Adjustment Administration* (Washington, D.C.: Brookings Institution, 1937).
32. Stephen J. Frese, "Aldo Leopold: An American Prophet," *History Teacher* 37, 1 (November 2003): 100–103.
33. Ibid., 103–105; Kline, *First Along the River*, 62–63.
34. Aldo Leopold, *A Sand County Almanac; with Essays on Conservation from Round River* (New York: Ballantine Books, 1990 [1949]), 138–139. See also Philip Shabecoff, *A Fierce Green Fire: The American Environmental Movement* (Washington, D.C. and Covelo, Calif.: Island Press, 2003), 82–83.
35. Leopold, *A Sand County Almanac*, 239, 240, 243. See also Devall, "The End of American Environmentalism?" 159.
36. Frese, "Aldo Leopold," 99–100; Kline, *First Along the River*, 62–63; Max Oelschlaeger, "Ecological Restoration, Aldo Leopold, and Beauty: An Evolutionary Tale," *Environmental Philosophy* 4, 1 & 2 (Fall/Spring 2007): 149–161; Gavin Van Horn, "The (Religious) Naturalist's Eye: An Introduction to 'Aldo Leopold: Ethical and Spiritual Dimensions,'" *Journal for the Study of Religion, Nature, and Culture* 5, 4 (2011): 397–409; Barbara E. Willard, "Rhetorical Landscapes as Epistemic: Revisiting Aldo Leopold's *A Sand County Almanac*," *Environmental Communication* 1, 2 (November 2007): 218–235.
37. Bibi Booth, "Robert Marshall (1901–1939)," in *Environmental Activists*, John Mongillo and Bibi Booth, editors (Westport, Conn.: Greenwood Press, 2001), 174; Phil Brown, "Wilderness Advocate," *New York Conservationist*, 62, 1 (August 2007): 3–5; James A. Glover, *A Wilderness Original: The Life of Bob Marshall* (Seattle, Wash.: The Mountaineers Book, 1986), 89–90; Donald Dale Jackson, "Just Plain Bob Was the Best Friend Wilderness Ever Had," *Smithsonian* 25, 5 (August 1994): 92–94.
38. Quoted in Robert Gottlieb, "Reconstructing Environmentalism: Complex Movements, Diverse Roots," in *American Environmental History*, Louis S. Warren, editor (Oxford, U.K.: B. H. Blackwell, 2004), 246–247. See also Booth, "Robert Marshall (1901–1939)," 176; Brown, "Wilderness Advocate," 6; Glover, *A Wilderness Original*, 238.
39. Brown, "Wilderness Advocate," 5–6; Gottlieb, "Reconstructing Environmentalism," 247–248; Jackson, "Just Plain Bob Was the Best Friend Wilderness Ever Had," 96–100.
40. Brown, "Wilderness Advocate," 5–6; Glover, *A Wilderness Original*, 163; Gottlieb, "Reconstructing Environmentalism," 247–248.
41. Robert Marshall, *The People's Forests* (Iowa City: University of Iowa Press, 2002 [1933]), 22, 123.
42. Glover, *A Wilderness Original*, 244–245; Dennis K. McDaniel, "The First Congressman Martin Dies of Texas," *Southwestern Historical Quarterly* 102, 2 (October 1998): 130–161.

43. Quoted in Gottlieb, "Reconstructing Environmentalism," 248. See also Booth, "Robert Marshall (1901–1939)," 176; Brown, "Wilderness Advocate," 6.
44. Peter A. Appel, "Wilderness, the Courts, and the Effect of Politics on Judicial Decision-making," *Harvard Environmental Law Review* 35, 2 (2011): 290; Booth, "Robert Marshall (1901–1939)," 177; Brown, "Wilderness Advocate," 6; Gottlieb, "Reconstructing Environmentalism," 248; Jackson, "Just Plain Bob Was the Best Friend Wilderness Ever Had," 99–100.
45. Quoted in Gottlieb, "Reconstructing Environmentalism," 250, 251. See also William T. Moye, "BLS and Alice Hamilton: Pioneers in Industrial Health," *Monthly Labor Review* 109, 6 (June 1986): 24–27; Elizabeth Fee and Theodore M. Brown, "Alice Hamilton: Settlement Physician, Occupational Health Pioneer," *American Journal of Public Health* 91, 11 (November 2001): 1767.
46. Alice Hamilton, *Industrial Poisons in the United States* (New York: Macmillan, 1929), v. See also Gottlieb, "Reconstructing Environmentalism," 251.
47. Quoted in David K. Hecht, "Constructing a Scientist: Expert Authority and Public Images of Rachel Carson," *Historical Studies in the Natural Sciences* 41, 3 (Summer 2011): 286. See also Devall, "The End of American Environmentalism?" 159.
48. Charles T. Rubin, *The Green Crusade: Rethinking the Roots of Environmentalism* (Lanham, Md.: Rowman & Littlefield, 1998), 30–31; William Souder, *On a Farther Shore: The Life and Legacy of Rachel Carson* (New York: Crown Publishers, 2012), 24–25.
49. Linda J. Lear, "Rachel Carson's *Silent Spring*," *Environmental History Review* 17, 2 (Summer 1993): 24–25; Souder, *On a Farther Shore*, 86–87.
50. Rubin, *The Green Crusade*, 30–32; Souder, *On a Farther Shore*, 142–160.
51. Gottlieb, "Reconstructing Environmentalism," 253–254; Hecht, "Constructing a Scientist," 287–294; Souder, *On a Farther Shore*, 255–263.
52. Hecht, "Constructing a Scientist," 283–284; Lear, "Rachel Carson's *Silent Spring*," 39–40.
53. Rachel Carson, *Silent Spring*, 40th Anniversary Edition (Boston: Houghton Mifflin Harcourt, 2002 [1962]), 2. See also Souder, *On a Farther Shore*, 307–308.
54. Carson, *Silent Spring*, 22.
55. Gottlieb, "Reconstructing Environmentalism," 253–254; Lear, "Rachel Carson's *Silent Spring*," 32–34.
56. Quoted in Lear, "Rachel Carson's *Silent Spring*," 36.
57. Lear, "Rachel Carson's *Silent Spring*," 36, 38–40; Rubin, *The Green Crusade*, 32–33; Souder, *On a Farther Shore*, 3–4.
58. Lear, "Rachel Carson's *Silent Spring*," 30, 37–38; Souder, *On a Farther Shore*, 347–348; Michael B. Smith, "'Silence, Miss Carson!' Science, Gender, and the Reception of *Silent Spring*," *Feminist Studies* 27, 3 (Fall 2001): 736–737.
59. Quoted in Kline, *First Along the River*, 74–75.
60. Quoted in Smith, "'Silence, Miss Carson!'" 738, 739.
61. David K. Hecht, "How to Make a Villain: Rachel Carson and the Politics of Anti-Environmentalism," *Endeavour* 36, 4 (December 2012): 149–155; Kline, *First Along the River*, 73–74; Lear, "Rachel Carson's *Silent Spring*," Souder, *On a Farther Shore*, 16.
62. Lear, "Rachel Carson's *Silent Spring*," 39–40; Souder, *On a Farther Shore*, 389.
63. Kline, *First Along the River*, 73–74; Shabecoff, *Earth Rising*, 177–178; Souder, *On a Farther Shore*, 3–7.
64. Lewis Mumford, *Technics and Civilization* (Chicago and London: University of Chicago Press, 1934), 172.

65. Udall, *The Quiet Crisis*, 188. See also Devall, "The End of American Environmentalism?" 159.
66. René Jules Dubos, *So Human an Animal: How We Are Shaped by Surroundings and Events* (New Brunswick, N.J.: Transaction Publishers, 1998 [1968]), 3.
67. Barry Commoner, *Science and Survival* (New York: Ballantine Books, 1971 [1966]), 3, 157.
68. Barry Commoner, *The Closing Circle: Nature, Man, and Technology* (New York: Bantam, 1980 [1971]), 29–42.
69. Barry Commoner, *Making Peace with the Planet* (New York: Pantheon, 1990), 99. Italics found in the original.
70. Alvin Toffler, *Future Shock* (New York: Bantam, 1984 [1970]), 2.
71. Devall, "The End of American Environmentalism?" 160–161; Kline, *First Along the River*, 77–78; "David R. Brower," *Earth Island Journal* 16, 1 (Spring 2001): 21–24.
72. Brown, "Wilderness Advocate," 6; Devall, "The End of American Environmentalism?" 161; Jackson, "Just Plain Bob Was the Best Friend Wilderness Ever Had," 99–100; Kline, *First Along the River*, 76; Walter A. Rosenbaum, *Environmental Politics and Policy*, 4th ed. (Washington, D.C.: Congressional Quarterly Press, 1998), 310–312.
73. Richard N. L. Andrews, "The EPA at 40: An Historical Perspective," *Duke Environmental Law and Policy Forum* 21, 2 (Spring 2011): 226–231; Susan J. Buck, *Understanding Environmental Administration and Law*, 2nd ed. (Washington, D.C. and Covelo, Calif.: Island Press, 1996), 21–22; Devall, "The End of American Environmentalism?," 161; Kline, *First Along the River*, 92; Robert V. Percival, "Checks Without Balance: Executive Office Oversight of the Environmental Protection Agency," *Law and Contemporary Problems* 54, 4 (Fall 1991): 129–135; Rosenbaum, *Environmental Politics and Policy*, 82.
74. Andrews, "The EPA at 40," 253–258; Rosenbaum, *Environmental Politics and Policy*, 107–112.
75. Andrews, "The EPA at 40," 235–253; Buck, *Understanding Environmental Administration and Law*, 23–24; Kline, *First Along the River*, 102–104; Rosenbaum, *Environmental Politics and Policy*, 107–112; Shabecoff, *Earth Rising*, 122–123.
76. Andrews, "The EPA at 40," 253–258; Rosenbaum, *Environmental Politics and Policy*, 108–212.
77. Bill Christofferson, *The Man from Clear Lake: Earth Day Founder Senator Gaylord Nelson* (Madison: The University of Wisconsin Press, 2004), 273, 284; Devall, "The End of American Environmentalism?" 161; David J. Webber, "Earth Day and Its Precursors: Continuity and Change in the Evolution of Mid-Twentieth-Century U.S. Environmental Policy," *Review of Policy Research* 25, 4 (July 2008): 314.
78. Gaylord Nelson and Earth Day Website: "Meet Gaylord Nelson, Founder of Earth Day," http://www.nelsonearthday.net/nelson/index.htm (Accessed October 10, 2012).
79. Gaylord Nelson and Earth Day Website: "Meet Gaylord Nelson, Founder of Earth Day," n.p.; Gaylord Nelson and Earth Day Website: "Modest Beginnings: Nelson's Idea on a Plane," http://www.nelsonearthday.net/earth-day/beginnings.htm, (Accessed October 10, 2012); Webber, "Earth Day and Its Precursors," 318.
80. Christofferson, *The Man from Clear Lake*, 531; Gaylord Nelson and Earth Day Website: "Meet Gaylord Nelson, Founder of Earth Day," n.p.; Mark Hertsgaard, "Save Earth Day," *The Nation* 294, 19 (May 7, 2012): 3; Webber, "Earth Day and Its Precursors," 318–319.

81. Christofferson, *The Man from Clear Lake*, 517; Gaylord Nelson and Earth Day Website: "Choosing Not to Be in Charge," http://www.nelsonearthday.net/earth-day/not-incharge.htm (Accessed October 10, 2012).
82. Christofferson, *The Man from Clear Lake*, 8; Gaylord Nelson and Earth Day Website: "Choosing Not to Be in Charge," n.p.; Hertsgaard, "Save Earth Day," 3; Shabecoff, *Earth Rising*, 186.
83. Quoted at Gaylord Nelson and Earth Day Website: "Choosing Not to Be in Charge," n.p.
84. Douglas Brinkley, *Cronkite* (New York: Harper, 2012), 431–434; Christofferson, *The Man from Clear Lake*, 2–8; Finis Dunaway, "Gas Masks, Pogo, and the Ecological Indian: Earth Day and the Visual Politics of American Environmentalism," *American Quarterly* 60, 1 (March 2008): 67–99. Gaylord Nelson and Earth Day Website: "April 22, 1970," http://www.nelsonearthday.net/earth-day/4-22-1970.htm (Accessed October 10, 2012); Hertsgaard, "Save Earth Day," 3; Kline, *First Along the River*, 81; Webber, "Earth Day and Its Precursors," 318.
85. Quoted at Gaylord Nelson and Earth Day Website: "April 22, 1970," n.p.
86. Quoted in Kline, *First Along the River*, 81. See also Gaylord Nelson and Earth Day Website: "April 22, 1970," n.p.; Hertsgaard, "Save Earth Day," 3–4; Shabecoff, *A Fierce Green Fire*, 105–112; Webber, "Earth Day and Its Precursors," 319.
87. Riley E. Dunlap and Angela G. Mertig, "The Evolution of the U.S. Environmental Movement from 1970 to 1990: An Overview," *Society & Natural Resources: An International Journal* 4, 3 (1991): 209–218; Gaylord Nelson and Earth Day Website: "The Living Tradition of Earth Day," http://www.nelsonearthday.net/earth-day/living-tradition.htm, (Accessed October 10, 2012); Denis Hayes, "Earth Day! A Call to Unite in Defense of Our Planet," *Mother Earth News* 309, 1 (April/May 2005): 24–31; David Rejeski, "Any Big Ideas Left?" *The Environmental Forum* 28, 5 (September/October 2011): 36–41; Shabecoff, *A Fierce Green Fire*, 107–112.
88. Environmental Action Website, "About EA," http://www.environmental-action.org/content/about-ea (Accessed September 13, 2012).
89. League of Conservation Voters Website, "Mission," http://www.lcv.org/about/mission/ (Accessed September 13, 2012). See also Shabecoff, *A Fierce Green Fire*, 113.
90. Gaylord Nelson and Earth Day Website: "'The Wrong Kind of Pollution': Critical Responses to Earth Day," http://www.nelsonearthday.net/earth-day/critics.htm (Accessed October 10, 2012); Hertsgaard, "Save Earth Day," 3–4; David Rejeski, "Any Big Ideas Left?" 36–41.
91. Paul R. Ehrlich, *The Population Bomb* (Cutchogue, N.Y.: Buccaneer Books, 1995 [1968]), 3.
92. Kline, *First Along the River*, 101–104; Shabecoff, *Earth Rising*, 7–9, 122–123.
93. Kline, *First Along the River*, 110–113; Rosenbaum, *Environmental Politics and Policy*, 12–13.
94. Kline, *First Along the River*, 116–117; Rosenbaum, *Environmental Politics and Policy*, 14, 15; Shabecoff, *Earth Rising*, 112–113.

III

POSTMODERN AMERICAN ENVIRONMENTAL POLITICS AND POLICY

Chapter 8

Environmental Interest Groups

> Ecological problems won't be solved by special interest groups spreading pop hysteria and merchandising fashionable panic. Genuine hard-got knowledge is required.
>
> **—P. J. O'Rourke, *Parliament of Whores*[1]**

The founders of the American regime recognized that organizations existing outside of government would play a crucial role in the formulation and implementation of public policy. Although they feared the destructive power of factions—self-interested persons who join with other like-minded individuals to pursue a goal based on a common interest or passion—they recognized, in the immortal words of "Federalist 10," that the "latent causes of faction are thus sown in the nature of man." If they could not prevent factions from forming, they must mitigate the effects of self-interest and passion lest such forces hobble or even destroy the republic. Despite their concerns, the Founders envisioned the centrality of group participation in debate and governance.[2]

The Interest-Group Theory of Democracy

The classical understanding of life inside a political association assumed that governments should promote virtue, thereby lifting up human beings from a depraved condition. This vision of perfectible human nature, with slight variations and differing nuances across the centuries, dominated the western intellectual tradition for almost two millennia. The ancients assumed that human nature could be changed. Through education and training, human beings could be made to see the light, although whether that light emanated from truth or God or monistic "Nature" depended on one's worldview. This faith in the concept of perfection ultimately was unsustainable. The grand traditions of classical thought and early Christian doctrine gradually eroded during the sixteenth and seventeenth centuries owing to empirical and metaphysical assaults from modern political philosophers, especially Niccolo Machiavelli, Francis Bacon, René Descartes, Thomas Hobbes, and John Locke. These new visionaries contended that classical and medieval Christian concepts were misguided because they were grounded in utopian visions. The ancients saw the world as they wished it to be, not as it existed in all its squalor and depravity.

Beginning with the Renaissance and the Enlightenment, thinkers reassessed the idealistic view of virtuous human nature. The old politics sought to lift up humans, to instill virtues in the unenlightened soul, but modern philosophers were not nearly so ambitious. They reduced the scope of politics. In lieu of the lofty virtues heralded by the ancients, the new thinker accepted the self-interested, passionate nature of humans as unquestionably dominant. As discussed in Chapter 5, the "new science of politics" highlighted the efficacy of means rather than the nobility of ends. Commodious living and the protection of individual liberties were the legitimate ends of a republican form of government. Politics could not change human nature; therefore, a regime must govern a man not as we would want him to be, but as he exists with all his flaws and imperfections. This view could be seen most clearly when James Madison argued in "Federalist 10" that "the principal task of modern legislation" is the regulation of competing interests, and the enterprise "involves the spirit of party and faction in the necessary and ordinary operations of government."[3]

In writing *The Federalist Papers*, Madison, Alexander Hamilton, and John Jay sought to explain how American political science could address the inherent pathology of democracies, those governments that Madison characterized as "spectacles of turbulence and contention." Unlike the ancients, who envisioned a small, relatively homogenous polity, Madison argued that only a large republic could accommodate the diverse interests of multiple, presumably heterogeneous, and competing groups. When a republic is large, the likelihood of a majority faction seizing power and potentially oppressing smaller groups through a tyranny of the majority diminishes because the playing field is so big and the multiplicity of interests so vast that no one group can gather too much power in too few hands. The task is simple to articulate if difficult to implement: Governments must police the "common impulse" of passion or interest so that minority rights are protected.[4]

Passionate man always has been, and always will be, an ever-present feature of the political landscape. If the passions and interests of individuals cannot be reduced to uniformity absent coercion or brainwashing, they must be channeled into activities that do not destroy the regime. The solution is to nurture the conditions for politics by pitting one group of like-minded, self-interested individuals against another. Madison and his brethren were satisfied to follow the tendency of western Enlightenment thought, which reduced the scope of politics. For all of their celebrated wisdom and judgment, the ancients could not appreciate the contentious nature of modern, enlightened self-government. Madison rejected the notion that harmony would displace discord. As he explained in "Federalist 51," human nature is such that there are not enough "better motives" to go around, not enough citizens and politicians who are animated by ideals that rise above self-interest and the gratification of one's own passionate desires. This "defect of better motives" can be overcome by establishing a system of "various and interfering interests" that pits "rival ambition against rival ambition." Although Madison and the Founders did not argue on behalf of the modern idea of interest group politics—which sometimes travels under the name of the "interest group theory of democracy"—they set the stage for the subsequent development of this concept. In attempting to transfer the "necessary evil" of factions into a positive attribute—pitting one group against another in hopes of preventing a powerful faction from tyrannizing the populace—the Founders unequivocally supported the notion that acquisitive man could not be trusted to protect the common good without oversight and political competition. It is a lesson that has never strayed far from the American political consciousness.[5]

Alexis de Tocqueville, a Frenchman who traveled to the United States during the 1830s and became a perspicacious observer of the democratic ethos, was impressed with the American tendency to join voluntary associations that enriched political life (Figure 8.1). It is customary in the modern world to point out the greed that predominates when a commercial regime flourishes, but Tocqueville distinguished acquisitiveness from greed. The former emphasizes the propensity of human beings to devote their passions to participating in the give and take of society as a means of acquiring the "good" things in life—material possessions, the respect of one's neighbors, the honor of country, the love of another. The acquisitive man must accommodate himself to the interests of others so he may learn what they desire. When he sees what they desire and compares it with what he desires, he finds that tastes reflect subtle differences. Recognizing these differences, he works with others to negotiate and bargain, ideally so that all parties can engage in a transaction that makes everyone better off—satisfying some, but probably not all, of their desires. This is the fundamental premise of neoclassical economics, as discussed in Chapter 2. It is an early expression of Pareto efficiency.[6]

Tocqueville marveled at the extent to which Americans were self-governing. They actively engaged with each other on an individual level when it suited their interests. They also joined other like-minded individuals to pursue common goals. In this manner, Tocqueville believed that voluntary associations—what today would be labeled "interest groups"—could be beneficial to the perpetuation of a republican form of government. When groups of like-minded individuals transact

Figure 8.1 The Frenchman Alexis de Tocqueville, author of *Democracy in America*, argued that Americans' propensity to join voluntary associations was a beneficial development in the American political system. (Courtesy of the Library of Congress)

business with other groups of like-minded individuals, Tocqueville observed, they are maximizing human preferences and enriching community and political life in ways unknown in European societies that do not enjoy such convivial voluntary associations. Where the Federalists had feared that factions were a necessary evil that must be transformed into a positive good through the mechanisms of government, Tocqueville saw groups as beneficial. They ensured that civic virtue could be advanced outside the boundaries of government regulation.[7]

The interest-group theory of democracy sprang from these early sources. David Truman's groundbreaking 1951 book *The Governmental Process* summarized the modern view in favor of interest-group politics. Rather than view such organizations as unwanted appendages to the American political system, Truman contended that groups were an integral part of the policy process. Because human beings, especially Americans, tend to join groups, just as Tocqueville observed, they can be counted on to express their views to the group. When a group grows large enough to become concerned about government action on an issue important to its membership, the group will send a representative to influence the policy process. The age-old problem of governance where it is difficult to discern the "will of the people" can be overcome, to some extent, when robust interest groups participate in the political arena. Assuming that people "put their money where their mouth is," the presence of particular groups and their desire to "duke it out" in the policy process will provide a reasonably reliable approximation of public sentiment. Pluralism, as the Federalists and Tocqueville understood it, reinvigorates American politics.[8]

Truman's pro-interest group theory triggered an enormous body of literature during the second half of the twentieth century. Aside from its many adherents, the theory also attracted its share of detractors, some of whom focused on the premises of the theory itself and some of whom questioned the continued vitality of interest groups. According to critics of the theory, Truman and his progeny accept several fundamental assumptions that may or may not be accurate. First, they assume that group membership is an accurate reflection of citizen preferences. In *The Logic of Collective Action*, Mancur Olson argued that some groups become elite and overpower lesser groups because elites possess a disproportionate share of money and influence. Elite groups may therefore be overrepresented in the political process. Smaller, less-powerful groups also may struggle with the free rider problem where some members enjoy the benefits but do not bear the burden of group membership.[9] In *The Semisovereign People: A Realist's View of Democracy in America*, political scientist E. E. Schattschneider argued that the conventional perspective on pluralism as a multiplicity of competing interests is a myth. The "range of organized, identifiable, known groups is amazingly narrow," he wrote. "There is nothing remotely universal about it." In fact, the U.S. political system "is skewed, loaded and unbalanced in favor of a fraction of a minority."[10] Theodore Lowi, in *The End of Liberalism*, was even more suspicious of the efficacy of interest group theory. In his opinion, with the beginning of Franklin Roosevelt's New Deal and the rise of the administrative state in the United States, Congress delegated an inordinate amount of authority to unelected public administrators so elected officials could avoid making unpopular decisions. With the loss of popular control by elected officials, interest groups arose to lobby in favor of their interests. Consequently, the power in the American public policy process shifted to the realm of unelected bureaucrats and the interest groups with which they frequently deal. Far from being beneficial, interest groups represent a threat to the democratic process because their influence has increased in inverse proportion to a transparent process of decision-making by elected officials who are directly accountable to the electorate.[11]

Other critics have contended that even if interest groups support the institutional structure of the American republic, group membership is dwindling and therefore the health of the polity is imperiled. Tocqueville's observations may have been true in the 1830s and in a later period, but in a world where globalization has eroded national borders, people do not join groups to the same extent they did in years past. With the loss of their political identity, citizens no longer feel a strong sense of community. In his widely discussed 1995 tract *Bowling Alone: America's Declining Social Capital*, Harvard University political scientist Robert Putnam suggested that "social capital" was declining because Americans were becoming more isolated and less likely to participate in groups. He used the declining membership in bowling leagues as a metaphor for a marked decrease in civic engagement. If interest groups are beneficial to the development of American public policy, Putnam observed, declining group membership portends negative consequences for the policy process.[12] In *Jihad vs. McWorld: How Globalism and Tribalism Are Reshaping the World*, Benjamin Barber, a Rutgers University political scientist, contended that two cultural and economic trends also threaten the policy process. As globalization increases—he referred to this process as "McWorld," the idea that corporations (such as McDonald's) frequently cross borders and are not answerable to particular governments—citizens become worried about the loss of their traditional identities. When traditional identities are threatened, individuals retreat into their tribal cultures, which Barber called "jihad." These competing forces undermine the traditional nation-state and its policymaking apparatus.[13]

Not everyone agrees with the observation that Americans increasingly are opting out of group membership. *Washington Post* columnist Robert J. Samuelson criticized Putnam's "intellectual ambition and journalistic superficiality" in a devastating critique of *Bowling Alone* published shortly after the book appeared. As Samuelson explained, "Putnam argues that civic life is collapsing—that Americans aren't joining, as they once did, the groups and clubs that promote trust and cooperation. This undermines democracy, he says." Samuelson countered that Americans have shifted their membership to different types of clubs than they joined in the past, but their civic engagement is as strong as it has ever been. Referring to *Bowling Alone*, Samuelson dismissed the book as "mostly bunk" because "although Americans may be sour, the reason is not that civic life is vanishing." In his opinion, "Americans mingle across racial, sexual and ethnic lines more now than ever." Samuelson was not alone in his criticism. Putnam's central thesis became one of the most hotly disputed issues in American political science during the late twentieth and early twenty-first centuries.[14]

While competing theories about overall interest group effectiveness and membership have been debated, the power of interest groups devoted to environmental concerns has increased during the early years of the twenty-first century. Modern American environmentalism illustrates the Founders' and Tocqueville's prescience about the prevalence of factions and voluntary associations. A 2006 research note in the academic journal *Environmental Politics* identified 92 environmental organizations that maintained an office in Washington, D.C., to lobby for their preferred positions on Capitol Hill. Even if, for the sake of argument, the overall number of groups is declining and these groups are not representative of citizen perspectives and preferences, the power and influence of environmental associations are undeniable.[15]

The environmental groups discussed here are non-governmental organizations (NGOs), an imprecise term referring to voluntary entities that work with individuals, multinational corporations, and government officials to effect changes in public policy. Whether the NGO model represents democracy at its finest or reflects the supremacy of mostly unaccountable private interest groups has been rancorously argued. What cannot be disputed is that environmental NGOs have become a ubiquitous presence in the U.S. policymaking arena.[16]

Well-Known Environmental Groups in the American Landscape

A detailed profile of every environmental organization that operates in the American political arena is far beyond the scope of this book. Nonetheless, it is instructive to narrow the list to a dozen prominent environmental organizations to understand their objectives, strategies, and tactics for influencing public policy. The groups vary widely in their mission, organizational structure, and operations as well as their histories, yet each has professed to be a steward for the natural environment. The list is admittedly subjective, and some well-known and important environmental groups such as Defenders of Wildlife, the Earth Island Institute, and the Environmental Working Group as well as regional organizations such as the Surfrider Foundation and Heal the Bay in California have not been included. The environmental groups profiled here represent some of the most influential lobbying organizations in the American polity, but they are by no means the only active associations. The groups are discussed in the order in which they were established.[17]

Sierra Club

No list could omit John Muir's organization, the Sierra Club, one of the oldest, continuously operating environmental groups in the United States. Founded in 1892, the group counts itself among the preservationist associations that Muir so admired. The idea of creating an organization originated when Muir and Robert Underwood Johnson, editor of a well-regarded publication, *The Century Magazine*, were sitting around a campfire in the Tuolumne Meadows of central California in 1889. The two men shared a passion for the Yosemite Valley and had spent much of their time strategizing on methods for convincing the federal government to protect the area from industrialization. During the campfire discussion, they debated the merits of establishing a club comprised of concerned environmental stewards.

The following year, their lobbying succeeded when Congress officially established Yosemite National Park. Despite this initial triumph, Muir and Johnson harbored no illusions: If they did not monitor federal efforts and ensure ongoing enforcement of the law, the area might well lose its pristine quality. As he had explained during the 1889 trip, Muir recognized that a permanent organization would be the most effective method of overseeing the activities of both private developers as well as federal government officials. Nothing should be left to chance.

Muir and Johnson were not alone in their concerns. Other parties were interested in protecting the Yosemite Valley, among them J. Henry Senger, an environmentalist from the University of California. Senger had met informally with several people who were interested in promoting the superior recreational features found in the area. When Muir learned of their interest, he agreed to meet with Senger as well as prominent men in the San Francisco Bay area, including Stanford University President David Starr Jordan, attorney Warren Olney, artist William Keith, and several professors from California universities. On May 28, 1892, the men assembled in Olney's office and incorporated the Sierra Club "to explore, enjoy, and render accessible the mountain regions of the Pacific Coast; to publish authentic information concerning them" as well as "enlist the support and cooperation of the people and government in preserving the forests and other natural features of the Sierra Nevada." As an article celebrating the group's centennial birthday explained, "The Sierra Club was born of the passion a small group of people held for the mountains of California. We envy them their simple love, untroubled by our own dark fears for the planet's future." It was an idyllic time shortly before the dawn of the twentieth century, an age when American environmentalism focused primarily on wilderness preservation. The club founders "marked the path, scouted the passes, and discovered the springs; they taught us to honor the wilderness. They are our parents, literally and figuratively. We celebrate what they did for us, rejoice in what they saved, and hope to do as much for those who follow." In the final tally, 182 charter members signed on to the club's mission.[18]

The founders did not seek to create a purely social club. They strongly believed their mission encompassed scientific research and exploration. If they intended to protect the region from industrial encroachment, it was necessary to map out the area as well as identify the species of plants and animals along with the geological features that must be included. To support the mission, the group began publishing *The Sierra Club Bulletin* in 1893. After a name change to *Sierra* in 1977, the publication continued to review noteworthy books about the environment, provide accounts of scientific expeditions into the wild, and sound urgent alarms about crucial environmental issues. As the editors noted in an article on the bulletin's history, "Any similarity between *Sierra* and the venerable journal of a hundred years ago is proudly admitted; the name has changed, but the purpose is the same."[19]

Figure 8.2 The Sierra Club is one of the oldest and best-known environmental groups in the United States. (Courtesy of the Granger Collection, New York)

Early group endeavors included lobbying government officials to preserve forest resources, especially California's famous redwood and sequoia trees. In fact, the group adopted the Sierra sequoia as part of its official seal (Figure 8.2). Beginning in 1901, club members began to participate in weeklong "high trip" outings to introduce them to the splendors of outdoor life among the spectacular sites of the Sierra Nevada. Muir's celebrated high trip with President Theodore Roosevelt occurred in 1903. In later years, the group focused on preserving lands outside of California.

By the early years of the twenty-first century, the club could boast of more than 750,000 members with state and local chapters scattered across the United States. The chapters were home to special-interest sections, task forces, and committees that emphasized environmental issues important to local communities. A 15-member board of directors oversaw the groups' activities. From the national headquarters offices in San Francisco, an executive director coordinated the activities of a paid staff of more than 500 and the efforts of scores of volunteers. The club maintained an active national presence through numerous state and local offices. To assist in lobbying government officials in Washington, D.C., as well as in state capitals, the Sierra Club Activist Network coordinated projects in a variety of areas, including global warming, water quality, coal mining, and nuclear power production.[20]

National Audubon Society

The National Audubon Society began as the result of a long train of events, although most historians credit Ms. Harriet Lawrence Hemenway as the driving force behind the modern organization. She hailed from a prominent Boston family. Her grandfather had become wealthy as a textile magnate while her father had once run for governor of Massachusetts and had served as treasurer of Harvard University. Her father also had been a devoted anti-slavery man in the antebellum era, founding Lawrence, Kansas, the famous abolitionist stronghold in the middle of a deeply divided state, and hence his daughter's middle name. For his part, husband Augustus had amassed a fortune from South American mining operations and had served in the state legislature.

Throughout her life, Harriet Hemenway continued the family's record of public service. When black leader Booker T. Washington could not secure suitable accommodations on a visit to Boston, she welcomed him into her home. In 1896, outraged at the indiscriminate slaughter of birds to manufacture women's hats, she set out to find a workable solution. Following in the footsteps of the famous naturalist and *Forest and Stream* publisher George Bird Grinnell, who had long lamented the frivolous, yet destructive, habit of decorating hats with bird feathers, Hemenway and her cousin, Minna Hall, both committed ornithologists, organized their friends and acquaintances to create the Massachusetts Audubon Society, named for the famous bird-lover. The organization's mission was straightforward: to "further the protection of our native birds" and to "discourage buying and wearing for ornamental purposes the feathers of any wild birds."[21]

The society was an immediate success. By October 1897, 111 chapters had sprung up throughout Massachusetts, 105 of which were headed by women. Similar groups appeared in neighboring states. Twenty states hosted a state organization with approximately 40,000 members by 1900. The chapters attracted a variety of people in all walks of life—high-society matrons, journalists, educators, naturalists, and professional lobbyists—all of whom criticized women's late-nineteenth-century fashions. They took their cues from Grinnell, a Yale-educated conservationist who had founded a precursor Audubon Society in 1886. He used the group as well as the pages of his magazine *Forest and Stream* to attack hunters who avariciously killed birds to supply the ghastly industry. In one memorable article, Grinnell remarked on the beauty of a living bird moving in its natural habitat, "but ten thousand dead birds with glass or waxen eyes, withering beaks and shrunken skins, perched above the noses of ten thousand women are anything but a source of joy."[22]

State Audubon Societies lobbied state legislatures to enact tough measures prohibiting the killing of non-game birds. The American Ornithologists' Union even developed model legislation to assist sympathetic lawmakers in enacting effective provisions. The quest to preserve fowl was a serious business: In November 1905, poachers ambushed and killed Guy Bradley, a game warden hired to protect bird habitats on the first bird sanctuary at Pelican Island in the Florida Everglades. Audubon Society members and game wardens threatened the livelihood of numerous hunters who would not cede the ground without a figurative and literal fight. If non-game birds were to be protected, tough laws and sufficient resources were needed to curb malicious behavior.[23]

In 1900, the movement gained momentum when the U.S. government passed the Lacey Act, sponsored by Congressman John B. Lacey of Iowa, outlawing the interstate shipment of birds killed in violation of state laws. In 1918, largely as a result of the state and national chapters of the Audubon Society, the federal government approved the Migratory Bird Treaty Act, an agreement with Great Britain (acting on behalf of Canada) to protect birds migrating across the Canadian-U.S. border. The measures might have passed without pressure from the state Audubon Societies, but their participation probably hastened the bills through the legislative process.[24]

A national committee of Audubon Societies emerged in 1902. Three years later, on January 5, 1905, William Dutcher filed papers with the state of New York creating the National Association of Audubon Societies for the Protection of Wild Birds and Animals. Three weeks afterward, the new organization held its inaugural meeting at the American Museum of Natural History. George Bird Grinnell attended. In the March 18, 1905, issue of *Forest and Stream*, he enthusiastically endorsed the Society as a natural extension of his decades-long effort to protect birds. Frank Chapman, founder of two magazines, *Audubon* and *Bird Lore*, also attended the first meeting and announced his support. The national organization was on its way to a long and storied existence.[25]

In the more than a century since its creation, the National Audubon Society has engaged in partnerships with government leaders and undertaken projects on its own. When federal officials have been unwilling or unable to preserve bird sanctuaries, the Society has stepped in, purchased the land, and established critical habitats without assistance. Since 1934, when NAS President John H. Baker published the first field guide during the height of the Great Depression, the organization has produced the *Audubon Field Guides* to assist serious ornithologists and amateur bird-watchers alike.[26]

The Audubon Society of the twenty-first century expanded its purview beyond birds, as the mission statement on the group's website explained. The organization's purpose was "To conserve and restore natural ecosystems, focusing on birds, other wildlife, and their habitats for the benefit of humanity and the earth's biological diversity." The NAS achieved the goal by mobilizing a strong network of chapters—more than 500 at last count—throughout the country to lobby legislators on the importance of protecting wildlife habitats. It also partnered with groups such as BirdLife International to acquire and develop more than 2,500 Audubon-designate Important Bird Areas (IBAs) as well as other wildlife sanctuaries and educate students and members of the public on scientific issues. During the first years of the twenty-first century, the NAS proudly touted its successes in protecting the Arctic National Wildlife Refuge and other fragile habitats, recovering the imperiled California condor and brown pelican, and restoring land in the Everglades and on Long Island Sound.[27]

Izaak Walton League

When Englishman Izaak Walton died at the age of 90 on December 15, 1683, he had spent much of his long life as a reasonably well-to-do businessman residing in burgeoning, thriving London. Fortunately for generations of outdoor enthusiasts, Walton had earned enough money through his judicious business dealings to retire comfortably from the commercial life by the age of 50. In his twilight years, he indulged in a life of outdoor contemplation and sport fishing, not to mention his hobby of writing about his thoughts and experiences. As early as 1619, he had put pen to paper, producing poems and biographies, but his most famous work became a 1653 book titled *The Compleat Angler: Contemplative Man's Recreation*, an idiosyncratic mix of reflections on travel, natural history, and fishing. In one memorable passage, Walton waxed eloquent on the pleasures of nature:

> I in these flow'ry meads would be:
> These crystal streams should solace me;
> To whose harmonious bubbling noise,
> I, with my angle, would rejoice.
> Sit here, and see the turtle-dove,
> Court his chaste mate to acts of love.[28]

The book went through several printings during the seventeenth century and proved to be popular among country gentlemen who wished to acquire the latest works on outdoor recreation. The publication also received criticism owing to Walton's politics. He was a Royalist living in the Cromwellian era when the book initially appeared. Following Restoration of the English monarchy, his political enemies still nursed old wounds, which they expressed by bitterly denouncing his work. It is difficult to fathom that a book on fishing and philosophical reflections on nature could attract vituperative political commentary, but it was a tumultuous age. At the time of Walton's death 30 years after its initial publication, *The Compleat Angler* had fallen out of favor. Eventually, it went out of print. It was only after nineteenth-century Englishmen fell in love with the outdoor life that a new generation discovered the delights of the whimsical, witty, contemplative fishing manual. With a new appreciation for the book came a realization that Walton's tome was not merely an instructional manual on the intricacies of hooking one's supper. It was a paean to nature and her adherents—and a well-written work of literature to boot. Izaak Walton thereafter became a symbol of the successful man who recognizes the virtues of pursuing a reflective, philosophical life in harmony with the natural environment.[29]

Fast-forward 238 years after the great man's demise. A group of sportsmen from Chicago, Illinois, gathered to create an organization dedicated to promoting and protecting fishing opportunities for themselves and their friends. As they searched for a name, the sportsmen naturally fixed on this classic, symbolic personality renowned for his love of nature and fishing. A new environmental group was born. As the organization's website stated in 2012: "Founded in 1922, the Izaak Walton League is one of the nation's oldest and most respected conservation organizations. With a powerful grassroots network of more than 250 local chapters nationwide, the League takes a common-sense approach toward protecting our country's natural heritage and improving outdoor recreation opportunities for all Americans. We invite you to learn about our work and to join us in supporting important conservation initiatives in your community."[30]

Under the leadership of the first president, Will Dilg, the Izaak Walton League of America championed policies to ensure pristine waters were available and cleaned up areas where the water already had become polluted. One early success included the creation of the Upper Mississippi River National Wildlife and Fish Refuge in 1924. In later years, the group urged legislators to establish a national clean water policy. Legislation enacted in 1948 and 1956 was marginally effective, but the League's crowning achievement was enactment of the Clean Water Act of 1972. Although other factors contributed to passage of the statute, the League's indefatigable advocacy on behalf of clean water resources was instrumental in creating the new law.[31]

As of 2012, the organization boasted of 38,000 members in the United States. Each member received periodic advocacy/action alerts highlighting important environmental issues and legislation involving wetlands protection, federal agricultural policy, regional conservation initiatives, wildlife protection, and sustainable community development. A quarterly magazine, *Outdoor America*, the League's annual report, and a periodic blog kept interested parties informed on the latest issues affecting the environment. Although it had a small number of active members compared with other mainstream environmental groups, the Izaak Walton League remained one of the most important advocacy organizations in the conservationist school of American environmentalism.[32]

Wilderness Society

The Wilderness Society was the organization that Bob Marshall, Aldo Leopold, and Robert Sterling Yard created in 1935 along with Benton MacKaye, the "Father of the Appalachian Trail," Ernest Oberholtzer, Harvey Broome, Bernard Frank, and Harold C. Anderson. Each of these men was a committed, life-long environmentalist with a love of wilderness areas. When the independently wealthy Bob Marshall died unexpectedly at a young age in 1939, a trust he left behind financed many of the group's activities in the ensuing years. Yard became the first secretary of the Wilderness Society and edited the organization's magazine, *The Living Wilderness*.[33]

The Wilderness Society proved adept at forming coalitions of interest groups to work toward achieving common goals, even when pursuing those goals initially appeared daunting. Representing the culmination of almost three decades of advocacy, enactment of the Wilderness Act of 1964 was no accident. Working with representatives from the Sierra Club, the General Federation of Women's Clubs, the American Planning and Civic Association, and the American Federation of Labor-Congress of Industrial Organizations (the leading labor organization in the United States), Wilderness Society lobbyists persuaded Congress to protect 9.1 million acres of forestland. Howard Zahniser, executive secretary of the Wilderness Society, wrote the statutory language prior to his untimely death a few months before President Lyndon Johnson signed the bill into law on September 3, 1964. Zahniser, a well-known adherent of the preservationist school of environmentalism, explained in his draft that wilderness lands included as part of the National Wilderness Preservation System (NWPS) must be "undeveloped by man's mechanical tools and in every way unmodified by his civilization." The measure secured bipartisan support in both houses of Congress, winning approval by a margin of 73-12 in the Senate and 373-1 in the House of Representatives.[34]

Pleased with the victory, Wilderness Society members were under no illusions that the battle had ended. Preserving more than 9 million acres of wilderness was a significant victory, but the total acreage represented a small fraction of the 800 million acres of forestland that existed in the United States in 1964. If additional lands were to be added to the NWPS in the future, the Society must keep up the pressure on federal officials to increase the acreage as well as enforce protectionist laws already in existence. Owing in no small part to the Wilderness Society efforts, the federal government increased the amount of protected lands every year for the next four decades. By 2005, 677 separate wilderness areas in 44 states encompassed more than 105 million acres—approximately 4.4 percent of the land mass of the United States.[35]

In the years following passage of the Wilderness Act, the group pushed for additional lands to be protected under the auspices of the NWPS, including the Arctic National Wildlife Refuge, a 1.5-million-acre coastal plain that serves as home to 250 animal species. The Society also urged Congress to prohibit oil and gas leases in wilderness areas, curtail road building in "roadless" portions of the National Forest

System, and adopt policies associated with the American Voice for Conservation, Recreation, & Preservation (AVCRP) to invest in green spaces in urban and suburban areas.[36]

By the early twenty-first century, the Wilderness Society was led by a governing council, a president's council, a president, and six vice presidents. Headquartered in Washington, D.C., as of 2012 the organization maintained nine regional offices and a staff of over 150 people. According to the Wilderness Society website, the organization actively pursued three goals: "Completing a system of protected wildlands in the United States; driving energy development away from sensitive wildlands and toward compatible places; and engaging an increasingly urban nation with the wildlands they own and inspiring them to care for wild places." Their methods included partnering with other associations to lobby legislators, publishing the *Wilderness* newsletter three times a year, and publicizing the need for forest preservation with assistance from influential members such as actress Betty White. Each year, the group presents the Ansel Adams Award, named for the celebrated wilderness photographer, to current or former federal officials who have served as effective conservation advocates. In addition, beginning in 1981, the Wilderness Society presents the annual Robert Marshall Award to a deserving individual who has championed wilderness protection and preservation.[37]

National Wildlife Federation

By the dawn of the twenty-first century, the National Wildlife Federation (NWF) had become one of the most recognizable environmental organizations in the United States. With an annual budget of $80 million, more than 4 million members, and a staff of 700 people, it was also the largest non-profit environmental organization in the country. By offering a variety of programs aimed at educating young people about protecting animals and enjoying the outdoor life, the group hoped to accomplish its mission of "inspiring Americans to protect wildlife for our children's future." The NWF worked with its national affiliates to address issues such as global warming, the loss of natural animal habitats, and the need to decrease dependence on fossil fuels.[38]

The Federation began in the 1930s at the behest of Jay Norwood "Ding" Darling, a cartoonist, self-proclaimed conservationist, and "man who saved ducks." As a newspaper writer and frequent traveler, he lived in Sioux City, Iowa, and spent much of his time in nearby South Dakota. He grew fond of the wildlife he encountered in his journeys, but he was worried that no one in the halls of power was looking for them. Despite lobbying from several environmental organizations, the federal government's commitment to nature had diminished in the years since Theodore Roosevelt, John Muir, and Gifford Pinchot had influenced public policy. To do his part, Darling stepped up and agreed to serve as chief of the Bureau of Biological Survey, the predecessor agency of the U.S. Fish and Wildlife Service. During his tenure there, he reduced the waterfowl bag limits to repopulate diminishing waterfowl and helped to set aside 3 million acres of public land as wildlife refuges. It also did not hurt that he could count President Franklin D. Roosevelt as a friend.

In fact, Darling convinced President Roosevelt to invite more than 2,000 anglers, conservationists, and wildlife enthusiasts to the first North American Wildlife Conference in 1936. Among other achievements, the session highlighted the need for an organization to unite and speak for the diverse individuals and groups that sought to protect wildlife and wild places. Darling came away from the meeting with an idea for a new organization, the General Wildlife Federation, a forerunner of the National Wildlife Federation. As the founder of the group, Darling assumed the role of first president. The Federation would serve as the lead coordinator for an active legislative agenda. Darling observed in 1935, "Wildlife interests remind me of an unorganized army, beaten in every battle, zealous and brave, but unable to combat the trained legions that are organized to get what they want." If Ding Darling got his way, the General Wildlife Federation would change all that.[39]

During his long, eventful life, Darling was a man of many accomplishments. Shortly after he stepped into his new role as president of the General Wildlife Federation, he successfully persuaded Congress to enact the Federal Aid in Wildlife Restoration Act of 1937, also known as the Pittman-Robertson Act, providing funds to states so they could purchase game habitats as well as finance wildlife research through a tax on firearms and ammunition. In his career as a cartoonist, he won the Pulitzer Prize in 1924 and again in 1943. He also designed the first duck stamp for the U.S. Postal Service. His "flying goose" drawing became the symbol for all federal wildlife refuges in the country. In 1978, 16 years after his death, the federal government dedicated the J. N. "Ding" Darling National Wildlife Reserve on Sanibel Island, Florida, in his honor.[40]

In the years since Darling served at the helm, the NWF has been involved in numerous legislative battles. The group's list of successes is long and impressive. Among the laws passed with Federation assistance were the 1956 Water Pollution Control Act, stipulating that federal grants may be given for the construction of water treatment plants; the Fish and Wildlife Reorganization Act, dividing the Fish and Wildlife Service into two bureaus—Sport Fisheries and Wildlife, and Commercial Fisheries; the Endangered Species Preservation Act of 1966, directing the Interior Department to carry out a land acquisition program to "conserve, protect, restore, and propagate selected species of native fish and wildlife"; the 1968 Wild and Scenic Rivers Act, segmenting seven major rivers into free-flowing states for recreational and conservation purposes; the Coastal Zone Management Act of 1972, designed to "preserve, protect, develop, and, where possible, to restore or enhance the resources of the nation's coastal zone"; and the Fish and Wildlife Conservation Act of 1980 to encourage comprehensive conservation planning.[41]

In 2005, Jerome Ringo, a former petrochemical worker from Louisiana's "Cancer Alley," became NWF chairperson. In an interview he granted the year following his accession, Ringo explained how he envisioned the organization moving forward. "I am the first African-American to head a major conservation group," he said. The problem was that many people of color had not viewed environmentalism as important, but times were changing. In Ringo's opinion, environmental groups of yesteryear "were founded by people who fished to put fish on the wall, not by people who fished to put fish on the table. And for poor people, issues like ozone depletion have not been a priority, compared with next month's rent. But I tell people in Cancer Alley, what good is next month's rent if you're dying of cancer?"[42]

As of 2012, the NWF was lobbying at the federal and state levels to enact laws preserving wildlife habitats, protecting endangered species, preventing water pollution, and calling for greater protection of public lands. The Federation also worked with Native American tribes. To build on past successes, the group promoted its message through a magazine, *National Wildlife*, and two publications, *Ranger Rick* (for children ages 7 to 14) and *Ranger Rick, Jr.* (for children ages 4 to 7), aimed at introducing young readers to wildlife protection.[43]

Nature Conservancy

The Nature Conservancy was founded in 1951 to save threatened species and habitats by buying land and setting it aside in the public interest. A group of scientists from the Ecological Society of America recognized that a pragmatic, incremental approach would accomplish more goals than all the grandiose plans and ambitious promises emanating from more radical environmental organizations. From modest beginnings—in 1956, the Conservancy had a staff of one (the indefatigable George Fell) and a $15,000 budget—by the early twenty-first century the group could claim to have compiled an impressive track record. With 1 million members worldwide and assets totaling more than $5 billion, the Nature Conservancy had protected 119 million acres of land and 5000 miles of rivers worldwide, and operated more than 100 marine conservation projects globally. Its reach was far and wide: The Conservancy maintained a presence in all 50 states of the United States and in more than 30 countries. Based on the totality of its financial assets, it was the wealthiest nonprofit environmental group in North America. Most donations came from the private sector.[44]

From its inception, the organization's staffers adopted a scientific approach to the issues, scouting out land in areas with a fragile ecosystem, an endangered animal species, or plant life that could not survive without protection. The ranks increased dramatically during the 1970s as the conservation movement expanded in conjunction with the enactment of new environmental laws and regulations. With a growth in donations and membership, the Conservancy also grew in sophistication. During a 1995 interview, then-chief executive officer John Sawhill explained the evolving approach over the years. "For four decades, the Conservancy focused almost exclusively on setting aside critical habitats and endangered species," he said. "In practice, that meant buying the specific piece of wetlands, forest, or prairie that supported a particular species or natural community. Like Noah, the

Conservancy was intent on building an ark—or, more accurately, building a lot of little arks." This approach was hands-on, measurable, and deeply satisfying to the group's membership. As the years progressed, however, the leadership recognized flaws in the effort. "Here was the problem," Sawhill explained. "We thought we could buy a piece of land, fence it off, and thereby protect whatever was in that preserve. But that thinking proved mistaken, which meant that our old performance measures—such as how much land we have acquired for conservation—weren't valid indicators of institutional progress."[45]

Because pollution spreads over large swaths of land, it is impossible to fence out all threats to a critical habitat. Instead of solely focusing on specific parcels of land, the Nature Conservancy eventually developed the concept of Last Great Places, protecting larger landscapes by touting prudent land-use management practices and partnering with federal government agencies, such as EPA, the Department of Agriculture, the U.S. Agency for International Development, the Department of the Interior, the National Park Service, and the Department of Defense as well as state and local agencies. The Conservancy also worked with private companies to discuss methods for incorporating sound environmental practices into their decision-making processes. Most notably, the Conservancy reached an agreement with Coca-Cola to determine the corporation's "water footprint" so that water would remain available for people, plants, and animals in areas where scarcity existed. Another company, Crystal Light, a popular Kraft Foods brand, worked with the group to protect and conserve freshwater supplies. In one of the most highly publicized partnerships, Dow Chemical Company announced a $10 million, five-year agreement to join forces with the Conservancy "to advance the incorporation of value of nature into business, and to take action to protect the earth's natural systems and the services they provide people, for the benefit of business and society."[46]

As of 2012, Mark Tercek, a former Goldman Sachs managing director, was at the helm. He recalled the day in May 2008 when he first learned that he had been tapped to lead the Conservancy. Giddy with excitement, he jumped into his Jeep Grand Cherokee, started the engine, and promptly backed into a tree, shattering the vehicle's rear window. Fearful of the negative omen, he was pleased to find that the damage to his vehicle was much worse than the damage to the tree. Tercek had good reason to be fearful. Similar to the rest of the nation, the Nature Conservancy was suffering through hard times in 2008. As the stock market declined and the economy slid into a deep recession, donations shrank by 27 percent. The Conservancy's normally robust $1.6 billion investment portfolio lost $321 million by the end of the 2009 fiscal year. To add insult to injury, group members expressed deep reservations about bringing in a banker to head the country's largest environmental group. Indeed, the disdain for Wall Street financiers was so strong and bitter that Tercek had a doubly difficult task of making unpleasant, difficult decisions in the face of internal dissension.

As a "numbers guy," Tercek recognized he had to place the organization on solid financial footing if the 60-year-old Conservancy was to survive the economic turmoil. He trimmed $80 million from the budget, reducing the 3,700-member staff by 9 percent. Despite these painful cuts, staffers begrudgingly conceded that the new executive director had served the group well. Peter Kareiva, the Conservancy's chief scientist, reflected the view of many colleagues when he ruminated on the dark days of 2008 and 2009. "I had to lay off a large number of people that I really respected. Yet everybody realized that Mark was right—that he could prioritize, see this unemotionally, that his Goldman experience was a credit."[47]

Under Tercek's leadership, the Conservancy continued its collaborative efforts with government entities and private companies. Supporters believed it was preferable to avoid labeling opponents as "bad actors" as a means of soliciting funds among true believers. Instead, sitting down with corporate officers to figure out solutions was more beneficial to the natural environment. As co-chair of the Nature Conservancy, Teresa Beck, explained in a 2012 interview, the group is "more likely in the boardroom than the courtroom." Her perspective mirrored Tercek's vision. "We're collaborators. We pursue non-confrontational, pragmatic solutions to conservation challenges, so you don't hear of the Nature Conservancy suing people to get things done. We try to work as partners with people to get things done."[48]

It was exactly this non-confrontational attitude that led Greenpeace to label the group a "right-wing environmental organization." After the Conservancy brokered a deal to provide tradable pollution credits to three large U.S. companies to protect Bolivian forests, Greenpeace denounced the enterprise as "a carbon scam." When asked about the criticism, Tercek was philosophical and, as always, pragmatic. "Why wouldn't we want to work with companies with the biggest environmental footprints? It would be irresponsible not to try."[49]

World Wildlife Fund

In 1961, 16 conservationists, including biologist and African-wildlife expert Sir Julian Huxley, International Union for the Conservation of Nature and Natural Resources (IUCN) Vice President Sir Peter Scott, and director-general of the British Nature Conservancy, E. M. Nicholson, were mulling over fundraising strategies for wildlife conservation. The group eventually produced the Morges Manifesto (named for a city in Switzerland) explaining that the expertise existed to protect the global environment, but funding was lacking. The World Wildlife Fund (WWF) was born as a result of the manifesto. It was designed as an international fundraising organization to collaborate with existing conservation groups and provide financial resources to areas in desperate need of support. The group's office opened in Morges on September 11, 1961.[50]

As stated in the manifesto, the WWF originally raised money to provide grants to deserving NGOs, especially those dedicated to protecting endangered species. As donations increased and the budget grew, the organization's purview expanded to encompass issues such as preserving biological diversity, promoting sustainability in natural-resource use, pollution reduction, and managing global climate change. The group's power and influence increased tremendously during the 1970s. In 1973, the WWF was instrumental in securing authorization of the Endangered Species Act. In 1979, the organization created the Trade Records Analysis of Flora and Fauna in Commerce (TRAFFIC) to monitor the wildlife trade. Any data collected on illegal activity is shared with the appropriate government sources, including law enforcement agencies in the countries where the activity occurs.[51]

The organization changed its name to World Wide Fund for Nature in 1986, although it retained the acronym "WWF" as well as its distinctive and widely recognized panda logo. The original name remained in place in the United States and Canada. That same year, the WWF celebrated its 25th anniversary by meeting in Assisi, Italy (home of St. Francis, well known for his love of animals). To mark the auspicious occasion, HRH Prince Philip, the Duke of Edinburgh, asked religious authorities from many faiths, including Buddhism, Christianity, Hinduism, Islam, and Judaism, to join the WWF in reaffirming the spiritual bond between their followers and nature. The Assisi Declarations, as they came to be known, promoted a connection between religion and conservation.[52]

In 1996, the group obtained general consultative status with the United Nations Educational, Scientific, and Cultural Organization (UNESCO). In essence, this status recognizes the legitimacy of an NGO contributing to "peace and security by promoting international collaboration through education, science, and culture." By the twenty-first century, owing in no small part to the UNESCO-approved status, the WWF continued to fulfill its mission with a great deal of success. According to its website, the group was protecting and restoring species and their habitats; strengthening local communities' ability to conserve their natural resources; transforming markets and policies to reduce the impact of the production and consumption of commodities; ensuring that the value of nature is reflected in decisions made by individuals, communities, governments, and businesses; and mobilizing hundreds of millions of people to support conservation. "We seek to be the voice for those creatures who have no voice," the organization's directors explained. "We speak for their future. We seek to apply the wealth of our talents, knowledge, and passion to making the world wealthier in life, in spirit, and in living wonder of nature." The WWF claimed to be the world's leading conservation organization, working in 100 countries and supported by 1.2 million members in the United States and close to 5 million around the world.[53]

Experienced natural-resource managers on the WWF payroll typically assist governments and private companies in developing national parks and protected lands. The level of assistance varies. Sometimes funding is all that is required, but on other occasions, the managers provide technical advice, supply equipment, or assist in constructing visitor centers and onsite facilities. These efforts have been especially valuable in developing nations. As then-chief executive director Greg Smith noted in 2008, "We've been doing a lot of things to benefit the Earth and the environment for more than 40 years. WWF has overall responsibility to be green for the planet and to set an example for the rest of the world."[54]

Environmental Defense Fund

The Environmental Defense Fund (EDF) has been one of the best-known and most influential environmental groups in the United States since its founding in 1967. The organization's stated objectives are clear and concise: "We are passionate, pragmatic environmental advocates who believe in prosperity *and* stewardship Grounded in science, we forge partnerships and harness the power of market incentives." Unlike some environmental groups that refuse to work with private companies in developing corporate "green" strategies, EDF prides itself on partnering with corporations in the private sector to develop alliances. Although many observers applaud these "market-based" solutions, radical environmentalists charge that long-time EDF president Fred Krupp is a "corporate stooge" while some corporate officers view the group's "partnerships" as thinly veiled strong-arm tactics designed to threaten weak-kneed company executives with negative publicity and product boycotts if they refuse to capitulate to EDF's demands that they fundamentally alter their purchasing decisions and supplier relations.

The EDF model evolved from the organization that was born decades earlier. The impetus for creating the group came from fears over the use of the pesticide dichlorodiphenyltrichloroethane (DDT) highlighted in Rachel Carson's 1962 best-seller *Silent Spring*. Founders Art Colley, Dennis Puleston, Robert Smolker, George Woodwell, Charles Wurster, and Victor Yannacone were lawyers living on Long Island in Suffolk County, New York. Sitting around a kitchen table one day as they discussed the loss of osprey and other wildlife at alarming rates, the men decided to do what they did best: litigate. Focusing on the area around their homes, they filed a lawsuit against the Suffolk County Mosquito Commission setting forth the dangers of DDT use. The first such suit against DDT had been filed a decade earlier—news coverage of the 1957 case had influenced Rachel Carson as she wrote her book—but the chemical remained in widespread use. After years of frustrating legal wrangling, the elated environmentalists eventually helped to ban DDT use during the 1970s. Following the victory, they recognized the value in taking their efforts to a national level and creating a permanent organization. "Sue the bastards" became their unofficial motto.[55]

Environmental groups reveled in the dramatic increase in environmental laws and regulations during the heady days of the 1970s. By the mid-1980s, however, the good times appeared to have ended. The Reagan era had dawned and the nation had moved to the right of the political spectrum as conservative commentators disparaged "big government" solutions to challenges that could best be addressed in the private sector or through state governments that operated closer to the citizenry. Donations declined while many environmental groups struggled to justify their continued existence. Concerned about EDF's stagnation, the group hired a dynamic young environmentalist to take the helm. His name was Fred Krupp.

Krupp refused to be dispirited by changing sensibilities. He had devoted his life to environmental causes. Even as a child, he later claimed, he could remember the devastating effects of chemical spills and suburban sprawl on the land adjacent to his middle-class neighborhood. "I can literally remember seeing wildlife scurrying away from the bulldozers," he once remarked in an interview. As an idealistic undergraduate at Yale University, he initially shared his fellow students' antipathy for big corporations and profit-driven business enterprises. He credited an engineering professor with broadening his perspective. "He talked about people lowering their voices and understanding each others' goals," Krupp recalled years later. "I just found his philosophy so positive and optimistic. That you could solve problems if people could really talk to each other."[56]

After graduating from Yale and earning a law degree at the University of Michigan, Krupp practiced law briefly before joining the EDF at the age of 30 in 1984. He found an organization in disarray. With the momentum from the environmental movement waning, the old approaches no longer worked. If EDF was going to persevere in a brave new world, new strategies and tactics were needed. Krupp went to work as though he were a business guru hired to turn around a floundering business, which in a manner of speaking he was. The old motto of "Sue the bastards" was softened so that it became the more convivial "What bastards?" Eschewing the old confrontational methods, Krupp reached out to large donors for charitable contributions. He cultivated friendly relationships with the media. He extended a hand to corporate executives, offering partnerships where litigation once had been the first response. He did not hesitate to use force when necessary—a gloved fist is still a fist—but he saw a better way. Rather than haul companies into court as a first resort, he resolved to meet with their leadership to show them how to reform their wicked ways. If they could see the economic and public relations value inherent in waste reduction, recycling, and "greening" their businesses, they would change their behavior of their own accord.[57]

He signaled his intentions in a 1986 *Wall Street Journal* editorial heralding a new, "third stage" of American environmentalism where non-profit environmental organizations would meet with corporate leaders to develop voluntary strategies for improving their business practices. Everyone would benefit if litigation costs were reduced or avoided altogether, to say nothing of the advantages from avoiding nasty boycotts and protests in the news media. Detailing his position in one of the most politically conservative, business-friendly publications in the country, Krupp assured his readers that his proposal was "in no sense a move toward compromise, a search for the in-between position." Instead, it was a move toward the "increased use of market-oriented incentives" which, in his view, "holds promise for greater environmental and economic benefits at a lower social and economic cost." This market-based approach to environmentalism suggested that modest, incremental change might be in the offing. Krupp's measured suggestions did nothing to endear him to radical environmentalists who denounced his timidity and willingness to be co-opted by the enemy. For radical environmentalists who rejected compromise and made a name (and encouraged contributions) using high-profile scare tactics, Krupp's approach undermined their efforts to portray private companies as greedy, villainous fiends who could not be trusted to negotiate a satisfactory solution to environmental problems.[58]

In the ensuing years, Krupp directed his staff scientists to focus on big, multifaceted issues such as global warming—an issue that was just emerging during the 1980s. Yet he also understood the importance of symbolic victories involving smaller issues that had captured the public imagination. One high-profile battle occurred in 1990. McDonald's Corporation had been packaging its sandwiches in polystyrene foam clamshell containers. During the late 1980s, polystyrene foam—sometimes mistakenly identified as "Styrofoam," Dow Chemical Company's trademarked name for its polystyrene foam insulation material used in the housing industry—became the "poster child" of the nation's solid waste "crisis." When pressed to explain their intense hatred for the product, few opponents could accurately pinpoint the reasons, but clearly, public opinion had identified an environmental villain. Much to its consternation, McDonald's, arguably the world's most recognized quick-service restaurant, had become the target of a high-profile campaign of protests and boycotts over many issues, including the company's packaging waste. Most environmentalists were satisfied with the ongoing denigration of the hamburger chain in the press—after all, prominent scapegoats incessantly pilloried in a public forum will call attention to issues that continually fill the coffers of "white knight" organizations—but Krupp recognized an opening for a collaborative effort.

Seeking to test the precepts of his new environmentalism, Krupp initiated contact with a letter to McDonald's outlining his plan. Desperate to alleviate growing public pressure to manage its packaging waste, McDonald's executives had been mulling over the costs and benefits of instituting a polystyrene foam recycling program. Responding favorably to Krupp's entreaty, the company established an eight-member McDonald's-EDF joint task force on waste reduction. On August 1, 1990, the group, led by Krupp and Shelby Yastrow, general counsel and senior vice president at McDonald's, paraded before an audience at the National Press Club and announced the formation of the task force to study available options for six months. Two months later, McDonald's expanded its polystyrene foam clamshell recycling program on the West Coast as a show of good faith that the hamburger chain was serious about responsibly handling its packaging. Although the EDF was pleased with the company's cooperation, accommodating the polystyrene industry would not present an environmental group with the incontrovertible victory that Krupp and his staffers required. He had risked his credibility in the environmental community to work with McDonald's. Krupp could not settle for a subtle, nuanced accommodation without appearing to be the corporate stooge his radical brethren had always claimed he was. Thus, on November 1, 1990, the joint task force reversed course and announced that McDonald's would no longer use clamshell packaging made from polystyrene foam. Instead, the company would deliver sandwiches in a paper wrapping that was perceived to be environmentally preferable to polystyrene foam. In later discussions, the task force identified 42 initiatives to reduce waste in the company's operations.[59]

Despite criticism from environmentalists who believed that EDF should have pushed the company to abandon paper along with polystyrene foam and supporters of polystyrene foam recycling who had been pleased that McDonald's was going to lend its heft to a nationwide program, the joint partnership achieved its goals. McDonald's reduced the level of public protests by showing its willingness to work with a major environmental group to reduce its environmental footprint. For its part, EDF demonstrated that it could stand up to a major corporation and forge a victory where other environmentalists had failed. Fred Krupp's reputation as a pragmatic champion of market-based reforms was solidified. Both parties enjoyed mostly favorable media coverage. The fact that McDonald's continued to serve coffee in polystyrene foam cups, that it had set back the state of polystyrene foam recycling for years, perhaps decades, and that credible scientific evidence strongly indicated that polystyrene foam was less damaging to the environment than the alternative packaging were never seriously addressed. Why would they be? In the absence of public pressure, virtually everyone assumed the matter was resolved. Whatever else the episode indicated, one point was clear. The EDF had demonstrated a crucial insight in the modern environmental wars: Scoring a win and improving the group's public perception were the important variables. Whether change that genuinely benefits the natural environment occurs is a secondary consideration. It is desirable to clean up the environment, but such goals—clouded as they are by conflicting, iterative, complex scientific data—must never supersede the public relations triumphs in cases where perception and reality diverge.[60]

During the decades that followed, EDF expanded its reach. By the early twenty-first century, the organization, which was headquartered in New York City, operated regional offices in Austin, Texas; Boulder, Colorado; Washington, D.C.; Raleigh, North Carolina; Boston, Massachusetts; and California offices in San Francisco, Los Angeles, and Sacramento. The group was involved in a variety of issues from climate change, ocean pollution, toxic releases from chemicals leading to air pollution, and the protection of endangered species, among others. In 2011, U.S. Secretary of Energy Steven Chu asked Krupp to serve on a working group of experts assembled to make recommendations on safety and environmental standards for natural gas hydraulic fracturing from shale formations.[61]

In the twenty-first century, EDF is most closely associated with the global warming and climate change debate. During a visit to Princeton University in 2003, Fred Krupp explained its importance. "Global warming is the largest environmental threat we face," he told the audience. "As citizens in a democracy, we carry the heavy responsibility of prompting action against this danger early enough to head off the most dangerous consequences. Global warming is particularly insidious because, unlike some environmental ills, it does not manifest itself in ways that are plainly obvious to policymakers or the public. Its symptoms creep up on us slowly, over many decades." He urged his listeners to become involved in the fight to make a difference and spark a green reform movement across the country and around the world.[62]

Friends of the Earth

Friends of the Earth (FOE) was the brainchild of the renowned environmentalist David Brower. After he resigned as executive director of the Sierra Club in 1969, in part because he opposed the development of the California coastline and he did not believe the organization adequately supported his position, Brower searched for a new outlet for his activism. It was an era devoted to civic engagement, and he recognized the rising tide of environmentalism that would lead to the first Earth Day a year later. Riding the crest of this wave of Sixties enthusiasm, Brower created FOE to be a more aggressive organization than the Sierra Club had been. FOE could not be called "radical"—violence was not in the offing—but it was devoted to confrontational tactics that the more establishment-oriented Sierra Club was reluctant to employ. The group set up shop in San Francisco and opened a second office in Washington, D.C. Brower later spun off two additional environmental organizations, the League of Conservation Voters and the Earth Island Institute. In those early years, FOE launched high-profile campaigns against a variety of development projects favored by industrialists, including construction of the Alaska pipeline, nuclear reactors, and the supersonic transport (SST) airplane. The group also voiced vehement opposition to continued use of the toxic defoliant herbicide Agent Orange. During the 1980s, FOE lobbied against the efforts of President Reagan's notorious secretary of the Interior Department, James G. Watt, to lease and sell public lands near western national parks.[63]

Brower realized that FOE could maximize its efficacy by expanding beyond the U.S. border. Fortunately for his supporters, he had many international connections, which he used to encourage the group's overseas development. The European chapter of FOE, formed in the United Kingdom in 1971, eventually helped the group to become one of the world's leading environmental networks, providing a home for organizations from 76 countries. In the early years, FOE's European branch pursued modest, incremental change. One initiative in the 1970s involved a campaign to persuade citizens to drop non-returnable bottles on the doorstep of a major beverage company, Schweppes, in London. In 1972, the group scored a significant victory by successfully lobbying the English government to ban the importation of skins from snow leopards, clouded leopards, cheetahs, and tigers. As the successes mounted over time, FOE expanded its purview to address a variety of issues, including efforts to save the whales, protect endangered species, prevent ozone depletion, reduce or eliminate rainforest destruction, block nuclear reactor construction, and curb industrial practices that lead to climate change.[64]

FOE raised its international profile during the 1980s and 1990s. The group's well-publicized public relations campaigns to prevent the importation and exportation of toxic wastes to and from the United Kingdom changed national policy. By threatening to engage in a noisy product boycott against companies that used ozone-depleting chlorofluorocarbons (CFCs) in their manufacturing processes, FOE pressured the manufacturers to phase out the use of those chemicals by the end of the 1980s. The group set an ambitious recycling agenda in 1989, pushing the United Kingdom to recycle 25 percent of household waste. Aside from efforts urging businesses and governments to act on environmental policies, the group also campaigned to prevent certain types of actions. In 1990, FOE coordinated a campaign among many diverse environmental groups to persuade the United Kingdom to abandon a road construction project.[65]

The Friends' organizational structure always distinguished it from competing environmental groups and allowed it to have an advantage in global environmental politics, especially compared with its two closest competitors, Greenpeace and the WWF. In the international environmental community, organizations from northern hemisphere nations usually have enjoyed greater attention and higher donations than organizations headquartered in southern hemisphere nations. The disparity has generated great resentment as leading international environmental groups have inadvertently failed to recognize how their policies contribute to continued inequality. Greenpeace International, for example, has centralized its operations with franchises in dozens of countries around the world. For most of its existence, Greenpeace has been controlled by a board of directors composed of members based on the level of financial contributions from the franchises. Thus, the Dutch, German, British, United States, and Australian branches have fashioned a majority coalition owing to their high level of funding. FOE, by contrast, allows each member group to vote regardless of the size of the financial contribution. This leveling process has ensured that typically marginalized groups with few resources, especially in developing southern-hemisphere countries, could exercise a stronger voice in group decisions. It also has meant that FOE addresses social and political inequality with a strong commitment to environmental justice. WWF, in the meantime, has focused on sustainable development through cooperative programs that have eschewed the confrontational politics of Greenpeace, preferring to avoid the pejorative "radical" label for fear that it would undermine collaborative efforts with businesses and governments. WWF's emphasis on providing grant monies and technical assistance to parties interested in preserving parks has ensured that poorer nations do not benefit from many of the organization's programs. Moreover, both Greenpeace and WWF, despite their philosophical differences, frequently have focused on headline-generating single-issue campaigns. FOE has tried to carve out a middle position between the two groups. On one hand, Friends has sought to be less aggressive and more representative of less financially-affluent members than Greenpeace. On the other hand, Friends has rejected WWF's cautious approach, neither seeking nor avoiding negative labels. In the words of one commentator, "For these reasons FOE is the only one of the three largest transnational environmental [non-governmental organizations] where it is possible to see the *negotiation* of a common identity between environmental groups from northern and southern countries."[66]

FOE has not been reluctant to use aggressive tactics, including litigation, when necessary to achieve its goals. The highest profile lawsuit arose during the 1990s, when the U.S. chapter challenged Laidlaw Environmental Services, a corporation that FOE alleged was dumping mercury into South Carolina's North Tyger River in violation of the U.S. Clean Water Act. The lawsuit was appealed all the way to the U.S. Supreme Court, which held in a 5-2 decision that citizens living near the North Tyger River had standing to sue Laidlaw without proving actual harm from the illegal dumping. The *Laidlaw* case and other litigation have solidified FOE's place as one of the most influential and effective environmental groups operating in the United States during the twenty-first century.[67]

Natural Resources Defense Council

The Natural Resources Defense Council (NRDC) began after a group of law students became concerned about a spate of environmental disasters that generated headlines in 1969. In January and February of that year, a huge oil spill off the coast of Santa Barbara galvanized the growing West Coast environmental movement. In June, an oil slick on the Cuyahoga River in Cleveland, Ohio, caught fire, sending plumes of dark smoke billowing into the atmosphere and alarming concerned citizens around the country. One person watching these and other tragic incidents was deeply affected and resolved to take action. John H. Adams, a Duke University Law School graduate and former assistant U.S. Attorney for the Southern District of New York as well as a special attorney with the Joint Task Force on Organized Crime and Racketeering, was searching for a new job at precisely the moment when he realized he wanted to handle environmental cases. "I loved being in court, thinking on my feet, the daily interaction with a multitude of people," he subsequently wrote. "However, it was a job most people do for only a few years, and I was ready to move on."[68]

Adams reached out to his circle of friends and acquaintances to discover whether any positions existed in environmental law. Through his networking efforts, Adams stumbled onto Whitney North Seymour, a lawyer who recently had agreed to create a new public interest law firm called the Natural Resources Defense League along with a colleague, Stephen Duggan. Shortly before he was to take the helm, Seymour accepted a position as the U.S. Attorney for the Southern District of New York—the same office where Adams had worked previously. Searching for a suitable replacement, Seymour asked Duggan to interview Adams. Adams and Duggan established an instant rapport. In their discussions, they decided to change the name of the nascent organization, substituting "council" for "league." Thus, it became the

Natural Resources Defense Council, or NRDC. The next step was to seek start-up funding. After several fruitless searches, the men approached the Ford Foundation.[69]

At about the same time that Adams was searching for a new challenge in his legal career, four law students from Yale University—Gus Speth, Dick Ayres, John Bryson, and Edward Strohbehn—and Harvard Law student and former Rhodes Scholar Tom Stoel joined forces to ruminate over how they could combat environmental problems through the courts. As Adams later explained, the impetus came from an article Speth read in the *New York Times* about the National Association for the Advancement of Colored People (NAACP) Legal Defense Fund. The idea struck him immediately: "Why not an NAACP-style legal defense fund for the environment?" With help from two well-known Yale faculty members, Charles Reich, author of the influential book *The Greening of America*, and Boris Bittker, a highly regarded expert on tax law, the students approached the Ford Foundation for funding. Foundation officers were interested in the idea, but they wanted to avoid duplicated efforts. They directed the students to meet with Adams, the seasoned attorney. When the groups met and discovered their mutual goals, they agreed to join forces. At a conference held in Princeton, New Jersey, in February 1970, the NRDC was created with Ford Foundation financing.[70]

Adams stepped in as executive director and held the position for 36 years. In a 1990 interview, he succinctly explained the group's fundamental approach to environmentalism. "We are here to represent the environment in the courts, before Congress, and the government agencies," he said. "Twenty years ago, we started out strictly as a law firm, but we soon realized that winning in the courts alone wasn't enough to protect the environment. We have to keep an eye on the entire process, from the moment legislation is introduced to its implementation by government agencies."[71]

Keeping an eye on the process involved filing lawsuits as well as working behind the scenes to draft landmark legislation that shaped the federal government's legal framework during the 1970s. Environmental protection historically had been handled by states, but the federal government's role increased throughout the decade, especially after Jimmy Carter moved into the White House in 1977. The new president reached out to the NRDC to fill key environmental policy positions within his administration. Gus Speth became chairperson of the White House Council on Environmental Quality. NRDC attorney David Hawkins was confirmed as the assistant administrator for air at the U.S. Environmental Protection Agency. In fact, so many NRDC staffers worked inside the Carter administration that critics complained that the law firm was a shadow EPA.[72]

Even after Carter lost his reelection bid, the NRDC remained an influential player in the policy process. In 1982, the group announced the Citizen Enforcement Program targeting corporate polluters that violated federal environmental statutes, especially the Clean Water Act, a law that allows citizen suits to compel compliance. The NRDC's program created a network for private environmental watchdogs to police corporate polluters, ostensibly in the public interest but also for private gain. With several environmental statutes providing for attorney fees, *qui tam* suits increased in frequency. For environmentalists anxious to use legal mechanisms to chase down offenders, such aggressive legal maneuvering was hailed as an effective method of protecting nature. For proponents of free-markets and smaller government, the overreaching breadth of environmental laws enacted beginning in the 1970s has hampered businesses, needlessly increased the size and cost of the federal bureaucracy, and encouraged "bounty hunter" law firms to file lawsuits against companies even when evidence of violations is scant or non-existent.[73]

The NRDC has experienced its share of victories and defeats throughout the years. Perhaps its best known lawsuit was *Chevron U.S.A., Inc. v. Natural Resources Defense Council*, a case that reached the U.S. Supreme Court in 1984 and became a landmark opinion in federal administrative law. The matter originated after the Clean Air Act of 1977 required states that failed to meet air quality standards to establish a permit program regulating "new or modified major stationary sources" of air pollution. The U.S. EPA initially interpreted a "source" broadly to include any device installed inside a manufacturing plant that produces air emissions. During the Reagan administration, however, the agency, reflecting the president's desire to relax stringent air quality standards, reinterpreted the term to allow currently operating plants to acquire permits for new equipment that did not meet the air quality standards as long as the total emissions from the plant did not increase. This became known as the "bubble" concept. In its role as an environmental watchdog, the NRDC filed suit to compel the agency to use the original interpretation and reject the "bubble" approach. The U.S. Court of Appeals for the District of Columbia ruled in the NRDC's favor, so EPA appealed to the Supreme Court. Writing for the court majority, Justice John Paul Stevens held that the high court would not intercede and second-guess EPA's interpretation. According to Stevens, courts must defer to administrative agency interpretations of the authority granted by Congress where the intent of Congress was ambiguous and the interpretation was reasonable or permissible. This opinion created a doctrine subsequently referred to as "*Chevron* deference," which represented a short-term defeat for the NRDC. In the long term, the doctrine expanded the power of administrative agencies during judicial review of administrative actions because it ensured that courts would be reluctant to impose their judgment on agencies in the absence of clear abuse of discretionary authority.[74]

Aside from the *Chevron* case, NRDC may be best known for its role in a controversial environmental health alert involving daminozide, a chemical sprayed on fruit to regulate growth, make it easier to harvest, and improve its color. In 1963, the Uniroyal Chemical Company registered daminozide under the trademarked name "Alar" for use on fruits consumed by human beings in the United States. Fifteen years later, the National Cancer Institute (NCI) published the results of a daminozide bioassay indicating that the chemical was a weak carcinogen. The measurement was so minuscule that EPA did not rely on the NCI data in preparing a quantitative risk assessment for daminozide. The U.S. Food & Drug Administration (FDA) approved the chemical for commercial use in 1968.

Throughout the years, new testing failed to indicate dangerous human health effects from using the chemical, but neither could testing demonstrate conclusively that daminozide was "safe," however the term was defined. The NRDC pressured EPA to ban daminozide use in the 1980s after a series of contested studies suggested that residue present in apple juice and applesauce could be harmful. When these behind-the-scenes negotiations failed to produce the desired result, the Council employed the ultimate weapon: the media. After launching the Children's Environmental Health Initiative early in 1989, the NRDC retained a public relations firm, Fenton Communications, to arrange an exclusive agreement with the popular, self-styled muckraking television show *60 Minutes*. Relying on an NRDC-produced report, "Intolerable Risk: Pesticides in Our Children's Food," the CBS television network aired a segment titled "'A' Is for Apple" on February 26, 1989. Presenting an alarmist theme, the segment featured an image of an apple with a skull and crossbones superimposed on top as *60 Minutes*-correspondent Ed Bradley issued a dire warning. The "most potent cancer-causing agent in our food supply is a substance sprayed on apples to keep them on the trees longer and make them look better," he said. Lest anyone lose the point owing to unintentional subtlety, the episode featured a scene from a pediatric cancer hospital. Minnesota Congressman Gerry Sikorski appeared in one frame lamenting, "See these bald, wasting-away kids."

The resultant hysteria succeeded in galvanizing public opinion far beyond the NRDC's wildest dreams. Empathetic television personality Phil Donahue aired an episode where he breathlessly informed his audience, "Don't look now, but we're poisoning our kids. I wouldn't lie to ya." School officials across the country destroyed apples and apple products found in their cafeterias owing to the precautionary principle, which states that it is better to be safe than sorry. One hysterical parent realized that a child had transported an apple to school; consequently, state troopers were dispatched to chase down the school bus before any harm could come to the children.

Although additional testing could not establish the dangers attributed to daminozide, apple growers were losing money owing to the negative publicity. They ceased using the chemical after EPA ordered an Alar phase out by July 31, 1990. Uniroyal stopped marketing Alar shortly thereafter. Although the chemical was used for only 15 percent of U.S. apple trees, the consequences for the apple industry were devastating. Apple-orchard owners lost $250 million and manufacturers of apple products posted losses of $125 million.

Experts from a variety of government agencies and independent laboratories later insisted that Alar had never posed a risk to human health, but their protests were too little and too late. For its part, far from being chagrined about the scare, the NRDC promoted the issue on its website as a case study of effective environmental management: "Much as Upton Sinclair's *The Jungle* led to passage of the

Pure Food and Drug Act of 1906, and Rachel Carson's *Silent Spring* triggered bans of DDT and other toxins, the airing in February 1989 of the *60 Minutes* broadcast, 'A Is for Apple'—based in large part on NRDC publication, 'Intolerable Risk: Pesticides in Our Children's Food'—sparked a chain of events that led to critical improvements in food safety policy."[75]

At the dawn of the twenty-first century, the Council had moved beyond the John Adams era following his 2006 retirement. It remained as bold and confrontational as ever—a fearless environmental hero to some, an opportunistic scaremonger to others. "NRDC is the nation's most effective environmental action group, combining the grassroots power of 1.3 million members and online activists with the courtroom clout and expertise of more than 350 lawyers, scientists and other professionals," the group's website proclaimed in 2012. "The *New York Times* calls us 'One of the nation's most powerful environmental groups.' The *National Journal* says we're 'A credible and forceful advocate for stringent environmental protection.'"[76]

Greenpeace

"There's an old joke that in any bar in Vancouver, Canada, you can sit down next to someone who claims to have founded Greenpeace," the Greenpeace website recounted in 2012. "In fact, there was no single founder, and the name, idea, spirit, tactics, and internationalism of the organization all can be said to have separate lineages." The website explained that a Greenpeace antecedent formed in 1970 when Irving and Dorothy Stowe (formerly Strasmich), Jim and Marie Bohlen, Ben and Dorothy Metcalfe, and Bob Hunter created the Don't Make a Wave Committee to protest nuclear weapons testing at Amchitka Island in the Aleutian archipelago. Alaska had been hit with a devastating earthquake in 1964, and anti-nuclear protesters feared that a nuclear test in the area might trigger another earthquake. A group of activists in the United States and Canada created signs warning "Don't Make a Wave. It's Your Fault if Our Fault Goes." Despite vehement opposition, the U.S. government carried out the testing as planned. Fortunately, no earthquakes occurred in the wake of the explosions.[77]

The men and women who eventually launched Greenpeace were members of Sierra Club Canada when the Amchitka episode erupted. They believed the organization, with its pragmatic, deliberative, methodological approach to natural-resource preservation, was not doing enough to protest the environmental damage from nuclear weapons. Jim Bohlen's wife, Marie, was the first person to argue that talk was not a strong enough response; a more "in your face" confrontational approach was necessary to drive the point home. She suggested that a group of activists should sail to Amchitka to protest the testing. As the small group shared its plans with others, word of their scheme leaked to the press, which ran a story linking the recommendation to the Sierra Club. After embarrassed club leaders disavowed the plan, the Stowes and Bohlens met with other disaffected environmentalists to create the Don't Make a Wave Committee.

The fledgling association received financial assistance from a benefit concert held on October 16, 1970, at the Pacific Coliseum in Vancouver by folk musicians Joni Mitchell, Phil Ochs, and surprise guest James Taylor. The funds they collected from the event allowed the activists to charter a ship, the *Phyllis Cormack* (which they later rechristened *Greenpeace*), for a journey to Amchitka in September 1971. The idealistic plan would now become a reality.

And what a reality it became. The expedition started as a disaster. Aside from braving rough seas for much of the journey, the travelers argued with each other incessantly and eventually turned back in the face of a challenge from the U.S. Coast Guard vessel *Confidence*. Fortunately for the activists, news coverage of their attempt engendered enormous sympathy for the group and its goals. Subsequent attempts to reach the testing site brought even more attention. The U.S. government eventually abandoned additional testing in Alaska in part owing to the negative publicity generated by the dedicated environmentalists. The group had snatched victory from the jaws of defeat.[78]

Emboldened by their unexpected success, Don't Make a Wave Committee members discussed forming a permanent organization to carry on the fight. They needed a new name. The origins of the term "Greenpeace" have been fiercely debated. Bob Hunter credited Canadian ecologist Bill Darnell as the serendipitous father of the term. Hunter later explained that the group had been plotting strategy in a church basement. The meeting broke up and everyone started to leave. "Somebody flashed two fingers as we were leaving the church basement and said 'Peace!'" Hunter recalled. "Bill said, 'Let's make it a Green Peace.' And we all went, 'Ommm.'" When Jim Bohlen's son, Paul, designed a button with the words "Green Peace," he realized the only way to fit all the letters on the button in a legible font was to combine the two words into one. In this unlikely fashion, the "Greenpeace" brand was born. In 1972, a year after the Amchitka voyage, the Don't Make a Wave Committee officially formed the Greenpeace Foundation.[79]

A larger group of affiliated activists, working under the name "Greenpeace," spent the 1970s perfecting an antagonistic style of confrontation reminiscent of the antiwar movement against the Vietnam War during the 1960s. This evolution was hardly surprising. Many Greenpeace participants hailed from the ranks of the counterculture hippie movement. Instead of engaging in the typical lobbying campaigns and seeking compromise solutions, as many mainstream environmental organizations did, this new wave of environmentalist would resist despoilers of the earth—whether they were governments or private companies—through aggressive confrontation. Their stage was global, and their intent was to shame their opponents into abandoning destructive practices.[80]

Also in 1972, a wealthy New Zealand businessman, David McTaggart, offered to use his yacht, the *Vega*, in a campaign against nuclear testing by the French government in the Mururoa exclusion zone in French Polynesia. Co-sponsored with the New Zealand branch of the Campaign for Nuclear Disarmament, this 1972 Greenpeace protest generated worldwide headlines when a French military ship rammed McTaggart's vessel, now renamed *Greenpeace III*, and beat him severely. A Greenpeace activist witnessing the assault photographed the incident. When the photographs circulated around the world, the adverse publicity was responsible to some extent for France's decision to forgo additional atmospheric testing of nuclear weapons.[81]

The Greenpeace model was firmly in place by decade's end: In lieu of bargaining and negotiation through the legislative process or legal challenges in the halls of power, group members would design audacious stunts to be carried out at the site of the environmental damage. The resultant media coverage and public outrage would force a change in the polluter's behavior in a way that traditional behind-the-scenes tactics never would. Literally dozens of grassroots groups, some of which had no formal ties to the original organization, used the name "Greenpeace" in their militant resistance to global environmental degradation. These zealous activists sponsored initiatives against nuclear weapons, whaling, climate change, chemicals such as chlorine and dioxins, pesticides, genetically modified crops ("FrankenFoods"), improperly disposed of electronics, plastic water bottles, ocean pollution, and dismantling and discarding oil rigs at sea. In 1979, the original organization established Greenpeace International to ensure a continued global presence.[82]

Greenpeace tactics frequently have caused controversy, sometimes triggering legal action against the group, and occasionally meeting with violent reactions. The most infamous incident occurred on July 10, 1985, when a French intelligence agency bombed the *Rainbow Warrior*, a Greenpeace vessel that had been protesting against French nuclear testing in the Mururoa exclusion zone, site of the altercation with David McTaggart 13 years earlier. As Greenpeace activists described the incident on their website, "an explosion aboard the *Rainbow Warrior* sank the ship, our hearts, and left one man dead." Photographer Fernando Pereira drowned when he was knocked unconscious by the blast and was trapped below deck as the vessel sank to the bottom of the ocean. The international brouhaha over the bombing pressured the French government to abandon its nuclear testing campaign in French Polynesia. As Greenpeace viewed the episode, "out of this tragedy eventually came triumph and an end to nuclear testing."[83]

As of 2012, Greenpeace International claimed to have a presence in 41 countries as well as 2.8 million active members around the globe. For its part, Greenpeace USA boasted of 250,000 members and an annual budget of approximately $10 million. Although these numbers were modest compared with some environmental groups, Greenpeace remained a high-profile activist organization because it devoted time and attention to publicizing worldwide environmental issues using aggressive, headline-generating antics. For detractors, the group has always overstated claims to terrify unsuspecting people using unconscionable scare tactics, but supporters believe that Greenpeace has taken the fight to the enemy, which accounts for the group's long record of success.[84]

Earth First!

Earth First! is a famous radical environmental group, an organization that has billed itself as a new kind of player in the policy arena. The group's website contained this provocative declaration in 2012: "Are you tired of namby-pamby environmental groups? Are you tired of overpaid corporate environmentalists who suck up to bureaucrats and industry? Have you become disempowered by the reductionist approach of environmental professionals and scientists? If you answered yes to any of these questions, then Earth First! is for you."[85]

David Foreman, the founder, was enamored of an infamous 1975 novel, *The Monkey Wrench Gang*, penned by a self-proclaimed anarchist, Edward Abbey. Abbey's book featured a Vietnam veteran, George Washington Hayduke III, who returned from his military service to discover, much to his horror, that the desert lands of his native southern Utah were threatened by industrial development. Abbey had worked as a seasonal park ranger in Arches National Monument (later a national park) near Moab, Utah, for a few years during the 1950s, so he was familiar with the landscape. As the novel's protagonist, Hayduke, journeyed down the Colorado River, he met a feminist saboteur, Bonnie Abbzug, a wilderness guide named Seldom Seen Smith, and a medical doctor who set fire to billboards, Doc Sarvis. The group groused about the sorry state of affairs in southern Utah, eventually deciding to engage in eco-terrorism to protect the pristine desert from developers by sabotaging their machinery.[86]

An environmentalist for most of his life, Foreman had been involved with the Wilderness Society and had served on the board of directors for the New Mexico chapter of the Nature Conservancy. By the late 1970s, as he approached his thirties, Foreman had become disillusioned with the mainstream environmental movement. He believed these groups were too cautious in their approaches and too willing to compromise their principles in bargaining and negotiating with corporations and other industrial interests. During a hiking trip with friends in 1980, Foreman first conceived of a new group that would abjure the traditional, incremental tactics of the established environmental organizations and actively encourage "monkey-wrenching," the term he borrowed from Edward Abbey to describe a campaign of strikes against logging and mining interests. Earth First! was born. Foreman was an alumnus of the 1960s counterculture movement; consequently, he adopted the tactics of anti-establishment protests and civil disobedience employed by the New Left during an earlier era. The *Earth First! Journal*, the organization's official publication, championed, and continues to champion, the deep ecology movement in provocative articles that challenge the reader to make radical lifestyle changes and resist the encroachment of industrialization. Foreman edited the journal from 1982 to 1988.[87]

Earth First! announced its presence in March 1981 during a demonstration at the Glen Canyon Dam, site of a much-hated construction project in northern Arizona. It was little wonder that Foreman and his cohorts chose the dam.

Characters from *The Monkey Wrench Gang* had discussed blowing up the dam. In the years since that first appearance in 1981, Earth First! supporters have avoided the fixed organizational structure and hierarchy of the mainstream environmental organizations. They have characterized themselves as a grassroots movement committed to coordinating environmental protection efforts. To that end, group members have protested the destruction of old-growth forests in Oregon and California and in other areas. They have always claimed not to be terrorists because they do not seek to hurt people. Instead, they have chained themselves to trees to prevent logging, strung cables to trees to increase the difficulty in timber cutting, and driven spikes into trees to foil chainsaws (although this tactic, which can harm workers, has caused deep divisions even among Earth First! activists). On some occasions, they have flung manure into the air conditioning units of buildings where timber industry activists worked in their corporate offices. Foreman later served time in jail for conspiring to sabotage U.S. nuclear power plants. As he described it, he was arrested in his bed at night by "a posse of gun-wielding FBI agents playing Dirty Harry."[88]

The most controversial incident occurred on May 24, 1990, when a car bomb exploded beneath the seat of a car occupied by Earth First! activists Judi Bari and Daryl Cherney in Oakland, California. Because she was driving, Bari was gravely injured, breaking her pelvis and suffering severe tissue damage. Cherney's wounds, an injury to his left eye, were comparatively minor. The Federal Bureau of Investigation (FBI) claimed that Bari and Cherney were carrying the bomb as part of the group's environmental protests and arrested them for transporting explosives. Two months later, the Oakland district attorney, citing insufficient evidence, refused to file criminal charges. The environmental community suspected that timber industry representatives had planted the device. Greenpeace even hired a private investigator to search for the assailants. Bari and Cherney filed a civil rights lawsuit against the FBI agents for false arrest. In 2002, the court awarded Cherney and Bari's estate—she had died of breast cancer in 1997—$4.4 million as compensation for the denial of their First and Fourth Amendment freedoms in the 1990 incident.[89]

Strengths and Weaknesses of the NGO Model

As discussed in this chapter, environmental NGOs exist outside of the formal institutions of a nation-state. They typically influence policymaking by engaging in public relations marketing to place salient issues on the national agenda, mobilizing grassroots support to publicize issues of importance to the organization, campaigning for candidates running for elective office, and lobbying legislators and regulators to adopt particular policy positions. Although Tocqueville heralded voluntary associations as integral features of civic life, he probably meant local civic groups; the term "NGO" was not used in his time. A multinational organization extending across national borders would have seemed unfathomable to him. Nonetheless, the concept of a private group influencing public policy has existed for centuries, even antedating the publication of *Democracy in America*.

In the waning years of the twentieth century and during the early years of the new millennium, NGOs became increasingly important in policymaking, especially because globalization of markets and international political concerns meant that effectively addressing issues such as the natural environment must transcend national boundaries. NGOs operate across borders by establishing an organizational presence in multiple countries and hemispheres. Greenpeace International, for example, has opened offices in more than 40 countries. A global presence allows an NGO to track issues from country to country as well as seek donations from partners around the world. By crossing borders and addressing problems on a broad, worldwide canvas, NGOs can influence international laws, treaties, and global relationships. Their reach is potentially limitless. For NGO enthusiasts, this transcendent character allows nonprofit organizations to compete effectively with multinational corporations that have long influenced international policies without an effective "check" on their power or influence. In short, the growth of NGOs is a healthy development because it levels the playing field among competing forms of conglomerated power.[90]

Critics of NGOs contend that such multinational organizations are accountable to no one—not even to their members. At least corporations must report to shareholders and a board of directors that can reign in objectionable activities, especially when government officials threaten to regulate their behavior or incensed customers boycott their products or services. Yet NGOs, while they generally have a board of directors, are staffed by true believers who have no incentive to curtail their behavior. Detractors note that Tocqueville and supporters of the interest-group theory of democracy envisioned civic organizations comprised of interested citizens working outside of the institutional structure of government to promote the interests of a small number of people in a local community. Because civic associations reflect community needs and can be modified in accordance with the wishes of community members relatively quickly, they do not pose a threat to the policy process. They are strictly accountable to members who can correct deficiencies with few impediments.

Detractors charge that powerful NGOs constantly petition current and potential members with vague requests for money—"Please donate $10 today to preserve a clean environment for your children and grandchildren!"—thereby amassing millions of dollars that can be channeled into a variety of accounts in multiple countries. With boards of directors and officers who do not have to stand for election or account for specific uses of the donations, these true believers in the cause are not effectively regulated. Although some groups are more conscientious than others are, an incentive exists for NGOs to use scare tactics, highlighting the crisis *du jour* to drum up additional support and dollars for pet causes. In the meantime, it may be career suicide for a political figure or corporate mogul to criticize the group because a public relations campaign can be launched to smear the would-be critic as someone willing to destroy the natural environment for profit. As conservative commentator Roger Scruton observed in his book *How to Think Seriously about the*

Planet, "While civic associations exist for the sake of their members, the big NGOs exist purely for the sake of their goals."[91]

The prescription for reform is to back away from NGOs and large, overarching solutions to global environmental problems. This insight may appear counterintuitive in an age of increasing globalization, but critics charge that large campaigns aimed at effecting massive changes seldom succeed in improving environmental quality because they do not garner genuine grassroots support. Rather than ask people to alter their fundamental way of life and embrace sweeping changes that may not work, critics suggest that small-scale civic associations, which are more in line with Tocqueville's original vision, would be more effective. When citizens acting in their own neighborhoods and communities seek their own solutions to problems, they achieve more success than when tens of thousands of people attempt to clasp hands across the world. An individual is more likely to take ownership of a problem and find solutions to pollution and environmental degradation in his own neighborhood than to worry about pollution in a faraway country.[92]

Because it offers root-and-branch, systemic change in lieu of incremental community reform, the NGO model is flawed, critics charge. The model ensures that problems are never solved. Citizens lurch from one crisis to another, eventually lapsing into cynicism and a malaise brought on by a feeling that the problems are so large that nothing can be done. Even an issue such as climate change, widely regarded as among the most pressing global environmental issues of the twenty-first century, can be solved, if it is solved at all, only through community approaches, according to anti-NGO commentators. To think otherwise is to fall victim to scare tactics that support a group's desire for self-perpetuation, but never solve the issues the group professes to consider. As political satirist P. J. O'Rourke wrote in his classic book *Parliament of Whores*, "Certain ecological doom-boosters are not only unreasonable in their attitude toward business; they're unreasonable in their attitude toward reason." Rather than seek common-sense solutions, which includes working with businesses and industrial concerns, radical NGOs spend their time demonizing opponents. "With the environmentalist movement, as with the homeless advocates, we see a special interest group more interested in marrying an evil than curing it," O'Rourke argued.[93]

Regardless of their strengths and weaknesses, NGOs remain an important part of the American environmental landscape, and they engage in a variety of sophisticated, effective activities to promote their goals. NGO representatives often use environmental laws and regulations, which provides an intricate web of remedies for potential and actual environmental degradation. In some cases, federal environmental law allows a private right of action for NGOs and other third parties to ensure that government officials enforce the law. Outside of the statutory and regulatory structure, environmental management systems and standards exist to reward corporations and other entities that consider environmental concerns in their decision-making. It is to these substantive features of American environmentalism that we turn in Chapter 9.

Notes

1. P. J. O'Rourke, *Parliament of Whores: A Lone Humorist Attempts to Explain the Entire U.S. Government* (New York: The Grove Press, 2003), 198.
2. James Madison, "Federalist 10," in Alexander Hamilton, James Madison, and John Jay, *The Federalist Papers,* Clinton Rossiter, editor (New York: New American Library, 1961), 79.
3. Madison, "Federalist 10," 79. The literature on the perfectibility, or lack thereof, of human nature is voluminous. See, for example, Harold Coward, *The Perfectibility of Human Nature in Eastern and Western Thought* (Albany, N.Y.: State University of New York Press, 2008); Kenneth Schmitz, "Towards the Reciprocity of Man and Nature: Receptivity, Normativity, and Procreativity," *Nova et Vetera* (English Edition) 10, 1 (Winter 2012): 81–94.
4. Madison, "Federalist 10," 81. See also, for example, Martin Diamond, "Democracy and *The Federalist*: A Reconsideration of the Framers' Intent," *American Political Science Review* 53, 1 (March 1959): 52–68; David F. Epstein, *The Political Theory of The Federalist* (Chicago: The University of Chicago Press, 1984), 4, 60–76; Gottfried Dietze, *The Federalist: A Classic on Federalism and Free Government* (Westport, Conn.: Greenwood Press, 1960), 316.
5. Madison, "Federalist 51," 322. See also Maynard Smith, "Reason, Passion, and Political Freedom in *The Federalist*," *Journal of Politics* 22, 3 (August 1960): 525–544.
6. See, for example, Ernst Fehr and Simon Gachter, "Reciprocity and Economics: The Economic Implications of *Homo Reciprocans*," *European Economic Review* 42, 3 (May 1998): 845–859.
7. Alexis de Tocqueville, *Democracy in America*, Arthur Goldhammer, translator (New York: Library of America, 2004), Book II, Chapter 4, 215–223. See also Sarah R. Adkins, "Democracy's Quiet Virtues," in *Ethics and Character: The Pursuit of Democratic Virtues*, William D. Richardson, J. Michael Martinez, and Kerry R. Stewart, editors (Durham, N.C.: Carolina Academic Press, 1998), 203–223; Arthur Goldhammer, "Tocqueville, Associations, and the Law of 1834," *Historical Reflections* 35, 3 (Winter 2009): 74–75; Henk E. S. Woldring, "State and Civil Society in the Political Philosophy of Alexis de Tocqueville," *Voluntas: International Journal of Voluntary and Nonprofit Organizations* 9, 4 (December 1998): 363–373.
8. David B. Truman, *The Governmental Process: Political Interests and Public Opinion* (New York: Knopf, 1971 [1951]).
9. Mancur Olson, *The Logic of Collective Action: Public Goods and the Theory of Groups* (Cambridge, Mass.: Harvard University Press, 1971).
10. E. E. Schattschneider, *The Semisovereign People: A Realist's View of Democracy in America* (Belmont, Calif.: Wadsworth, 1975), 30, 36.
11. Theodore J. Lowi, *The End of Liberalism: The Second Republic of the United States* (New York: W. W. Norton, 2009). For a case study illustrating the power of interest groups and the bureaucracy in political decision-making, see, for example, M. Christine Cagle and J. Michael Martinez, "Have Gun, Will Travel: The Dispute Between the CDC and the NRA on Firearm Violence as a Public Health Problem," *Politics & Policy* 32, 2 (June 2004): 278–310.
12. Robert D. Putnam, *Bowling Alone: The Collapse and Revival of American Community* (New York: Simon & Schuster, 2000).

13. Benjamin R. Barber, *Jihad vs. McWorld: How Globalism and Tribalism Are Reshaping the World* (New York: Ballantine Books, 1996).
14. Robert J. Samuelson, "Join the Club," *Washington Post National Weekly Edition* (April 15–21, 1996): 5. See also Steven N. Durlauf, "*Bowling Alone*: A Review Essay," *Journal of Economic Behavior and Organization* 47, 3 (March 2002): 259–273; Nicholas Lemann, "Kicking in Groups," *Atlantic Monthly* 277, 4 (April 1996): 22–26; Leslie Lenowsky, "Still 'Bowling Alone'?" *Commentary* 110, 3 (October 2000): 57–60.
15. Matt Grossmann, "Environmental Advocacy in Washington: A Comparison with Other Interest Groups," *Environmental Policy* 15, 4 (August 2006): 630.
16. Garry W. Jenkins, "Nongovernmental Organizations and the Forces against Them: Lessons of the Anti-NGO Movement," *Brooklyn Journal of International Law* 37, 2 (2012): 465–467.
17. See, for example, Defenders of Wildlife Website: "Home Page," http://www.defenders.org/ (Accessed November 1, 2012); Earth Island Institute Website: "Home Page," http://www.earthisland.org/ (Accessed November 1, 2012); Environmental Working Group Website: "Home Page," http://www.ewg.org/ (Accessed November 1, 2012); Heal the Bay Website: "Home Page," http://www.healthebay.org/ (Accessed October 26, 2012); Surfrider Foundation Website: "Home Page," http://www.surfrider.org/ (Accessed October 26, 2012).
18. "A Centennial Celebration, 1892–1992," *Sierra* 77, 3 (May/June 1992): 52. See also Douglas Brinkley, *The Wilderness Warrior: Theodore Roosevelt and the Crusade for America* (New York: HarperCollins, 2009), 541; Sierra Club Website: "Sierra Club History," http://www.sierraclub.org/history/ (Accessed October 15, 2012).
19. "The *Sierra Club Bulletin*," *Sierra* 78, 5 (September/October 1993): 54.
20. "A Centennial Celebration, 1892–1992," 52–73; Sierra Club Activist Network Website: "Sierra Club Activist Network: Organize, Share, Work, Win!" http://connect.sierraclub.org/ActivistNetwork/home (Accessed October 18, 2012).
21. Quoted in Kathy S. Mason, "Out of Fashion: Harriet Hemenway and the Audubon Society, 1896–1905," *Historian* 65, 1 (Fall 2002): 1. See also Frank Graham, Jr., *The Audubon Ark: A History of the Audubon Society* (Austin: University of Texas Press, 1992), 3–10; David J. Miller, "Rich History: Celebrating National Audubon Society's Centennial," *New York State Conservationist* 60, 2 (October 2005): 2–5.
22. Quoted in Mason, "Out of Fashion," 4. See also Brinkley, *The Wilderness Warrior*, 10–11, 184–189.
23. Brinkley, *The Wilderness Warrior*, 497–499.
24. Graham, *The Audubon Ark*, 107; Mason, "Out of Fashion," 10–11; Miller, "Rich History," 2–3.
25. Graham, *The Audubon Ark*, 46; Mason, "Out of Fashion," 13; Miller, "Rich History," 2.
26. Miller, "Rich History," 2–5.
27. BirdLife International Website: About BirdLife," http://www.birdlife.org/worldwide/index.html (Accessed October 23, 2012); Graham, *The Audubon Ark*, 134; National Audubon Society Website: "About Us," http://www.audubon.org/about-us (Accessed October 23, 2012).
28. Izaak Walton and Charles Cotton, *The Compleat Angler: Contemplative Man's Recreation, Being a Discourse on Rivers, Fishponds, Fish, and Fishing* (London: Henry Washbourne, 1842 [1653]), 111. See also John Lowerson, "Izaak Walton: Father of a Dream," *History Today* 33, 12 (December 1983): 28–29.

29. Raymond A. Anselment, "Robert Boyle, Izaak Walton, and the Art of Angling," *Prose Studies* 30, 2 (August 2008): 124–141; Lowerson, "Izaak Walton," 30–32.
30. Izaak Walton League Website: "Home Page," http://www.iwla.org/ (Accessed October 23, 2012).
31. Izaak Walton League Website: "Clean Water," http://www.iwla.org/index.php?ht=d/sp/i/1976/pid/1976/cat_id/217 (Accessed October 24, 2012); Izaak Walton League Website: "Waltonians (October 1924)," http://www.iwla.org/index.php?ht=d/ContentDetails/i/31132 (Accessed October 24, 2012).
32. Izaak Walton League Website: "Advocacy Page," http://www.iwla.org/index.php?ht=d/sp/i/196/pid/196 (Accessed October 24, 2012); Izaak Walton League Website: "News You Can Use," http://www.iwla.org/index.php?ht=d/sp/i/2246/pid/2246 (Accessed October 24, 2012); Walter A. Rosenbaum, *Environmental Politics and Policy*, 4th ed. (Washington, D.C.: Congressional Quarterly Press, 1998), 31, 35.
33. Tracy Marafiote, "The American Dream: Technology, Tourism, and the Transformation of Wilderness," *Environmental Communication* 2, 2 (July 2008): 155; Douglas Scott, *The Enduring Wilderness: Protecting Our National Heritage through the Wilderness Act* (Golden, Colo.: Fulcrum Press, 2004), 32–40; "The Wilderness Society 75 Years: 1935–2010," *Wilderness* 26, 1 (October 2009): 26; The Wilderness Society Website: "Founders," http://wilderness.org/bios/founders (Accessed October 24, 2012).
34. Michael Fromme, *Battle for the Wilderness* (Salt Lake City: University of Utah Press, 1997), xii–xix, 17, 228; Marafiote, "The American Dream," 155–156; Katherine Daniels Ryan, "Preservation Prevails over Commercial Interests in the Wilderness Act: *Wilderness Society v. United States Fish & Wildlife Service*," *Ecology Law Quarterly* 32, 3 (2005): 543–544; Scott, *The Enduring Wilderness*, xi–xii, 4; James Morton Turner, "'The Specter of Environmentalism': Wilderness, Environmental Politics, and the Evolution of the New Right," *The Journal of American History* 96, 1 (June 2009): 126.
35. Ryan, "Preservation Prevails over Commercial Interests in the Wilderness Act," 544; Turner, "'The Specter of Environmentalism,'" 126–135.
36. Ryan, "Preservation Prevails over Commercial Interests in the Wilderness Act," 539–573; "The Wilderness Society 75 Years: 1935–2010," 26–27.
37. The Wilderness Society Website: "Leadership," http://wilderness.org/leadership (Accessed October 24, 2012); The Wilderness Society Website: "Mission and Impact," http://wilderness.org/article/mission-and-impact (Accessed October 24, 2012); The Wilderness Society Website: "Publications," http://wilderness.org/publications (Accessed October 24, 2012).
38. Mark Hertsgaard, "Green Goes Grassroots: The Environmental Movement Today," *The Nation* 283, 4 (July 31/August 7, 2006): 11; National Wildlife Federation Website: "About Us—Annual Report," http://www.nwf.org/About/Annual-Report/Annual-Report-Archive.aspx (Accessed October 26, 2012).
39. Quoted in Chuck Berry, "National Wildlife Federation Is 75 Years Old," *South Dakota Wildlife Out of Doors* 51, 2 (February 2011): 2. See also National Wildlife Federation Website: "About Us—Our History and Heritage," http://www.nwf.org/About/History-and-Heritage.aspx (Accessed October 26, 2012).
40. Berry, "National Wildlife Federation Is 75 Years Old," 2–3; National Wildlife Federation Website: "About Us—Creation of National Wildlife Federation," http://www.nwf.org/About/History-and-Heritage/Creation-of-NWF.aspx (Accessed October 26, 2012).

41. National Wildlife Federation Website: "About Us—Legislative Victories," http://www.nwf.org/About/History-and-Heritage/Legislative-Victories.aspx (Accessed October 26, 2012).
42. Quoted in Hertsgaard, "Green Goes Grassroots," 11.
43. National Wildlife Federation Website: "What We Do," http://www.nwf.org/Wildlife/What-We-Do.aspx (Accessed October 26, 2012); Garrit Voggesser, "When History Matters: The National Wildlife Federation's Conservation Partnerships with Tribes," *Western Historical Quarterly* 40, 3 (Autumn 2009): 349–357.
44. Bill Birchard, *Nature's Keepers: The Remarkable Story of How the Nature Conservancy Became the Largest Environmental Organization in the World* (San Francisco, Calif.: Jossey-Bass, 2005), 3–5, 9–10; Alice Howard and Joan Magretta, "Surviving Success: An Interview with the Nature Conservancy's John Sawhill," *Harvard Business Review* 73, 5 (September/October 1995): 109; Angela G. Mertig, "The 'Nature' of Environmentalism: Nature Protection in the USA," in *Protecting Nature: Organizations and Networks in Europe and the USA*, C. S. A. (Kris) van Koppen and William T. Markham, editors (Cheltenham, U.K.: Edward Elgar Publishing, 2008), 251–252, 254; Nature Conservancy Website: "About Us," http://www.nature.org/about-us/index.htm (Accessed November 1, 2012); Nature Conservancy Website: "Celebrate 60 Years of Conservation," http://my.nature.org/nature/posts/public/celebrate-60-years-of.html (Accessed November 1, 2012).
45. Quoted in Howard and Magretta, "Surviving Success," 110. See also Dan Brockington, Rosaleen Duffy, and Jim Igoe, *Nature Unbound: Conservation, Capitalism, and the Future of Protected Areas* (New York and Oxford: Routledge, 2008), 155.
46. Quoted in Christina Asquith, "Dow Chemical and The Nature Conservancy Team Up to Ask, What is Nature Worth?," *Solutions Journal* 2, 6 (November-December 2011): 20. See also James J. Griffith and Charles R. Knoeber, "Why Do Corporations Contribute to the Nature Conservancy?" *Public Choice* 49, 1 (1986): 69–77; Nature Conservancy Website: "Our Partners in Conservation,"<http://www.nature.org/about-us/our-partners/index.htm (Accessed November 4, 2012).
47. Quoted in Ken Wells, "Nature's Own Hedge Fund," *Bloomberg Business Week* 4282, 1 (June 4–June 10, 2012): 50. See also Environmental Defense Fund Website: "Home Page," http://www.edf.org/ (Accessed October 29, 2012).
48. Quoted in Brice Wallace, "Nature Conservancy 'More Likely in the Boardroom than the Courtroom,'" *The Enterprise* [Salt Lake City, Utah], 41, 25 (January 16–22, 2012): 7.
49. Quoted in Wells, "Nature's Own Hedge Fund," 50. See also Glen Martin, "Lessons from the Mountain: Adapting Strategies for Conservation," *Bay Nature* 12, 3 (July-September 2012): 53; "The Nature of Parks," *Parks & Recreation* 45, 9 (September 2010): 19.
50. Sir Peter Scott, "World Wildlife Fund (1961)," *Creative Review* 31, 4 (April 2011): 43; David Walker, John Paul Jones III, Susan M. Roberts, and Oliver R. Frohling, "When Participation Meets Empowerment: The WWF and the Politics of Invitation in the Chimalapas, Mexico," *Annals of the Association of American Geographers* 97, 2 (June 2007): 428; World Wildlife Fund Website: "History," http://worldwildlife.org/about/history (Accessed November 5, 2012).
51. Randall Kremer, "Institutions: World Wildlife Fund," *Environment* 27, 4 (May 1985): 42–43; World Wildlife Fund Website: "History," n.p.
52. World Wildlife Fund Website: "History," n.p.

53. Wildlife Fund Website: "About Us," http://worldwildlife.org/about (Accessed November 5, 2012).
54. Quoted in Stacey Collett, "World Wildlife Fund," *Computerworld* 42, 8 (February 2008): 40. See also Kremer, "Institutions," 42; Robert Napier, "World Wildlife Fund: Making a Genuine and Lasting Difference to the Planet," *OECD Observer* 235, 1 (December 2002): 42–43; Yvonne M. Scherrer, "Environmental Conservation NGOs and the Concept of Sustainable Development," *Journal of Business Ethics* 85, Supplement 3 (April 2009): 562.
55. Environmental Defense Fund Website: "Home Page," n.p.; Environmental Defense Fund Website: "Our Mission and History," http://www.edf.org/about/our-mission-and-history (Accessed October 29, 2012); William H. Miller, "Fred Krupp: A Different Kind of Environmentalist," *Industry Week* 243, 15 (August 15, 1994): 93, 96, 98; James Verini, "The Devil's Advocate: Is Fred Krupp an Environmental Savior or a Corporate Stooge?," *The New Republic* 237, 6 (September 24, 2007): 24–29.
56. Quoted in Verini, "The Devil's Advocate," 27.
57. Miller, "Fred Krupp, 93; Bret Schulte, "Teaming Up with the Enemy," *U.S. News & World Report* 143, 18 (November 19, 2007): 54–56; Verini, "The Devil's Advocate," 26–27.
58. Fred Krupp, "New Environmentalism Factors in Economic Needs," *Wall Street Journal* (November 20, 1986): 34. See also Sharon M. Livesey, "McDonald's and the Environmental Defense Fund: A Case Study of a Green Alliance," *The Journal of Business Communication* 36, 1 (January 1999): 10–11; Verini, "The Devil's Advocate," 27.
59. Livesey, "McDonald's and the Environmental Defense Fund," 14–18; Miller, "Fred Krupp, 93, 96; Schulte, "Teaming Up with the Enemy," 28; Verini, "The Devil's Advocate," 28.
60. For information on the effects of the product de-selection on McDonald's and EDF, see Livesey, "McDonald's and the Environmental Defense Fund," 17–19; Miller, "Fred Krupp," 93, 96; Schulte, "Teaming Up with the Enemy," 28; Verini, "The Devil's Advocate," 28. For contrary evidence indicating that polystyrene foam is not as damaging to the natural environment as alternative products, see, for example, Judd H. Alexander, *In Defense of Garbage* (Westport, Conn. and London: Praeger, 1993), 54–55; *Final Peer-Reviewed Report: Life Cycle Inventory of Polystyrene Foam, Bleached Paperboard, and Corrugated Paperboard Foodservice Products* (Prairie Village, Kans.: Franklin Associates, Ltd., March 2006); Margaret Hardy and John A. Charles, Jr., *Sustainable Failure: Why Portland's Polystyrene Foam Ban Should Be Repealed* (Portland, Ore.: Cascade Policy Institute, November 2007); Martin B. Hocking, "Is Paper Better Than Plastic?," *Consumer's Research* 74, 10 (October 1991): 28–29; Martin B. Hocking, "Relative Merits of Polystyrene Foam and Paper in Hot Drink Cups: Implications for Packaging," *Environmental Management* 15, 6 (1991): 731–747; Martin B. Hocking, "Reusable and Disposable Cups: An Energy-Based Evaluation," *Environmental Management* 18, 6 (1994): 889–899; James Michael Martinez, "Packaging Legislation and Unintended Consequences: A Case Study on the Necessity of Life Cycle Management," in *Towards Life Cycle Sustainability Management*, Matthias Finkbeiner, editor (London and New York: Springer, 2011), 371–380; William L. Rathje and Cullen Murphy, "Five Major Myths about Garbage, and Why They're Wrong," *Smithsonian* 23, 4 (July 1992): 113–122; William L. Rathje and Cullen Murphy, *Rubbish! The Archaeology of Garbage* (New York: HarperCollins, 1992), 97–98.

61. Diane Brady, "Fred Krupp," *Bloomberg Business Week* 4284, 1 (June 18–June 24, 2012): 76; Environmental Defense Fund Website: "Home Page," n.p.; Fred Krupp, "Grab the Green Brass Ring," *Business Week* 4020, 1 (February 5, 2007): 104.
62. Fred Krupp, "Global Warming and the USA: Sparking Reengagement," *Vital Speeches of the Day* 69, 13 (April 15, 2003): 399. See also "The Environmental Defense Fund, Part 1," *Business & The Environment* 22, 5 (May 2011): 4–6; "The Environmental Defense Fund, Conclusion," *Business & The Environment* 22, 6 (June 2011): 3–5.
63. Bill Devall, "The End of American Environmentalism?," *Nature and Culture* 1, 2 (Autumn 2006): 161; Christopher Rootes, "Environmental Movements," in *The Blackwell Companion to Social Movements*, David A. Snow, Sarah A. Soule, and Hanspeter Kriesi, editors (Oxford, U.K.: B. H. Blackwell, 2004), 614–615.
64. Friends of the Earth Website: "Achievements," http://www.Foe.org/about-us/achievements (Accessed November 6, 2012); Duncan McLaren, "Spotlight on…Friends of the Earth," *Planning Practice & Research* 7, 3 (Winter 1992): 43–47.
65. Friends of the Earth Website: "Achievements," n.p.; Friends of the Earth Website: "Who We Are," http://www.Foe.org/about-us (Accessed November 6, 2012); McLaren, "Spotlight on…Friends of the Earth," 43–44.
66. Brian Doherty, "Friends of the Earth International: Negotiating a Transnational Identity," *Environmental Politics* 15, 5 (November 2006): 862. See also Michael S. Greve, "Friends of the Earth, Foes of Federalism," *Duke Environmental Law & Policy Forum* 12, 1 (Fall 2001): 167–182.
67. *Friends of the Earth, Inc. v. Laidlaw Environmental Services (TOC), Inc.*, 528 U.S. 167 (2000). See also David N. Cassuto, "The Law of Words: Standing, Environment, and Other Contested Terms," *Harvard Environmental Law Review* 28, 1 (2004): 79–128; Hudson P. Henry, "A Shift in Citizen Suit Standing Doctrine: *Friends of the Earth, Inc. v. Laidlaw Environmental Services*," *Ecology Law Quarterly* 28, 2 (2001): 233–252; Jeffrey G. Miller and James A. Hopkins, "The Standing to Enforce against Violations of Environmental Statutes in the United States *Friends of the Earth v. Laidlaw Environmental Services*," *Journal of Environmental Law* 12, 3 (2000): 370–380; "Proactively Fighting Pollution," *Friends of the Earth Magazine* 39, 2 (Summer 2009): 6C; Richard J. Pierce, Jr., "Issues Raised by *Friends of the Earth v. Laidlaw Environmental Services*: Access to Courts for Environmental Plaintiffs," *Duke Environmental Law & Policy Forum* 11, 2 (Spring 2001): 207–246.
68. John H. Adams and Patricia Adams with George Black, *A Force for Nature: The Story of NRDC and Its Fight to Save Our Planet* (San Francisco, Calif.: Chronicle Books, 2010), 12–13. See also Susan J. Buck, *Understanding Environmental Administration and Law*, 2nd ed. (Washington, D.C. and Covelo, Calif.: Island Press, 1996), 17.
69. Adams, Adams, with Black, *A Force for Nature*, 13–20; Paul J. Allen, "Institutions: Natural Resources Defense Council," *Environment* 32, 10 (December 1990): 2.
70. Adams, Adams, with Black, *A Force for Nature*, 12–21; Allen, "Institutions," 2; Philip Shabecoff, *Earth Rising: American Environmentalism in the 21st Century* (Washington, D.C. and Covelo, Calif.: Island Press, 2001), 38–39.
71. Quoted in "NRDC Executive Director Adams on the Environment," *Southern Energy Report* 38, 1 (January 1990): 2; See also Natural Resources Defense Council Website: "About NRDC," http://www.nrdc.org/about/jha.asp (Accessed November 7, 2012); "Ten Questions for RFK Jr.," *New Jersey CPA* 33, 1 (June 2012): 4.

72. Bonnie R. Cohen, "Natural Resources Defense Council—Weapons in the Environmentalist Arsenal: Lawsuits, Blacklists and Publicity," *Organization Trends* (Washington, D.C.: Capital Research Center, August 2003): 2.
73. Cohen, "Natural Resources Defense Council," 3; Richard MacLean, "Think Like a Lawyer?" *Environmental Quality Management* 20, 1 (Autumn 2010): 117–123; Kenneth Smith and Jack Raso, M.S., R.D., *An Unhappy Anniversary: The Alar "Scare" Ten Years Later* (New York: American Council on Science and Health, 1999): 4.
74. *Chevron U.S.A. Inc. v. Natural Resources Defense Council, Inc.*, 467 U.S. 837 (1984). Numerous articles have been written on the effect of the *Chevron* decision. See, for example, Jack M. Beermann, "End the Failed *Chevron* Experiment Now: How *Chevron* Has Failed and Why It Can and Should be Overruled," *Administrative & Regulatory Law News* 35, 2 (Winter 2010): 3–15; Eric M. Braun, "Coring the Seedless Grape: A Reinterpretation of *Chevron U.S.A. v. NRDC*," *Columbia Law Review* 87, 5 (June 1987): 986–1008; Thomas W. Merrill, "Justice Stevens and the *Chevron* Puzzle," *Northwestern University Law Review* 106, 2 (2012): 551–566.
75. Natural Resources Defense Council Website: "A Case Study: Alar," http://www.nrdc.org/health/alar/> (Accessed November 7, 2012). See also Cohen, "Natural Resources Defense Council," 3; Gregg Easterbrook, *A Moment on the Earth: The Coming Age of Environmental Optimism* (New York: Viking, 1995), 117, 249; Wallace Kaufman, *No Turning Back: Dismantling the Fantasies of Environmental Thinking* (New York: Basic Books, 1994), 80–84; Smith and Raso, *An Unhappy Anniversary*, 1; Jacqueline Vaughn Switzer, *Green Backlash: The History and Politics of Environmental Opposition in the U.S.* (Boulder, Colo.: Lynne Rienner Publishers, 1997), 85, 133.
76. Natural Resources Defense Council Website: "About NRDC," n.p.
77. Greenpeace Website: "Our History," http://www.greenpeace.org/usa/en/campaigns/history/ (Accessed October 18, 2012).
78. Rex Weyler, *Greenpeace: How a Group of Ecologists, Journalists, and Visionaries Changed the World* (Emmaus, Penn.: Rodale Books, 2004), 27–28, 52–60.
79. Greenpeace Website: "The Founders," http://www.greenpeace.org/usa/en/campaigns/history/the-founders/ (Accessed October 18, 2012).
80. Ibid., n.p.
81. Ibid., n.p.; Weyler, *Greenpeace*, 176–185, 339.
82. Greenpeace Website: "Greenpeace International," http://www.greenpeace.org/international/en/ (Accessed October 19, 2012); Weyler, *Greenpeace*, 10.
83. Greenpeace Website: "The Bombing of the *Rainbow Warrior*," http://www.greenpeace.org/usa/en/campaigns/history/The-Bombing-of-the-Rainbow-Warrior/ (Accessed October 19, 2012); Weyler, *Greenpeace*, 396–403.
84. ActivistCash.com Website: "Greenpeace," http://activistcash.com/organization_overview.cfm/o/131-greenpeace (Accessed October 19, 2012); Greenpeace Website: "Greenpeace USA," http://www.greenpeace.org/usa/en/about/ (Accessed October 19, 2012); Greenpeace Website: "Questions about Greenpeace in General," http://www.greenpeace.org/international/en/about/faq_old/questions-about-greenpeace-in/ (Accessed October 19, 2012); Michael Roose, "Greenpeace, Social Media, and the Possibility of Global Deliberation on the Environment," *Indiana Journal of Global Legal Studies* 19, 1 (Winter 2012): 347–364; Leslie Spencer and Jan Bollwerk, "The Not So Peaceful World of Greenpeace," *Forbes* 148, 11 (November 11, 1991): 174–180; Scherrer, "Environmental Conservation NGOs and the Concept of Sustainable

Development," 559–561; Kristin Stromsnes, Per Selle, and Gunnar Grendstad, "Environmentalism between State and Local Community: Why Greenpeace Has Failed in Norway," *Environmental Politics* 18, 3 (May 2009): 391–407.

85. Earth First! Website: "About Earth First!," http://www.earthfirst.org/about.htm (Accessed October 15, 2012).
86. Edward Abbey, *The Monkey Wrench Gang* (New York: Harper Perennial Modern Classics, 2000 [1975]). See also Dave Foreman, *Confessions of an Eco-Warrior* (New York: Crown Books, 1993), 4, 17, 114–115; Roger Scruton, *How to Think Seriously about the Planet: The Case for an Environmental Conservatism* (New York and Oxford: Oxford University Press, 2012), 360–361.
87. Foreman, *Confessions of an Eco-Warrior*, viii, 17. See also Charles T. Rubin, *The Green Crusade: Rethinking the Roots of Environmentalism* (Lanham, Md.: Rowman & Littlefield, 1998), 179–180.
88. Foreman, *Confessions of an Eco-Warrior*, 167; Paul K. Conklin, *The State of the Earth: Environmental Challenges on the Road to 2100* (Lexington: University of Kentucky Press, 2007), 276–277; Scruton, *How to Think Seriously about the Planet*, 360–361; Derek Wall, *Earth First! and the Anti-Roads Movement: Radical Environmentalism and Comparative Social Movements* (New York and Oxford: Routledge, 1999), 5, 18, 44.
89. Foreman, *Confessions of an Eco-Warrior*, 126; Kathlyn Gay, *American Dissidents: An Encyclopedia of Troublemakers, Subversives, and Prisoners of Conscience* (Santa Barbara, Calif.: ABC-CLIO, 2012), 47–50; Bron Taylor, "The Religion and Politics of Earth First!" *The Ecologist* 21, 6 (November/December 1991): 261; Wall, *Earth First!*, 143.
90. Frank Biermann, Olwen Davies, and Nicolien van der Grijp, "Environmental Policy Integration and the Architecture of Global Environmental Governance," *International Environmental Agreements* 9, 4 (November 2009): 351–369; Jenkins, "Nongovernmental Organizations and the Forces against Them," 465–467; Michael Mason, "Representing Transnational Environmental Interests: New Opportunities for Non-Governmental Organization Access within the World Trade Organization?" *Environmental Politics* 13, 3 (Autumn 2004): 566–589; Scherrer, "Environmental Conservation NGOs and the Concept of Sustainable Development," 555–571; Karla W. Simon, "International Non-Governmental Organizations and Non-Profit Organizations," *International Lawyer* 44, 1 (Spring 2010): 399–414.
91. Scruton, *How to Think Seriously about the Planet*, 28. See also Jon Entine, "NGO Openness is a Two-Way Street," *The Central New York Business Journal* 21, 52 (December 28, 2007): 23; Jenkins, "Nongovernmental Organizations and the Forces Against Them," 479–492.
92. Scruton, *How to Think Seriously about the Planet*, 248–252.
93. O'Rourke, *Parliament of Whores*, 197, 198.

Chapter 9

The Foundations of Twenty-First-Century Sustainability

We have met the enemy and he is us.

—Walt Kelly, "Pogo"[1]

The challenge for the twenty-first-century sustainability movement is to "operationalize the variable," in the parlance of scholars, in a manner that leads to effective action. Supporters of environmental stewardship must ensure that abstract, amorphous, vague concepts such as "sustainable," "green," and "environmentally friendly," among others, are translated into practical, "real world" solutions leading to demonstrable, measurable effects. The literature on environmentalism is filled with stirring, noble, eloquent pleas to save Mother Earth from humanity's thoughtlessness, yet the prescriptions for change are sweeping and unrealistic. Such unattainable objectives can demoralize the rational actor and embolden the skeptic. Alternatively, some environmental activists expend time and energy pursuing narrow, ineffectual goals or niche, "pet peeve" issues driven by emotionalism and political expediency that have no appreciable effect, positive or negative, on systemic environmental quality. What is needed in lieu of extreme positions is a thoughtful, realistic, incremental, science-based approach to environmentalism that balances competing values of freedom and authority, international, national, state, and local political power as well as environmentalism and industrialization.[2]

Consider an example of the noble, yet impracticable, call to action championing worldwide systemic change. In his well-known 1992 book *Earth in the Balance*, U.S. senator and later vice president Al Gore argued that environmentalism desperately requires a "global Marshall Plan" modeled on the U.S. strategy for rebuilding Europe's fractured economy following World War II. Gore argued that modern sustainability requires all nations to stabilize their populations, upgrade their technology, reform their economic systems, abide by uniform "rules of the road" in protecting the natural environment, negotiate enforceable international environmental agreements, and educate citizens around the world about the global environment. "For civilization as a whole," the senator concluded, "the faith that is so essential to restore the balance now missing in our relationship to the earth is the faith that we do have a future. We can believe in that future and work to achieve and preserve it, or we can whirl blindly on, behaving as if one day there will be no children to inherit our legacy. The choice is ours. The earth is in the balance."[3]

Most citizens concerned about global environmental problems would agree with Gore's sensible, if unrealistic, prescriptions. Yet the likelihood of developing and implementing a global Marshall Plan that satisfies substantially all of the parties, at least in the near future, is slim. Such calls for systemic changes to ensure worldwide environmental sustainability are relatively easy to make but difficult to achieve. Critics of American environmentalism have mocked pleas for wholesale change, especially change fostered through the imperfect mechanism of international agreements. Because the United States cannot exercise sovereignty over foreign countries, the time devoted to negotiating treaties with developing nations, especially Brazil, India, and China, is wasted owing to the lack of enforceable mechanisms to ensure compliance. Even when countries that sign international agreements attempt to act in good faith and fully intend to comply with burdensome environmental controls, they justify subsequent noncompliance by arguing that practicalities associated with caring for a burgeoning population necessitate regrettable violations. Detractors charge that weak, unenforceable treaties seldom effect systemic change. Instead, they exacerbate tensions and ratchet up cynicism.[4]

Environmentalists that support international mechanisms cite the European Union (EU) as an effective model for handling environmental problems because nations that participate under that governance structure agree to abide by common regulations established by EU regulators. Ideally, such a transnational network creates efficacious public policy while avoiding the enforcement difficulties inherent in international treaties because EU signatories voluntarily relinquish a measure of national sovereignty in the interests of the common good. Unfortunately, reality seldom conforms to an ideal. Since its creation, the EU has produced 180,000 pages of regulations and directives, a complex labyrinth of behavioral norms that do not satisfactorily account for unintended consequences. Environmental policies that apply to one European nation might prove wholly ineffectual in another nation or, in fact, aggravate existing problems.[5]

The EU response to carbon emissions provides an instructive example. Environmentalists worried about global warming have urged nations to implement voluntary limits on artificial carbon production. In the wake of the Kyoto Protocol, a 1997 agreement among more than 190 countries to limit greenhouse gas emissions, the environmental community lobbied individual EU countries to establish a cap-and-trade policy whereby private companies would purchase from government the right to emit carbon from a company's manufacturing process. By forcing private enterprise to internalize costs that traditionally have existed as negative externalities, government policy can influence behavior in the private marketplace. Some firms will reduce carbon emissions under the plan and sell their carbon credits in the open market while large carbon producers must purchase credits to ensure compliance. Such a market mechanism, at least in theory, will maximize economic efficiency while concomitantly ensuring that private corporations factor concerns about global climate change into their decision-making. Presumably, over time all parties will have incentives to reduce carbon emissions as they seek to reduce costs and thereby hone a competitive advantage.[6]

This seemingly logical regulatory scheme created a multitude of unintended consequences across the EU. Great Britain and Spain, for example, established stringent carbon reduction targets while nation states such as Italy declined to set tough requirements. Because not all nations started from a common point, economic inefficiencies were introduced into the EU cap-and-trade system. Private companies in Great Britain and Spain were strictly regulated; thus, the need for emissions credits was greater in those countries than in relatively lax jurisdictions. Countries forced to comply with stringent regulations scrambled to purchase emissions credits, which made those credits valuable, hence expensive. Not only were corporations in Great Britain and Spain placed at a competitive disadvantage because they were required to purchase expensive credits that companies in other countries could forgo, but noncompliant companies had an incentive to purchase credits in jurisdictions where targets were less stringent and permits were less expensive. Consequently, companies in a few EU countries bore the burden while companies in other EU countries enjoyed the benefits of an inequitable cap-and-trade policy. The net effect was that carbon emissions in Europe did not decline. Instead, economic resources were redistributed among participating nations.[7]

Apologists for transnational environmental policies contend that emissions trading schemes and international agreements are imperfect mechanisms but they can be improved in time. If all EU countries, for example, faced the same stringent regulations, the system would work. Cap-and-trade apologists counsel patience and plead for more cooperation among nation-states before assessing the effect of international cooperation in environmental policymaking.[8]

As an alternative to systemic change, perhaps addressing a narrow issue with clearly delineated parameters is preferable to pursuing an overly broad and ambitious program because incremental progress is eminently demonstrable with a single, clearly defined project. This recommendation holds merit, but care must be taken not to allow emotionally driven, politically expedient concerns to drive the initiative. The "low-hanging fruit" of whether polystyrene foam products should be used to package food and beverages in one major metropolitan area provides an example of the perils involved with focusing on a narrow concern. In February 2009, a group of concerned citizens in Chicago, Illinois, led by self-proclaimed activist Stacy Pfingsten, founded an environmental group called No Foam Chicago. "Styrofoam has always been somewhat of a pet peeve of mine," Ms. Pfingsten explained. "When I saw that there were many cities that had already banned or are going to ban Styrofoam, I thought, 'Why not Chicago? Why haven't we?'" Ms. Pfingsten's group created a website lambasting the plastics industry as well as chemical manufacturers for creating a product that had the misfortune of becoming her pet peeve.[9]

Although No Foam Chicago raised a legitimate public policy issue about packaging waste and the desirability of using specific types of materials, it was clear that the group primarily was interested in championing a polemic based on scare tactics and political correctness. A reasoned, science-based debate about the trade-offs among packaging choices requires a disciplined, nuanced comparison among and between life cycle management (LCM) data. It is hardly the stuff of screaming headlines or alarmed citizens contributing funds to combat the latest environmental menace. Ignoring data contrary to the group's position, No Foam Chicago urged concerned citizens to join the group in prohibiting polystyrene foam food service products and thereby saving the planet.[10] Group members had found their pet peeve and were determined to move forward regardless of other factors. In lieu of engaging in a reasoned dialogue, allowing all credible voices to be heard and all data-driven positions to be debated, No Foam Chicago drew a line in the sand: Either you are with us (heroes) or against us (villains).[11]

Yet even if No Foam Chicago succeeded beyond its wildest expectations and convinced policymakers to ban the offending products and materials, the repercussions would be negligible. The products packaged in polystyrene foam would be packaged in something else, and that something else would present its own set of environmental problems and challenges. This message was lost or ignored in the quest for political expediency. No Foam Chicago became emblematic of environmental organizations that expend resources pursuing narrow, ineffectual goals that have no practical effect but hold enormous appeal for like-minded pseudo-environmentalists who seek to take advantage of a gullible public over the latest trendy crisis *du jour*. If the issue had been framed in a broader context—involving a larger set of packaging materials or covering a wider geographic area beyond Chicago—and if scientific data had been the paramount consideration, additional voices probably would have been heard and more parties would have participated in the debate. Of course, this insight presumes that a reasoned debate rather than an emotionally driven polemic is the objective.[12]

This critique is not to suggest that concerned policymakers should avoid tackling large, systemic problems or, alternatively, avoid an incremental approach to addressing issues of local concern. Even in cases where the practical effect may be marginal, undeniable value exists in pursuing high profile, symbolic environmental campaigns. Symbolism always has been, and remains, important in American environmentalism. Because the issues often become complex and impenetrable to the novice, symbolism appeals to an environmentally conscious but technically unsophisticated citizenry. Yet a symbolic effort must never supplant genuine, measurable progress; a pyrrhic victory must never be confused with a hard-got environmental conservation or preservation initiative. In fact, when "pie-in-the-sky" schemes and "the-sky-is-falling" alarms proliferate, they crowd out science-based, practical, good-faith efforts to tackle genuine environmental problems in a thoughtful, intelligible manner.

The pragmatic sustainability proponent seeks a good-faith "Goldilocks" approach to environmentalism—a middle course of action. For modern sustainability to achieve genuine, measurable results, it is necessary to operationalize the variable by rejecting broad, amorphous abstractions as well as calls for radical, overnight changes to multiple sovereign governments, on one hand, and shrill, symbolic, politically correct, ineffective polemics using scare tactics, on the other hand. The ubiquitous windmill must remain a symbol of the best features of environmentalism and not become a monument to futility (Figure 9.1). Otherwise, the environmental community risks marginalization on the fringes of mainstream public policymaking. The time for tilting at windmills and feeling good while

Figure 9.1 Wind turbines have become an iconic symbol of the search for renewable energy sources in modern American environmentalism. (Courtesy of the Library of Congress)

accomplishing nothing must be banished from mainstream American environmentalism. Recognizing this fact, the pragmatic environmentalist invests time and resources in creating mechanisms for measuring and evaluating success. He employs a variety of tools and mechanisms to improve the natural environment through realistic means, including carefully drafted and negotiated international treaties and agreements, enforceable environmental laws and regulations, and voluntary environmental management systems and certifications. In operationalizing the variable using these tools, the scope of the issue must be framed properly, the objectives must be clearly defined, the programs must be carefully monitored, and the results must be rigorously evaluated. By these standards, some sustainability mechanisms are more effective than others are.

Famous International Conferences and Agreements

Environmentalists interested in promoting sustainability have recognized the value of working across national boundaries. With the growing globalization of markets, and therefore a concomitant increase in widespread pollution, environmental problems often do not affect a single nation-state. Beginning late in the twentieth century, non-governmental organization (NGO) representatives, academics, engineers, legislators, regulators, and other interested parties engaged with their peers through a series of conferences and symposia designed to develop transnational solutions to global environmental problems. Although the number of meetings and seminars continues to grow, four conferences and their resultant agreements in particular have become well-known symbols of the strengths and weaknesses of international environmentalism. Participants have pushed government officials to enter into binding treaties with representatives from neighboring nation-states, but the record of successes has been mixed. Despite the difficulty in negotiating and enforcing international agreements, such conferences likely will remain a mechanism for promoting environmentalism well into the future.

Montréal Protocol

The Montréal Protocol on Substances that Deplete the Ozone Layer (a protocol to the Vienna Convention for the Protection of the Ozone Layer), generally known as the "Montréal Protocol," was an international treaty that restricted the production and use of chlorofluorocarbons (CFCs), an industrial chemical shown to damage the stratospheric ozone layer. Owing to a diminishing ozone layer, increased ultraviolet radiation at the earth's surface could lead to a variety of human illnesses, including eye cataracts and skin cancer. In 1978, the United States unilaterally prohibited nonessential CFC aerosol propellants. Seven years later, a group of scientists published research concluding that a "hole" existed in the ozone layer over Antarctica. After decades of research on the effects of CFCs, by the 1980s it was reasonably clear that scientific consensus and political will existed to phase out products containing the substance.[13]

In the meantime, nation-states met under the auspices of the United Nations to exchange scientific information and discuss strategies for reducing or eliminating ozone-depleting substances. The Montréal Protocol occurred as an addendum to a March 1985 meeting in Vienna, Austria. Thirty-six countries signed the protocol beginning in September 1987. The treaty went into effect on January 1, 1989. Participating nations held a series of follow-up conferences resulting in revisions to the document throughout the 1990s. By 2008, more than 190 nations had ratified the agreement. Among other things, the protocol required an incremental phase-out of CFC production, culminating in complete elimination by the middle of the twenty-first century. In response, industries that had relied on CFCs found suitable alternatives.[14]

Champions of transnational solutions to global environmental problems cite the Montréal Protocol as a model for future agreements. Nobel Laureate Kofi Annan, former Secretary-General of the United Nations, remarked that it was "perhaps the single most successful international agreement to date." Environmentalists concerned about greenhouse gas emissions (GHGs) frequently ask whether the terms and conditions of the Montréal agreement can be retrofitted to address GHG concerns or at least serve as a model for an analogous international treaty.[15]

United Nations Conference on Environment and Development (UNCED)—The Rio de Janeiro Earth Summit

The United Nations Conference on Environment and Development (UNCED) or, simply, the Rio Earth Summit, was an unprecedented gathering in Rio de Janeiro, Brazil, in 1992 devoted to heightening public awareness of global environmental issues by providing a forum for world leaders to grapple with socioeconomic and development issues in emerging nations while also protecting the natural environment from despoliation. Fans of the summit tout the impressive statistics: 120 heads of state appeared; 172 nations participated; 9,000 journalists covered the event; 1,600 NGOs attended; and 450,000 visitors descended on the city. Never had so many resources been channeled together in one place to consider global environmental concerns.[16]

The Rio Earth Summit was not the first global meeting devoted to the environment. A smaller group met in Stockholm, Sweden, in 1972. The earlier meeting resulted in 26 principles adopted as part of the *Declaration of the United Nations Conference on the Human Environment (Stockholm Declaration)*. Principle 2 set forth the credo that would govern all the international conferences to follow: "The natural resources of the earth, including the air, water, land, flora and fauna and especially representative samples of natural ecosystems, must be safeguarded for the benefit of present and future generations through careful planning or management, as appropriate." The Rio Earth Summit would build on, and add to, the *Stockholm Declaration*.[17]

By the time the Rio Earth Summit convened in 1992, environmentalists from around the world had agreed that a new paradigm was necessary to address global environmental problems. Hoping to replicate the success of the Montréal Protocol, participants at the Rio conference strove to negotiate treaties covering global climate change, the loss of biodiversity, worldwide deforestation, and sustainable development. By the end of the convention, they had forged agreements that set the stage for global environmentalism during the ensuing two decades. The *United Nations Framework Convention on Climate Change (UNFCCC)* became a mechanism for convening developed and developing nations to discuss, in the words of Article 2, a "stabilization of greenhouse gas concentrations in the atmosphere at a level that would prevent dangerous anthropogenic interference with the climate system." The *UNFCCC* was a precursor to the Kyoto Protocol. The *Convention on Biological Diversity (CBD)* contained a set of forestry principles designed to curtail the extinction of plant and animal species owing to human settlements by proposing management practices for agricultural activities, land use, and pollution. *Agenda 21*, an action plan encouraging sustainable development by focusing on deforestation, ocean pollution, workers' rights, and poverty, arguably was the most famous and influential deliverable to emerge from the Rio Earth Summit. Maurice Strong, Secretary-General of the Rio Earth Summit, reflected the view of many participants when he concluded that organizers "got agreement beyond what anybody thought was possible." Follow-up summits were held in Johannesburg, South Africa (Earth Summit 2002, or "Rio + 10") and Rio de Janeiro, Brazil (Earth Summit 2012, or "Rio + 20").[18]

Kyoto Protocol

The Kyoto Protocol, to date the most comprehensive international treaty designed to manage global GHG emissions, was created during an environmental conference held in Kyoto, Japan, in December 1997. The protocol amended the *UNFCCC*; however, while the *Convention on Climate Change* encouraged participating nations to stabilize and eventually reduce emissions, the Kyoto agreement required a binding commitment from signatories to achieve those goals. The protocol established targets for 37 industrialized countries as well as the European Community. It went into effect in February 2005; as of 2012, 191 countries had endorsed the protocol.[19]

The Kyoto Protocol set a 1990 baseline for four greenhouse gases: carbon dioxide, methane, nitrous oxide, and sulfur hexafluoride. Two groups of gases, hydrofluorocarbons and perfluorocarbons, were covered by the agreement as well. Industrialized countries, because they produced more emissions than developing nations, bore the brunt of the burden, although the protocol provided some flexibility in meeting the targets. Market-based mechanisms, such as a carbon emissions trading market, joint implementation, and the clean development mechanism

(CDM), were designed to encourage capitalistic, highly industrialized nations to seek economically efficient solutions.[20]

From its inception, the Kyoto Protocol proved to be bitterly divisive. Owing to its failure to regulate developing countries, the protocol met with resistance in developed nations, such as the United States, which contended that the treaty unfairly penalized affluent signatories. The agreement especially irked politically conservative detractors who viewed it as a veiled mechanism for redistributing wealth. In 1998, the U.S. Senate unanimously refused to ratify the treaty unless it "also mandates new specific scheduled commitments to limit or reduce greenhouse gas emissions for Developing Country Parties." In 2001, President George W. Bush complained that the Kyoto Protocol would harm the U.S. economy and yet it exempted "80 percent of the world, including major population centers such as China and India, from compliance."[21]

As Bush suggested, some nations straddle the line between developing and developed nations. With China rapidly becoming the world's largest GHG emitter, allowing the heretofore-developing nation to avoid bearing a large burden for reductions undermined the protocol's effectiveness. Moreover, the lack of an effective enforcement mechanism meant that if rapidly developing nations agreed to cut GHG emissions but failed to honor their commitments, the penalties remained unclear. "Expert review teams" were supposed to assess whether the protocol had been implemented in accordance with the treaty terms, but compliance depended on the continuing good faith of signatories.[22]

Aside from these considerable difficulties, the Kyoto agreement also suffered by comparison with the Montréal Protocol, in part, because of differences in CFC and GHG emissions. CFCs are a relatively small group of chemicals. Prior to the 1990s, they were used primarily as refrigerants and propellants in aerosol sprays. The United States took the lead in phasing out CFCs in large part because alternatives were readily available and comparatively inexpensive. GHG emissions, in contrast, are a byproduct of combustion, and combustion is an integral feature of life inside a highly industrialized nation. Operating a manufacturing plant, driving an automobile, or powering a modern home with multiple appliances and electronic devices requires combustion somewhere along the line. Effectively reducing GHGs in such a society without severely hampering a national economy is extraordinarily challenging. As a result, the difficulty in implementing the Kyoto Protocol is orders of magnitude more complicated than curbing CFC production and usage under the Montréal Protocol. Nonetheless, environmentalists have contended that the Kyoto agreement, as imperfect as it is, remains "the only game in town," and therefore must be retained and modified. Time will tell if protocol defenders can convince skeptics of the agreement's ongoing value.[23]

Copenhagen Climate Change Conference

Despite the ubiquity of global environmental summits, twenty-first-century conferences have failed to match the successes of their predecessors. This assessment is especially true for the Copenhagen Accord, a three-page interim agreement produced in Denmark in December 2009. The original intent had been to correct the deficiencies in the Kyoto Protocol by carving out a legally binding treaty with aggressive, enforceable targets leading to significant global warming abatement. Owing to dissension among conference participants, the resultant accord failed to include specific goals or mandatory GHG emissions reductions targets. For disheartened environmentalists, the document seemed to be a retreat from almost two decades of environmental progress achieved under the UN umbrella.[24]

Despite its well-documented shortcomings, the Copenhagen Accord was not a complete failure. Participants agreed to set a goal of limiting GHG-induced global temperature rise to 2°C above preindustrial levels by 2050. In addition, the accord established a legal obligation to transfer hundreds of billions of U.S. dollars from developed nations to assist developing nations. Recognizing the continuing problem of global deforestation, conference attendees created a new mechanism to address the issue.[25]

For the first time, environmentalists pushed for "super-developing" nation-states such as Brazil, India, and China to be responsible for addressing GHG emissions. This was an important change. In Kyoto, the participants had embraced a principle of "common but differentiated responsibilities," which meant that industrialized countries would bear the major burden for GHG reductions. By placing responsibility on the largest GHG emitters, the Copenhagen Accord adopted the "polluter pays" principle, which does not distinguish between developed and developing nations when assessing the compliance burden.[26]

Disillusioned environmentalists left Copenhagen decrying the self-serving nature of modern nation-states, especially "isolationist and avaricious China." In a world where borders are less relevant than ever, heads of state must recognize the dangers of hiding behind national sovereignty. Climate change will wreak havoc on all nations of the world eventually, environmentalists charge. Irresoluteness and delay will only compound the difficulties faced by future generations. Weak-willed politicians and captains of industry simply were unwilling or unable to set aside their own petty, short-term self-interestedness in favor of the long-term collective good (Figure 9.2).[27]

International conferences and agreements remain standard features of global environmentalism despite evidence of their diminishing successes. Assuming *arguendo* that international agreements and transnational unions will come into their own at some future time—which is by no means an uncontested proposition—the question arises as to how individual countries can implement modern sustainability solutions in the meantime. The obvious answer is to use the mechanisms of law and regulation in a single nation to remedy problems associated with environmental

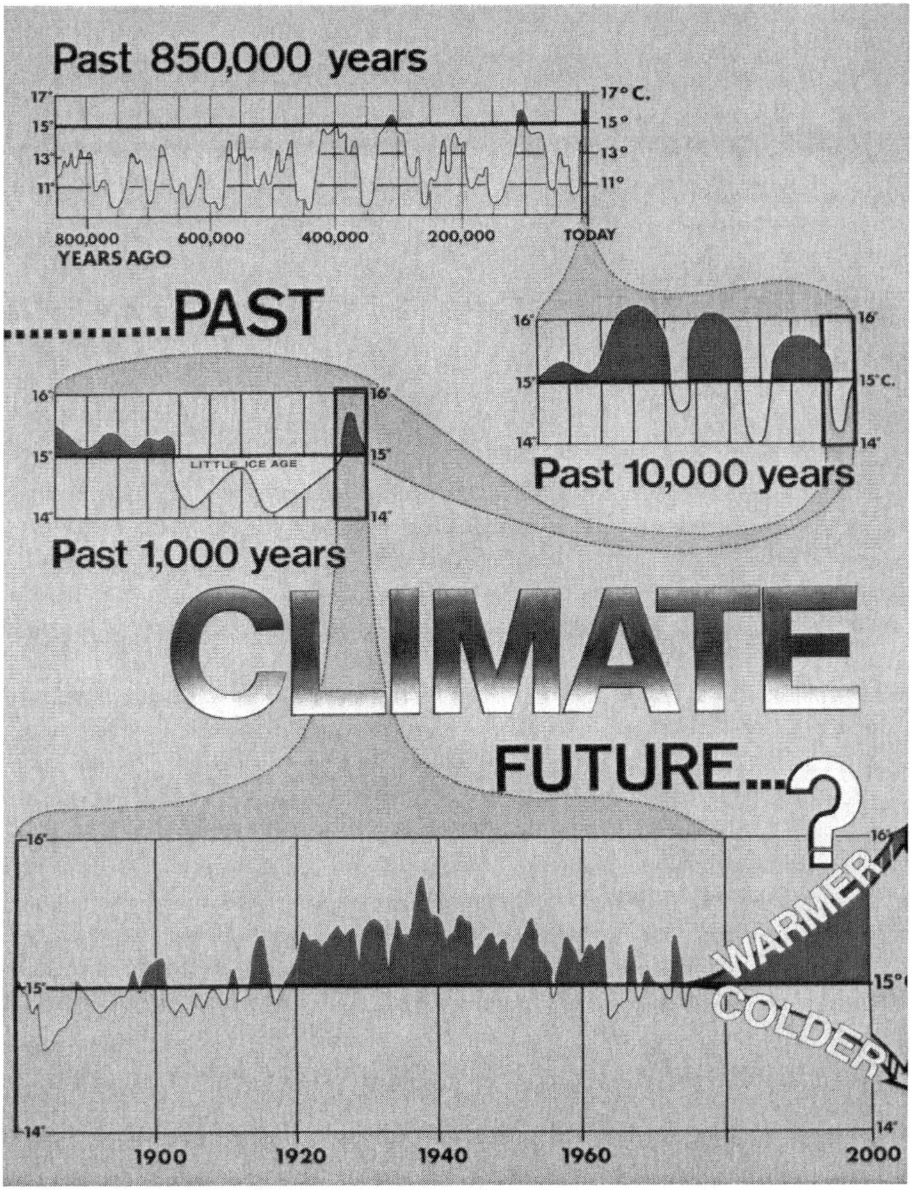

Figure 9.2 Climate change has been a controversial subject in recent decades, especially since scientists of the mid-twentieth century debated whether the climate was growing colder or warmer over time. Most major NGOs added climate change to their menu of critical environmental issues for the twenty-first century. (Courtesy of the Granger Collection, New York)

degradation. The United States, for example, has developed a system of federal environmental law that methodically addresses contamination of the air, water, and land as well as regulating hazardous materials, toxic chemicals, and the loss of plant and animal species and wetlands. Although a comprehensive review of all U.S. environmental laws is beyond the scope of this book, a brief discussion of federal environmental statutes enacted since the 1970s demonstrates how the gains of the twentieth-century American environmental movement were made permanent through legal mechanisms.

American Environmental Law in a Federal System

However overwrought and alarmist some aspects of the environmentalism of the 1960s and 1970s may seem when viewed with the benefit of hindsight, the movement undeniably helped to create the framework of American environmental law. The enactment of laws and regulations to preserve and protect the natural environment was perhaps the greatest legacy of American environmentalism during the latter half of the twentieth century. To translate activism from dire predictions into legislative victories was no mean feat—and grassroots activists could not take complete credit for the legal successes that followed—but it was possible because Americans had been awakened from their slumbers to realize that environmental issues had been ignored for far too long. In the aftermath of Earth Day celebrations as well as the publication of books and news stories about the devastation caused by clear-cutting of forests, toxic waste dumped near homes and communities, and polluters indiscriminately fouling the air and water, mainstream citizens who once had been oblivious to environmental issues became keenly aware of the old adage that they were either part of the problem or part of the solution.

At the beginning of the 1970s, as lawmakers and regulators grappled with methods for responding to increased public concerns about polluted air, water, and land, they expanded existing laws to erect a large, complex structure of environmental regulation. First, they based new laws on risk assessments by weighing the costs and benefits of their actions based on the hazard of a particular pollutant at a specific concentration; the human response in terms of carcinogenic, mutagenic, or pathogenic effects; the likely exposure through air, water, soil, or the food chain; and the likely incidence of disease or damage. They also looked to common law standards such as nuisance—the idea that no one may interfere with the property rights of another person, thereby causing substantial harm, unless the use of that private property right becomes a nuisance affecting the health, safety, welfare, comfort, or convenience of the general public. The long-standing trespass doctrine was also the source of environmental laws. The doctrine held that direct physical invasion or intrusion on a person's land was necessary to prevail in a civil suit for damages. A 1959 Oregon case, *Martin v. Reynolds Metal Company*, decided that the intrusion of invisible fluoride particles constituted trespass, the first step in a line of court opinions broadening this area of the law.[28]

Other common law doctrines were useful, but they had to be modified by statute. Negligence law, which governs personal injuries owing to the unintentional actions of a tortfeasor, holds that if a polluter has a duty to conform to a certain standard, violates that duty, causes damage as a consequence of the violation, and an actual injury results, the plaintiff can collect monetary damages. Because negligence was a difficult tort to prove, new environmental statutes relaxed stringent evidentiary requirements to increase the likelihood of holding polluters accountable. Strict liability, a tort that does not require the complaining party to demonstrate proof of fault, was useful in cases where the defendant engaged in an abnormally dangerous activity that subjected third parties to foreseeable harm resulting from the activity. Courts determined through case-by-case adjudication what constituted an "abnormally dangerous activity." Another common law principle, the public trust doctrine, dictated that some properties are held in trust by the sovereign on behalf of all people. As environmental law developed in the 1970s, courts expanded the definition of "properties" to include publicly held natural resources even if they did not have a measurable pecuniary value. The states and the federal government, the courts decided, have an affirmative duty to act as fiduciaries in managing public lands. Damage to a resource held in public trust is damage inflicted on all citizens.[29]

National Environmental Policy Act (NEPA)

As useful as the common law doctrines were in handling environmental matters, lawmakers realized that specific environmental problems required specific laws and regulations tailored to address those issues. The National Environmental Policy Act of 1970 (NEPA) was one of the earliest and best-known federal statutes enacted during the new era. Section 102(2)(C) of the statute requires a federal agency to develop a "detailed statement" when the agency proposes to undertake "a major federal action significantly affecting the quality of the human environment." The detailed statement, an "environmental impact statement (EIS)," must describe the proposed federal action, outline the expected environmental impacts, and consider the costs and benefits of alternative actions. Unlike subsequent environmental laws, NEPA was designed as an informational, not a regulatory statute. The agency is required to produce an EIS, but the preparation of such information does not preclude continuation of the activity even in cases when one or more alternatives might be deemed environmentally preferable. The law also established the Council on Environmental Quality (CEQ) to advise the president on environmental issues. Housed within the Executive Office of the President, the CEQ is responsible for overseeing NEPA implementation, resolving disputes between the U.S. Environmental Protection Agency (EPA) and federal agencies if a dispute arises concerning the adequacy of an EIS submission, and preparing an annual report summarizing the state of environmental conditions in the country.[30] Critics charge that NEPA involves a large volume of paperwork and triggers seemingly endless rounds of litigation requiring defendants to jump through a maze of "legal hoops" without necessarily improving environmental quality. EISs are seen as a legal requirement—a costly, time-consuming, and mindlessly rote *pro forma* exercise—but not as a tool for genuinely influencing agency decision-making or changing behavior. NEPA requires that an EIS be prepared, but the preparation of a statement does not mean the agency must perform any particular action because of the report—even if the proposed action results in destructive environmental consequences. NEPA allows for a number of exceptions to EIS preparation as well. Some proposed actions fall under a "category exemption," which is a predetermination that the proposed action has no significant environmental impact. In some instances, an agency can prepare an environmental assessment (EA), a shortened version of an EIS, to determine whether an EIS is needed in the first place. EPA may reach a finding of no significant impact (FONSI) based on the EA conclusions. For these reasons, detractors charge that NEPA is a paper tiger, a federal requirement with little practical effect.[31]

NEPA supporters contend that the statute, while far from ideal, has at least a marginal impact on agency decision-making. In considering the effects of a proposed federal action on the environment, agencies sometimes choose options that minimize the potential for environmental harm. Agency officials realize that the administrative record can be used to review decisions after the fact to discern whether the organization acted in an arbitrary and capricious manner; therefore, they are careful to document their decisions and reasoning. Activists have influenced and even modified federal action by threatening or, in some cases, filing lawsuits to ensure the agency has complied with NEPA's procedural requirements. In their view, the statute has altered the federal government's standard operating procedures by incorporating environmental considerations into the mix of crucial policy issues.[32]

Aside from NEPA, additional important environmental statutes enacted in the 1970s included the Clean Air Act (CAA), the Clean Water Act (CWA), the Toxic Substances Control Act (TSCA), the Resource Conservation & Recovery Act (RCRA), and the Comprehensive Environmental Response, Compensation, and Liability Act (CERCLA), often called "Superfund." These federal laws, along with other federal laws of a more limited scope as well as state analogues, have constructed a regulatory regime hailed by environmentalists as important (although not comprehensive) legal protections for the nation's environmental quality while private-sector industry representatives occasionally have denounced them as needlessly bureaucratic, an intricate web of complicated, confusing "rules for the sake of rules" that encourage litigation, drive up production costs, and place U.S. businesses at a competitive disadvantage when pitted against foreign firms in a global marketplace.[33]

Clean Air Act (CAA)

The CAA sports a long, complex history. After passing an initial law, the Air Pollution Control Act of 1955, Congress enacted the first Clean Air Act in 1963 to authorize the U.S. Department of Health, Education, and Welfare (HEW) to establish air quality criteria. Two years later, in the Motor Vehicle Pollution Control Act, the HEW secretary prescribed national standards to regulate emissions beginning with 1968 automobiles. In the Air Quality Control Act of 1967, Congress created Air Quality Control Regions, or AQCRs, a concept eventually supplanted by National Ambient Air Quality Standards (NAAQS). The concept was simple even if implementation proved to be challenging: States would set standards to limit the levels of pollutants identified in federal air quality control technology documents. Without enforceable national air quality standards, however, the law was not powerful enough to effect major change. Two competing schools of thought arose on how to strengthen the law. The emissions-standard school, largely with support from the environmental-activist community, argued on behalf of an absolutist, technology-based regulation with an emphasis on individual rights. Private-sector businesses favored the ambient-air school, a relativist approach that sought economically sound solutions and opposed technology-based regulation. This debate would continue through each version of the CAA.[34]

The Clean Air Act Amendments of 1970 represented a new method of regulating an old problem. Congress moved from a region-by-region approach to promulgate NAAQS for six criteria pollutants: particulate matter, primarily from coal mining processes; sulfur dioxide from fossil-fuel combustion; ozone emissions from smog, automobile exhaust, and chemical facilities; nitrogen oxides from smog and fossil-fuel combustion; carbon monoxide; and hydrogens. The law subsequently added lead to the list. The authors of the statute chose to focus on criteria pollutants—that is, chemicals that typically are not toxic when found in average concentrations in ambient air but which serve as precursors to pollution when combined with other chemicals—to ensure that substances contributing to atmospheric pollution would not slip below regulatory concern. To enforce NAAQS, the 1970 statute sought to bridge the gap between the creation of federal air quality standards and state enforcement. To that end, states were required to prepare and submit State Implementation Plans (SIPs) to EPA for review and approval. If a SIP failed to gain EPA approval, the agency could substitute a Federal Implementation Plan (FIP) in its stead. An approved SIP or FIP would be promulgated as a regulation. The original plan was to mandate NAAQS for criteria pollutants by 1975.[35]

Aside from NAAQS, the 1970 CAA established New Source Performance Standards (NSPS) to regulate emissions for new or modified sources of air pollution. Industries subject to the standards were required to use Best Available Control Technology (BACT), a stringent criterion that ensured additional air quality improvements. The statute also set forth National Emissions Standards for Hazardous Air Pollutants (NESHAPS), a standard that went beyond NAAQS criteria pollutants to regulate less-widely emitted, but highly dangerous, toxic air pollutants. New federal motor vehicle emissions limitations required automobile manufacturers to curtail hydrocarbon and carbon monoxide emissions as well as nitrogen oxide.[36]

In 1977, Congress revised the CAA to address existing as well as new pollution sources. Existing major stationary sources of pollution were subject to Reasonably Available Control Technology (RACT) requirements. As for new pollution sources, the amendments overlaid additional geographic-specific standards on top of the NSPS technology standards. For "non-attainment" areas, or so-called "dirty air" regions, a "new source review" was required to determine whether new or modified major sources of pollution were using the Lowest Available Emission Rate (LAER). In "attainment" areas, or "clean air" regions, Prevention of Significant Deterioration (PSD) plans were required.[37]

Another amendment, enacted in 1990, refined previous iterations of the statute. Primary NAAQS standards were set to protect public health while secondary standards protected public welfare. Five categories of non-attainment areas—marginal, moderate, serious, severe, and extreme—allowed resources to be devoted to parts of the country that were more polluted than others. The statute set target dates for cleaning up each area. A polluter designated as a "major source" of stationary emissions was required to obtain a permit under Title V.[38]

As one might expect with a complex law that has been amended repeatedly, the CAA suffers from no shortage of critics. Environmentalists decry the glacial speed of implementation, especially the constantly changing target dates. NAAQS are supposed to be revised every five years, but EPA seldom meets its goals. Industry leaders contend that compliance is expensive and does not necessarily result in cleaner air. Small businesses frequently experience difficulty keeping up with the requirements. Both sides argue vehemently about the scientific basis for the standards and the state of the latest research. EPA is assailed from all sides: Environmentalists believe the agency allows polluters to violate the law with impunity while private companies believe the agency's epidemiological findings do not enjoy sufficient support in the toxicological data.[39]

Clean Water Act (CWA)

The 1972 amendments to the Federal Water Pollution Control Act, commonly known as the Clean Water Act (CWA), were less complicated than the CAA. Yet the statute also envisioned a federal-state partnership and exposed fissures between the environmental and industrial sectors. When the law was enacted, the purpose was to eliminate pollutant discharges into navigable waters by 1985. The law did not meet the timeline, but the goal of ensuring water quality and safety remained unchanged. The CWA pledged to provide federal financial assistance for construction of publicly owned treatment works (POTW). Prior to 1977, many states relied on existing state systems, but this piecemeal approach led to variable water quality throughout the country. The CWA was designed to rectify the problem and allow for consistent, uniform regulation across the United States.[40]

The statute prescribed primary and secondary treatments for point-source water pollution from biological and chemical sources. Treatments were required for POTWs, combined commercial and residential sewer systems, and industrial facilities. In addition, the act addressed the problem of non-point-source pollution, which meant everything not characterized as point-source pollution, such as farms, construction sites, mining operations, and urban runoff, as well as oil and hazardous waste spills.[41]

The National Pollutant Discharge Elimination System (NPDES) became a central feature of the CWA. As was the case with Title V CAA permits to emit pollution, the CWA recognized that pollution could not be banned overnight without decimating the nation's industrial capacity. Instead, sources that intended to discharge effluents into navigable waters were required to apply for an NPDES permit, which tightly controlled the nature and quality of the discharges. To obtain storm water permits, POTWs were required to apply to EPA and provide detailed information such as a topographic map showing drainage areas, where chemicals were used, and waste material storage and sampling data for effluent guidelines and permits.[42]

As with the CAA, the U.S. EPA was designated as the responsible federal agency for CWA enforcement. The administrator was empowered to issue notices of violation (NOVs) for statutory or regulatory violations. A cease-and-desist order required the polluter to halt the discharges until an administrative penalty order or a consent agreement finalized a plan to correct deficiencies. Citizens were authorized to file suit under the statute either to force the EPA administrator to perform non-discretionary functions or to force emitters to comply with statutory provisions. This last feature resulted in many citizen suits, especially after 1983, as environmental groups became savvy in using litigation to promote their goals. Because dischargers were obligated to file discharge monitoring reports (DMRs) under the NPDES—and the DMRs became a matter of public record—and because strict liability was the standard of legal liability, litigators found CWA cases to be winnable and profitable. A party litigating under the CWA could recover attorney fees, which became an attractive bonus for the plaintiff's bar. Suspected violators and EPA administrators had to be afforded at least a 60-day notice to correct ongoing violations prior to commencement of the suit, but afterward the "bounty hunter" could pursue the case as zealously as time and resources permitted. Not surprisingly, citizen suits became a source of deep and bitter divisiveness between environmentalists and business leaders.[43]

Toxic Substances Control Act (TSCA)

Aside from cleaning up the air and water, the modern regulatory regime has focused on managing toxic substances and hazardous waste. The Toxic Substances Control Act (TSCA) was enacted in 1976 to regulate the manufacture, processing, distribution, use, and disposal of chemical substances in the United States. The act requires a risk assessment to be performed. First, a hazard, or chemical of concern, has to be identified. Next, a dose response assessment is necessary to determine the magnitude of exposure to the hazard. An exposure assessment reveals the extent of human contact with the hazard. A risk characterization describes the nature, magnitude, and uncertainty of the risk that a health effect will result from the dose and exposure. A full-fledged risk assessment thus combines all of these factors to provide a comprehensive overview of the causal nexus between use of the chemical and a specific outcome.[44]

Congress intended to develop adequate data on the effects of chemical substances on human health and the environment, but the authors also chose to consider economic factors. In lieu of developing specific treatment requirements, the statute allowed for the possibility of future technological innovations. The objective was to quantify risk and reduce, to the extent possible, "unreasonable" risk to the public; it was not to vilify chemicals or stymie chemical manufacturers from producing necessary and desirable products, even if risk was involved.[45]

Notable provisions in the 1976 statute included Section 4, which allowed the EPA administrator to order testing of chemicals where inadequate data existed, although the issue of selecting an appropriate testing method subsequently became controversial. Section 5, a premanufacture review provision, required chemical manufacturers to provide a 90-day notice to EPA before producing a new chemical or using an existing chemical in a "significantly new way." Afterward, it was incumbent upon the agency to approve or deny the chemical use or distribution. If EPA denied the application, the manufacturer had the right to seek an administrative review. Sections 6 and 7 provided enormous authority to the EPA administrator to determine whether a chemical presented an unreasonable risk to human health or the environment. The administrator could prohibit the manufacture of an offending substance and, in egregious situations where a substance was deemed "imminently hazardous," pursue a civil action to force compliance.[46]

Section 8 required manufacturers to report and retain information on the chemical, including adverse reactions and health and safety studies. New studies had to be reported to EPA through a mechanism known as a Section 8(e) letter. A letter had to be submitted as soon as preliminary results were available. Citizen suits were allowed under Section 20. Similar to provisions in the Clean Air Act and the Clean Water Act, Section 20 was designed to ensure that agency officials perform "non-discretionary" duties. Attorney fees were allowed.[47]

By the early twenty-first century, it was clear that TSCA was outdated and ripe for reform. As the American Chemistry Council (ACC), an industry trade association, assessed the situation, "While the law created a robust system of regulations, over time, confidence in EPA's regulation of chemicals has eroded." In the industry's view, EPA had not adequately considered new scientific data on chemicals introduced into the marketplace since TSCA was enacted. Performing credible risk assessments and relying on sound science required independent judgment all-too-lacking in the agency because EPA scientists were too beholden to scientists from

NGOs. "The lack of confidence has created pressure on individual state legislatures to create their own chemicals management laws and on retailers to pull products from their shelves, often based on the claims of activists rather than scientific conclusions," ACC complained.[48]

For activists, EPA had failed to perform its duties under TSCA, but their perspective was diametrically opposed to the industry's view. EPA was not too wedded to NGOs; it was too accommodating to the chemical industry. More importantly, a structural concern drove activists to push for TSCA reform. Rather than government scientists failing to consider credible data, the problem involved the burden of proof. As Richard Denison, senior scientist with the Environmental Defense Fund (EDF), explained in a 2009 article, "Current policy hinders government's ability to generate information and to act on such information when it indicates significant risk." Although EDF could not claim to speak for all NGOs, Denison's argument was emblematic of activist concerns over TSCA. "For the last several decades, government policy has granted tens of thousands of industrial chemicals already in commerce a strong 'presumption of innocence,'" he wrote. "In the absence of clear evidence of harm, companies have largely been free to produce and use such chemicals as they see fit." In Denison's opinion, this scenario was the major problem with the original statute. "Government—and, hence, the public—shoulders the burden of proof." Instead of requiring chemical companies to produce data proving safety, TSCA required EPA to "prove beyond all reasonable doubt that a chemical poses a risk in order to take any regulatory action to restrict its production or use." Denision recommended that TSCA reformers adopt a standard used by the European Union when it developed its chemical regulatory system known as REACH (Registration, Evaluation, Authorisation, and Restriction of Chemicals). "REACH is underpinned by the precautionary principle, which the European Commission indicates applies 'where scientific evidence is insufficient, inconclusive, or uncertain and there are indications through preliminary objective scientific evaluation that there are reasonable grounds for concern that the potentially dangerous effects on the environment, human, animal or plant health may be inconsistent with the chosen level of protection.'"[49]

The debate over the desirability, or lack thereof, of the precautionary principle went to the heart of the debate between industry groups and environmental NGOs over acceptable risk. Rather than risk calamitous consequences resulting from decision-making under conditions of uncertainty, precautionary principle advocates argued they would rather be safe than sorry. They assumed that a risk was unacceptable until proven otherwise. Therefore, they urged EPA to deny a chemical company the authority to manufacture or distribute a chemical if the company could not document the safety of the substance. In a typical case of decision-making under conditions of uncertainty, the burden falls on persons challenging the decision to demonstrate why a course of action is not safe, but the precautionary principle places the burden on the manufacturer to demonstrate why a chemical is safe.

The challenge for scientific decision-makers is that the precautionary principle presumes that uncertainty can be substantially reduced or eliminated altogether, and this assumption may be fundamentally flawed. A proponent of the principle might call for testing of a certain drug or chemical to ensure its safety, which sounds reasonable, but at what point should the drug or chemical be pronounced "safe"? The precautionary principle can serve as the basis for opponents to push for additional testing *ad infinitum*. Yes, a thousand tests suggest that the product is safe, but what about the potentially negative results from the one thousand-and-first test? We need to be safe rather than sorry—we need to wipe out uncertainty before we proceed—so researchers should perform another test followed by another and another. Demonstrating safety is tantamount to proving a negative, which is virtually impossible.[50]

Resource Conservation & Recovery Act (RCRA)

Prior to 1965, state and local governments were responsible for disposing of most types of discarded waste. Recognizing the need for a more centralized, consistent standard, Congress enacted the Solid Waste Disposal Act of 1965 to improve disposal methods. The statute provided grants to states and local governments to develop cutting-edge waste disposal techniques. Five years later, the Resource Recovery Act amended the 1965 statute to expand federal assistance to states and municipalities. Finally, in 1976, Congress created the Resource Conservation & Recovery Act (RCRA) as a management system to track hazardous waste from the cradle to the grave so the federal government could eliminate "the last remaining loophole in environmental law—unregulated land disposal of discarded materials and hazardous wastes." In 1984, Congress amended the statute to phase out land disposal of toxic chemicals and encourage industry to develop new technologies for detoxifying hazardous waste. RCRA also contained subsidiary regulatory programs governing underground storage tanks (USTs) and medical waste. Since its inception, the act has become one of the most pervasive and controversial statutes in American environmental history.[51]

RCRA defined "solid waste" as a broad category of concern. According to Section 1004(27), waste could be deemed "solid" even if it was a liquid or a gas. A series of inquiries established the appropriate characterization: Is it semi-solid, liquid, or does it contain gaseous material that is discarded or has served its intended purpose? Is the waste abandoned? Is it being recycled by being placed in the ground, incinerated for energy recovery, reclaimed, or accumulated for more than one year? Is it inherently waste-like? If the answer to any of these questions was "yes," the material met the RCRA definition of "solid waste." Domestic sewage, industrial wastewater charges, irrigation return flows, mining wastes not removed from the ground, nuclear material, household waste, fertilizer used in agriculture, and certain high-volume wastes were excluded so EPA could study them separately. These wastes were regulated elsewhere and therefore exempted from RCRA.[52]

The next RCRA trigger in the original act was whether the solid waste was characterized as "hazardous." Under RCRA, all hazardous waste is solid waste, but not all solid waste is hazardous waste. The statute designated a waste as "hazardous" because its quantity, concentration, or physical, chemical, or infectious characteristics "may cause or contribute to an increase in mortality, or an increase in serious irreversible or incapacitating reversible illness."[53]

After determining what material was regulated, RCRA identified who was regulated. A "generator" would be any person or entity that produced hazardous waste or caused a waste to be subject to regulation. A generator was required to obtain an identification number, appropriately store or package the waste, prepare a manifest, maintain records of manifests and reports, and dispose of waste at a permitted facility. The statute also defined a "small-quantity generator" and established procedures for transporting waste to a licensed treatment, storage, and disposal (TSD) facility. The 1984 RCRA amendments specified that because landfills are designed to handle solid waste, hazardous waste must be treated before it can be disposed of on land.[54]

RCRA contained a provision requiring TSD facilities to obtain a permit prior to operating. The permit application was required to include a factual description of site plans, a description of the process to be used at a particular facility, and a list of the hazardous wastes involved. In addition, the application had to include analytical information on the type of waste, plans for security and inspections, emergency procedures and contingency plans, relevant geological information, personnel training plans, closure plans, groundwater monitoring plans, and financial assurances. For its part, EPA was required to review the permit application to ensure that it contained the necessary information. In some instances, the agency could allow for public participation through a hearing process. After EPA was satisfied that the permit application was complete, the agency would issue a draft permit subject to appropriate conditions. It was incumbent upon any facility operators holding a RCRA permit to request a modification if they materially altered their operations, discovered new information about their processes, experienced problems complying with the permit terms, or found themselves subject to new regulatory requirements.[55]

One of the more far-reaching and therefore controversial statutory provisions concerned corrective action. RCRA provided EPA with authority to conduct a facility assessment if the agency reasonably believed that a potential for hazardous-waste release existed. The agency could order the owner or operator of a facility or site to undertake monitoring, testing, analysis, and reporting, as necessary, to satisfy EPA personnel that public health and safety were not endangered. Section 7002 of the 1976 statute allowed for any person to commence a civil action alleging a violation of any requirement or prohibition under RCRA. If the EPA administrator failed to perform a non-discretionary function, a citizen suit could compel the administrator to act in accordance with the statutory provisions. Citizen-plaintiffs were required to provide 60-days' notice of the intent to litigate or 90-days' notice for an "imminent hazard" action. If the citizen-plaintiff succeeded in the lawsuit, he or she could recover litigation costs, reasonable attorney fees, expert witness fees, and civil penalties of up to $25,000 a day as well as an injunction or permit revocation. A citizen suit could not be pursued to enjoin the siting or licensure of a new TSD facility nor could it be used to sue EPA if the agency was diligently pursuing an enforcement action. Moreover, citizen suits were barred if a state agency was diligently engaged in an enforcement action. Violations could not be wholly in the past.[56]

As one might expect, critics on all sides of the debate assailed the statute. Regulated industries protested the onerous record-keeping requirements as well as intrusive EPA inspections, to say nothing of the implications for federalism. With EPA assuming increased authority for regulating wastes traditionally handled by state and local governments, politically conservative critics bemoaned the growth of federal authority. Environmentalists, in the meantime, charged that the statute, while it improved the safe handling and containment of wastes when they were generated, did not address problems associated with inactive or abandoned dumpsites or chemical spills that required immediate attention. Contaminated sites at abandoned properties or in places where the previous owner was insolvent or no longer on the scene presented a multitude of problems that had not been addressed.[57]

Comprehensive Environmental Response, Compensation, and Liability Act (CERCLA)

The Comprehensive Environmental Response, Compensation, and Liability Act (CERCLA), first enacted in 1980, rectified problems that RCRA did not address. While the latter statute was designed to handle and manage hazardous and solid waste prospectively, CERCLA, colloquially known as "Superfund," was drafted to regulate spills and releases of hazardous substances retrospectively. That is, RCRA sought to manage waste so a release would never occur while CERCLA dealt with spills and releases that had already occurred.[58]

The infamous Love Canal fiasco was the impetus, at least in part, for enacting Superfund. From 1942 until 1953, Hooker Chemicals, a chemical and plastics manufacturing facility, buried 21,900 tons of industrial wastes on a 16-acre site in Niagara Falls, New York. In accordance with industry custom at the time, Hooker covered the waste with a thin layer of clay. In 1953, the company sold the land to the Niagara Falls Board of Education for $1. The deed explicitly revealed the existence of the waste and contained a liability limitation clause protecting Hooker from subsequent lawsuits. By 1970, the land had been subdivided for residential use and sported both a housing development as well as neighborhood schools. Unfortunately, heavy rainstorms had washed away the clay barrier, exposing toxic chemicals that seeped into household basements. Throughout the 1970s, residents complained about foul odors. A local newspaper, the *Niagara Falls*

Gazette, investigated the situation for two years beginning in 1976. *Gazette* staff writer Michael Brown first reported on potential health problems beginning in July 1978. A subsequent report indicated that Love Canal homeowners had "suffered higher than normal rates of miscarriages, birth defects, epilepsy, and liver ailments." The New York Commissioner of Health declared a public emergency. In 1979, he ordered all families with children less than two years of age as well as pregnant women to relocate. As the scandal attracted national attention, "Love Canal" became synonymous with environmental degradation and the need for a comprehensive federal regulatory program to ensure that citizens would be kept safe from the debilitating effects of improperly discarded toxic chemicals.[59]

Cognizant of RCRA deficiencies as well as the Love Canal saga, members of Congress enacted CERCLA to fill the void in federal environmental law. The statute called for the investigation, identification, and cleanup of abandoned sites by establishing a fund to finance federal remedial action and a mechanism for imposing liability on "potentially responsible parties" (PRPs). CERCLA empowered the federal government to clean up a contaminated site using federal monies (the "Superfund") and later sue PRPs for reimbursement to replenish the fund as well as provide for permanent remediation. Alternatively, the agency could issue administrative orders requiring the PRPs to clean up the sites with federal oversight.[60]

In many ways, CERCLA was a unique environmental statute. While the CAA, CWA, and RCRA regulate people, CERCLA regulates places. It was designed as a comprehensive, multimedia statute covering land, water, and air. No private rights of action were built into the statute. Unlike some environmental laws, which rely on the common-law doctrine of negligence, CERCLA imposes liability without requiring proof of negligence.[61]

The original statute contained four fundamental program elements. The first element involved information gathering, establishing EPA as the central authority for reporting a release of hazardous substances. Under Section 103 and Section 104, any person in charge of a "facility" who knows of the "release" of a reportable quantity of a hazardous substance must notify EPA's National Response Center, a federally administered information clearinghouse, immediately. Failure to make the required notification could result in a fine of up to $10,000 or a sentence of up to a year in prison. The National Response Center must notify the governor of an affected state as well as any appropriate government agencies about the release. EPA compiles a list of problem sites throughout the country and can enter the property, conduct inspections, and take samples. The agency possesses eminent domain authority.[62]

In addition, Section 104 and Section 105 authorize federal authorities to clean up sites where hazardous substances have been released. As part of this authority, the response can take two forms: short-term removal or emergency response activities as well as long-term remedial activities. The statute originally allowed only $2 million of federal funds to be expended on short-term removal or emergency response activity prior to listing the site on the Superfund National Priorities

List (NPL), a ranking of sites that require additional cleanup activities. All federal activities are governed by the National Contingency Plan (NCP), a set of regulations that specify the steps in cleaning up contaminated sites. As part of the NCP, EPA created the National Hazardous Substance Response Plan requiring that cost-effective measures be pursued and a community-relations plan be established if remedial action is necessary. All remedial actions must include a site investigation and analysis of remedial alternatives, compliance with NCP regulations on developing, screening, analyzing, and selecting cleanup methods, and a consideration of cost at every step in the process.[63]

Although the acronyms can be confusing, CERCLA's NCP process is relatively straightforward. The initial step is to perform a preliminary assessment to determine environmental and health risks and priorities among waste sites. Afterward, EPA quantifies the potential risks of each site to determine a numerical score on the Hazard Ranking System (HRS), which evaluates toxicity, quantity, and waste concentrations. The HRS allows the agency to establish a single risk value reflecting exposure potential and degree of danger. Based on the ranking, EPA will assign personnel and resources to oversee cleanup activities. The NPL designation can be thought of as a queue. After the queue is established, EPA officials conduct a remedial investigation (RI) to ascertain the contamination level of the particular site. A feasibility study (FS) evaluates a menu of remediation options. EPA issues a record of decision (ROD) to solicit public comment prior to commencing long-term cleanup activities. Legal challenges to the agency's cleanup remedies are confined to the administrative record established in the ROD. An EPA-maintained database, CERCLIS (Comprehensive Environmental Response, Compensation and Liability Information System), allows the agency to store information on the status of cleanup efforts, milestones, and the amount of liquid and solid media treated at NPL sites.[64]

Throughout the process, EPA shares responsibility with the federal Agency for Toxic Substances and Disease Registry (ATSDR). ATSDR establishes and maintains a national registry of serious diseases and illnesses of persons exposed to toxic substances and studies the health effects of exposure. Part of the agency's responsibility is to create a list of areas restricted or closed to the public owing to toxic-substance contamination.[65]

A crucial feature of the original law was the creation of a hazardous substance response trust fund, the so-called "Superfund," developed to provide monies for cleaning up orphan sites if the original owner or operator could not be found or held accountable in an expeditious manner. Sections 111 and 112 of the 1980 version of CERCLA specified that the trust fund would be established by sales taxes on chemical companies, although those provisions were subsequently amended. The statute did not anticipate that the trust funds would finance all site cleanup activities. Indeed, the cost of cleaning up contamination at most locations far exceeds the amount of money available from the trust fund. Instead, CERCLA anticipated that Superfund would provide initial financing until PRPs could be identified and held accountable.[66]

The final program element was a mechanism for imposing liability on parties directly or indirectly responsible for a hazardous-substances release. One of the difficulties with traditional common law standards in environmental cases was the onerous burden of proof placed on a plaintiff. In many situations involving environmental contamination, the party directly responsible for the release is missing, insolvent, or unknown. Locating the responsible party and proving negligence or intent was virtually impossible, especially when decades had elapsed between the time the contamination occurred and the time it was discovered. To overcome this evidentiary impediment, CERCLA established joint-and-several liability for restitution of response costs incurred by the government or a private party because of the potential or actual release of a hazardous substance. This provision, found in Section 107, was among the most controversial sections of the original statute.[67]

In 1986, Congress amended CERCLA to revise statutory standards, strengthen the settlement and enforcement provisions, and specify a larger revenue base for financing and replenishing the trust fund. The Superfund Amendments and Reauthorization Act of 1986 (SARA) set forth a stringent standard, requiring that remediation must be "protective of human health and the environment" as well as cost-effective and afford "permanent solutions and alternative treatment technologies to the maximum extent practicable." This provision sounded innocuous, but it empowered EPA to search for innovative technologies even if the technologies had not been implemented in previous situations. Such technology-forcing provisions placed the burden on private companies to seek out improved methods for site remediation or risk additional agency intervention.[68]

One defect in the original statute was the relatively small funding mechanism available through the Superfund. SARA established a broader source of revenues by levying an "environmental" tax on corporations. Petroleum companies would be required to provide funding along with chemical manufacturers that produced feedstocks or imported chemical derivatives.[69]

Corporations had complained bitterly that the original CERCLA statute encouraged costly and time-consuming litigation that did not necessarily result in improved environmental quality. To address this concern, SARA set forth extensive provisions allowing for contribution rights among PRPs and established settlement rules for *de minimis* parties. The amendments also mixed EPA and private PRP funding to improve financial flexibility. After SARA was enacted, a PRP had an incentive to join other PRPs to establish a working group dedicated to allocating costs and responsibilities lest a single recalcitrant party bear the full financial burden.[70]

Other Federal Environmental Statutes

Aside from NEPA, CAA, CWA, TSCA, RCRA, and CERCLA—all of which are comprehensive federal statutes—Congress has enacted many smaller, "niche" laws to address specific environmental concerns not covered elsewhere. The Federal Insecticide, Fungicide and Rodenticide Act (FIFRA), as an example, set forth legal requirements to register, classify, label, and distribute pesticides. Although pesticides could have been regulated under TSCA, RCRA, or CWA—and some aquatic pesticides were regulated under the latter statute—Congress chose to create a separate law specifically addressing such products. Under most environmental statutes, the objective is to protect animals, plants, land, air, or water from contamination. FIFRA was an unusual law because it was designed to destroy or control the growth of nonhuman animals and plants. By definition, a pesticide is a type of poison aimed at reducing or eliminating a target organism. A pesticide is engineered to eradicate a life form deemed a "pest." FIFRA, therefore, was intended to control the spillover effect; that is, the statute ensured that target organisms were affected and the damage to nontarget organisms was minimized or eliminated.[71]

Numerous uncertainties exist in pesticide regulation. The precautionary principle dictates that a chemical must be proven safe before it can be used; however, some chemicals, especially pesticides, contain incomplete data and scientific uncertainty. Pesticide manufacturers have long contended that indeterminacy is an unavoidable factor in pesticide production and use, especially at low levels where data on dose response are inconclusive. Because the chemical is specifically designed to eliminate a target organism, demonstrating the effects of the nonexistence of a target population is tantamount to proving a negative. Uncertainty always exists in instances where the lack of evidence is the primary starting point for assessing the efficacy of the law.[72]

If FIFRA was designed to govern how target organisms are reduced or eliminated, the Endangered Species Act (ESA) of 1973 was drafted to protect organisms deemed too valuable to lose through extinction. The purpose of the ESA was to preserve critical habitats on which designated species depend for survival. Stretching back to the Lacey Act of 1900, the ESA enjoyed an illustrious pedigree, although earlier statutes proceeded on a species-by-species basis. By the 1970s, the heyday of American environmental law, federal officials had come to realize that piecemeal federal regulation often addressed the problem after it had become a full-blown crisis. Consequently, the ESA focused on protecting species that might become extinct in the future.[73]

The law required the Secretary of the Interior or the Secretary of Commerce to develop a list of endangered or threatened species based on the "best scientific and commercial data available." Citizens could petition to modify the list as well. Afterward, the U.S. Fish and Wildlife Service (FWS) and the National Marine Fisheries Service (NMFS) were assigned responsibility for administering the act and protecting the species designated on the list. The ESA made it unlawful to import, export, possess, sell, deliver, transport, or ship an endangered species. Civil and criminal penalties could be assessed for violations. Citizen suits could be filed against any person allegedly violating the act or against the Secretary of the Interior for failing to perform a nondiscretionary duty.[74]

The business community was concerned that the statute granted too much authority to the federal government and could hinder economic growth. Their worst fears were realized in the snail darter case. Beginning in 1967, the Tennessee Valley Authority (TVA) broke ground on a new project, the Tellico Dam on the Little Tennessee River. Incensed environmentalists searched for a means to halt the project. They initially filed suit to require TVA to prepare an EIS as required by NEPA. Shortly after the agency prepared the EIS, a University of Tennessee ichthyologist, Dr. David A. Etnier, discovered a previously unknown species of perch, the snail darter, or *Percina (Imostoma) tanasi*. The small fish had been added to the Endangered Species List in 1975. Environmentalists noted that the snail darter only existed in the Little Tennessee River, and therefore the Tellico Dam might drive the snail darter to extinction. Relying on the ESA, opponents filed another lawsuit, this time to enjoin completion of the dam. When the case was appealed to the U.S. Supreme Court, the justices held by a 6-3 vote that the ESA prohibited completion of the construction owing to the detrimental impact to the snail darter—even though TVA had spent $100 million on the project.[75]

Conservative members of Congress as well as business leaders were aghast that a multimillion-dollar public-works project would be abandoned because it might damage a species of protected fish. Beginning in the spring of 1977, legislators searched for a mechanism to introduce "common sense" flexibility into the ESA. The following year, Congress amended the statute to create the Endangered Species Committee, a small group comprised of seven executive-branch decision-makers: the administrator of the U.S. EPA, the administrator of the National Oceanic and Atmospheric Administration (NOAA), the chairman of the Council of Economic Advisers (CEA), the Secretary of Agriculture, the Secretary of the Army, the Secretary of the Interior, and a representative of the affected state. The committee was empowered to provide exemptions from the ESA if five of the seven members determined that no reasonable alternatives existed, the benefits of the action outweighed the benefits of alternative courses of action, the exemption was in the public interest, and neither the agency nor the exemption applicant already had made an irreversible or irretrievable commitment of resources. Even when an exemption was granted, reasonable mitigation efforts had to be undertaken to reduce the negative effects on an endangered species.[76]

Owing to the committee's authority to determine the fate of endangered species, it earned the nickname the "God Squad." The God Squad was most famously put to the test in 1991 after a federal judge halted a logging operation in a Pacific Northwest national forest. The Northern Spotted Owl, another protected species, depended on old-growth forests for survival. Logging potentially harmed the animals. The U.S. Bureau of Land Management (BLM), which sought to sell 4400 acres of timber land on 44 tracts to timber companies, filed for an ESA exemption in due course. The God Squad convened and, in a 5-2 vote, approved exemptions in 13 cases where no reasonable alternatives existed. Eventually, the Clinton administration reached an accommodation with environmentalists where a Forest

Ecosystem Management Assessment Team would create a program to protect the spotted owl while also allowing some timber cutting in national forests.[77]

Aside from protecting endangered species and their habitats, wetlands regulation is another area where the federal government has taken a lead in managing and protecting critical landmasses. Although no single federal wetlands management statute exists, several laws have provided the requisite federal authority. In 1899, Congress passed the Rivers and Harbors Act empowering the Army Corps of Engineers to regulate construction activities involving dredging, filling, or obstructing navigable waters. In later years, NEPA expanded the Corps' authority to determine whether a federal agency could be permitted to undertake an action that might negatively affect wetlands. Section 404 of the CWA did not specifically mention the term "wetlands"; however, it was interpreted as the key provision for protecting wetlands from "creeping" encroachment during the late twentieth and early twenty-first centuries. The section detailed the procedure for applicants to obtain permits for discharging dredged or fill material into navigable waters, an action that generally results in a loss of wetlands. As with so many federal environmental laws, the CWA wetlands provisions proved to be controversial. Opponents complained that the law was overly burdensome and expensive while environmentalists argued that the law did not provide an adequate level of protection and needed to be modified accordingly.[78]

Management Systems, Standards Organizations, and Certifications

In addition to environmental statutes, regulations, and cases that promulgate compulsory legal requirements, voluntary management systems and standards organizations have played an important role in encouraging effective sustainability practices during the twenty-first century. The appeal of voluntary standards, especially when coupled with a certification, lies partially in the cover it provides to embattled corporations seeking to beef up their "green" bona fides. A voluntary requirement avoids the costly, time-consuming, sometimes contentious chore of complying with laws and regulations imposed with a threat of sanctions. In addition, a voluntary management standard that comes with a certification may prove beneficial in the marketplace. A company touting the environmental attributes of its product line can proudly point to a certification as proof of corporate good faith. Even if the certification is not designed as a "Good Housekeeping seal of approval," it is often interpreted that way by a public not attuned to the quality or legitimacy of various organizations and standards. Corporate officers find it difficult to promote the public-relations value of complying with laws and regulations, but they realize the value of enumerating the various "green" certifications of their products. For their part, lawmakers, regulators, and environmentalists recognize the utility of partnering with companies in lieu of always employing adversarial processes. Convincing a potential adversary to settle an issue generally is more effective than forcing compliance, assuming the result of environmental protection is more desirable than engaging in a high-profile public confrontation. A "green-certified" business, service, or product—if top management undertakes good-faith actions to avoid greenwashing—will adhere to measurable criteria theoretically indicating wise environmental stewardship.[79]

International Organization for Standardization (ISO)

The International Organization for Standardization (ISO) is perhaps the best-known private, voluntary standards association in the world. Created in 1947, ISO exists as an international federation comprised of representatives from 157 countries with a mission to develop common standards for private businesses. Organization leaders chose "ISO" as the acronym because the Greek word "iso" means "equal," thereby evoking the notion of equivalent standards. Unlike intergovernmental organizations such as the United Nations or NGOs such as the environmental groups discussed in Chapter 8, ISO members hail from private industry. The American National Standards Institute (ANSI), a private entity, represents the United States at ISO meetings.

Headquartered in Geneva, Switzerland, the organization schedules periodic meetings of its technical committees, subcommittees, and working groups. Although ISO standards are not binding under international treaties, they are afforded great weight. Each proposed standard undergoes a rigorous six-stage review-and-adoption process that provides interested parties with numerous opportunities for comment and review.[80]

ISO is renowned for its quality-management standard for manufactured products and services, the ISO 9000 standard, which provides guidelines for businesses to receive certification. An ISO 9000 certification does not guarantee that a particular company manufactures a superior product or provides quality service; instead, the organization's "seal of approval" recognizes that a company has formalized an effective management process that maximizes quality, to the extent possible. The ISO 9000 series served as a model for the ISO Technical Committee (TC) 207 to develop environmental management standards beginning in the mid-1990s.[81]

ISO 14000 is a term used to describe environmental management standards for labeling, auditing, and environmental assessments (EAs). ISO 14001 is the first element in the 14000 series, and it sets forth requirements for creating an environmental management system. Unlike environmental laws and regulations promulgated by authoritative governmental bodies, ISO 14001 is not a substantive standard. It is a voluntary procedural goal designed to improve a corporation's management processes. ISO 14001 focuses on inputs and throughputs in the expectation that an effective process will lead to improved outputs. Because the standard specifically calls attention to a company's environmental practices, it elevates the environment as a matter of importance in corporate decision-making. Adhering to ISO 14000 standards does not ensure that a company will comply with applicable environmental laws and regulations, but it indicates that corporate managers are committed to factoring environmental issues into their decision-making processes. In addition, as ISO standards have become well known in the marketplace, a company that gains ISO 14000 certification may enjoy an ancillary marketing advantage over competitors, although ISO standards are not designed as marketing tools.[82]

Critics charge that ISO standards do not necessarily result in substantive results. As is the case with any procedural requirement, a standard that focuses on process inadvertently can lead to "hoop jumping." Companies may focus on meeting the bare requirements necessary to become ISO certified without changing the basic corporate philosophy. A company may be so focused on fulfilling the letter of the standard that it ignores the spirit of the requirement. Because the ISO 14000 series contains numerous procedures that can be expensive and time-consuming to implement, some companies may choose to forgo a certification owing to a lack of resources or the inability to shut down operations for the time needed to inspect the facility. Other companies may be reluctant to allow third parties to examine the corporation's internal operating systems or proprietary equipment—generally the first step involved when an external certification body agrees to audit a prospective ISO participant. Even when a company agrees to provide access to third-party certifiers, the ISO certification process is not transparent. Environmentalists and members of the public who question the reasons for certifying a particular company or industry have no direct recourse for challenging a certification. U.S. environmental laws and regulations provide for citizen lawsuits as well as participation in the regulatory process, but ISO certification need not provide for public participation and therefore remains opaque.[83]

American Society for Testing and Materials International (ASTM)

The American Society for Testing and Materials International (ASTM), known simply as ASTM until 2001, is another well-publicized standards organization, although it does not directly focus on environmental management systems. Founded in 1898 as the American Section of the International Association for Testing and Materials, the venerable organization is headquartered in West Conshohocken, Pennsylvania, five miles northwest of Philadelphia. ASTM develops voluntary, consensus technical standards for a wide array of systems, materials, products, and services—more than 12,000 total. Each year, ASTM publishes the *Annual Book of ASTM Standards* to provide guidance on the latest global standards. The organization is supported by literally thousands of volunteer technical committees that draw expertise from members scattered around the world.[84] Similar national standards organizations include the American National Standards Institute (ANSI), a non-profit organization that sets voluntary standards for many products and services and represents the United States at ISO meetings;[85] Association Française de Normalisation (AFNOR), the French analogue, which sets national standards for France and represents the country at ISO sessions;[86] British Standards, produced by BSI Group, formally known as the National Standards Body;[87] and Deutsches Institut für Normung e.V. (known by the acronym DIN, or, in English, German Institute for Standardization), Germany's national standardization organization and ISO-member entity.[88]

Leadership in Energy and Environmental Design (LEED) Certification

Unlike international and national organizations that promote a wide range of technical standards, LEED (Leadership in Energy and Environmental Design) is a voluntary environmental management system specifically designed for buildings. Operated by a non-profit organization, the U.S. Green Building Council (USGBC), LEED certification requires building architects, engineers, and maintenance professionals to incorporate "environmentally friendly" features into designing, constructing, operating, and maintaining modern office buildings throughout the life cycle of the building. The process offers 110 points in categories that include sustainable sites, water efficiency, energy and atmosphere, materials and resources, and indoor environmental quality. Rating systems exist for new construction, existing buildings, homes, schools, and neighborhoods.

The process, as one might expect, has been controversial. LEED supporters argue that incorporating "green" features into buildings represents a positive step in American environmentalism. Since energy inefficiency frequently occurs in the construction industry as well as the operation of commercial buildings, multifamily dwellings, and single-family residences, developing a points system to encourage "green" attributes for all stages of a building will result in demonstrable environmental benefits. Moreover, because a significant amount of material sent to U.S. landfills consists of construction debris, the benefits of greening all aspects of building design and construction will extend far beyond the life of the structure itself.

Critics argue that the USGBC is simply another NGO operating on the fringes of science and political correctness. Because the organization charges an application fee to apply for certification, tuition for courses allowing attendees to become LEED-certified, and certification expenses, the USGBC is little more than a self-certifying, nontransparent, profit-making firm hiding behind the thin guise of a nonprofit, "do-gooder" organization. Even among parties that agree in principle with the concept of a "green"-building certification, the USGBC rating system has come under attack. Points are added or deducted for activities that may or may not improve the environmental quality of the structure. Plastics industry representatives, for example, have taken issue with the organization's decision to subtract points for the use of certain plastic products. Because plastics generally are lighter and less expensive than alternative materials, plastics industry supporters contend that the USGBC decision to penalize specific polymers was driven by politics, not sound science and defensible engineering principles. Other critics suggest that LEED points in categories fail to weigh important factors such as indoor air quality and aesthetics.[89]

A Sampling of "Green"-Certifying Organizations

Differing opinions on the efficacy of the USGBC initiative highlight a major difficulty in modern American environmentalism. Since the late twentieth century, a dizzying array of organizations has emerged to promote "greenness" certifications. The average consumer quickly can become confused and skeptical of these organizations even if, in many instances, the organizations perform a much-needed and legitimate service. As an example, consider the following organizations (listed and discussed in alphabetical order) and their claims about certifying "greenness." This discussion is not meant to denigrate or promote the work of these groups; rather, it is to underscore the difficult task for the novice who seeks to separate information that illuminates from information that obfuscates. Differentiating between white knights and charlatans is no easy task.

Consider first the Biodegradable Products Institute (BPI), a nonprofit organization that educates manufacturers, legislators, and the public about scientific standards for biodegradable and compostable products. According to the Institute's website in 2012,

> The BPI certifies compostability test results from BPI-approved testing laboratories, then licenses your company to use our trademarks (the Compostable Logo) on your products and marketing materials. To be BPI certified, you will need to first obtain the appropriate ASTM test (D6400 or D6868), plus other tests to assure compliance with municipal composting standards.[90]

Cedar Grove Composting, Inc., according to the website SustainablePlastics.org,

> is a family-owned waste management firm that has served the Puget Sound area of Washington and Oregon since 1938. In 1989, Cedar Grove began large-scale composting operations, and started including compostable products in 2003. Before they can be accepted for composting, biobased products must be submitted to Cedar Grove for testing at its site. To qualify for on-site testing, products must already be BPI certified, meet ASTM D6868 or meet ASTM D6400. Testing trials start during the first week of each month.[91]

Cedar Grove executives wrote in 2012 that the organization "plays an important role in diverting organic waste from landfills. From two local facilities, we compost over 350,000 tons of residential and commercial yard and food waste annually." The group has long promoted itself as a leader in the composting field, hosting more than 5,000 visitors from 17 countries throughout the years—more than any other composting facility in North America.[92]

ENERGY STAR is a joint program sponsored by U.S. EPA and the U.S. Department of Energy (DOE) to assist individuals and businesses in making decisions about energy efficient homes and products. The program began in 1992 when EPA developed a voluntary labeling program promoting energy efficiency, which presumably leads to a reduction in GHGs. Three years later, the agency expanded the product categories to include residential heating and cooling equipment as well as many office products previously excluded. In 1996, EPA partnered with DOE. Twenty years after its inception, the ENERGY STAR label was affixed to numerous appliances, office equipment, home electronics, and other common products. According to the organization's website,

> ENERGY STAR delivers the technical information and tools that organizations and consumers need to choose energy-efficient solutions and best management practices. ENERGY STAR has successfully delivered energy and cost savings across the country, saving businesses, organizations, and consumers about $18 billion in 2010 alone. Over the past decade, ENERGY STAR has been a driving force behind the more widespread use of such technological innovations as efficient fluorescent lighting, power management systems for office equipment, and low standby energy use.[93]

The Forest Stewardship Council (FSC), according to its website, has a mission

> to promote environmentally sound, socially beneficial and economically prosperous management of the world's forests. Our vision is that we can meet our current needs for forest products without compromising the health of the world's forests for future generations. To achieve our mission and vision, FSC has developed a set of 10 principles and 57 criteria that apply to FSC-certified forests around the world.

A group of environmentalists created the organization following the 1992 Earth Summit in Rio de Janeiro, Brazil. Concerned that the summit failed to produce an agreement on methods and processes for reducing or eliminating deforestation, these concerned foresters decided that the FSC could serve as an effective mechanism for calling attention to the plight of the world's forests. They met in Toronto, Canada, in 1993 and established a voluntary, market-based approach to improve forestry practices around the globe. The FSC initially established its headquarters in Oaxaca, Mexico, but later moved the FSC Secretariat to Bonn,

Germany. In 1995, the FSC created a U.S. chapter, which is headquartered in Minneapolis, Minnesota.[94]

The FSC certification process requires a forest owner to request that an independent certifier inspect the forest to ensure it meets the organization's criteria. Only an FSC-accredited certification entity can perform the inspection. Accreditation must be renewed each year and an annual fee must be remitted to the FSC. The organization requires certificate holders to adhere to a number of guiding principles, including compliance with all applicable laws and international treaties, recognition and respect for the rights of indigenous peoples, equitable use and sharing of benefits derived from the forest, and a reduction of environmental impact from logging activities.[95]

As is the case with many certifying organizations, the FSC has attracted its share of supporters as well as detractors. The former contend that providing a mechanism for judging the quality and health of forestry resources is a valuable tool in the global environmental struggle to promote sustainable practices. Critics worry whether the FSC has inadvertently become involved in "greenwashing." According to one commentator, "At a particular level, the FSC faces clear issues regarding its legitimacy, and hence its future as a global governance organization." Other commentators are more blunt. "Debates remain whether certification has improved sustainable forest management, environmental, economic, or social practices; whether its benefits are worth the cost required; whether there are significant differences among certification systems; and, indeed, whether it leads to green practices or if it is just 'greenwash.'"[96]

Reminiscent of the USGBC, Green Advantage Environmental Certification is an organization devoted to ensuring "that buildings throughout the world are constructed in an environmentally sensitive manner that supports social and economic sustainability." To that end, the organization seeks "to be an exemplary certifier of building-related practitioners, nationally and internationally." To receive certification, a candidate must pass a Green Advantage examination demonstrating proficiency in "green" construction concepts, materials, and practices. Successful candidates receive a three-year, renewable certification attesting to their "ability to meet or exceed environmental goals." A Green Advantage environmental certification "offers individuals a meaningful way to differentiate themselves within the construction industry" because a Green Advantage credential is a valuable addition "to your green tool box."[97]

Founded in 1982, Green America is a nonprofit organization that features the credo "growing the green economy for people and the planet." Before changing its name on January 1, 2009, Green America was known as "Co-Op America." According to the group's website, "Our mission is to harness economic power—the strength of consumers, investors, businesses, and the marketplace—to create a socially just and environmentally sustainable society." Green America's vision can be stated simply: "We work for a world where all people have enough, where all communities are healthy and safe, and where the bounty of the Earth is preserved for all the generations to come." The group focuses on economic strategies by mobilizing individual consumers, investors, workers, and business leaders. "We work on issues of social justice *and* environmental responsibility," the group explains. "We see these issues as completely linked in the quest for a sustainable world. It's what we mean when we say 'green.'"[98]

Green America accomplishes its mission through a certification program. The certification ensures that businesses are

> (1) Actively using their business as a tool for positive social change; (2) operating a "values-driven" enterprise according to principles of social justice and environmental sustainability; (3) environmentally responsible in the way they source, manufacture, and market their products and run their operations and facilities; (4) socially equitable and committed to extraordinary practices that benefit workers, customers, communities, and the environment; and (5) accountable for their work by continually improving and tracking their progress, and operating with radical transparency in every facet of their business.

Businesses earn a bronze certification "as they start their path to sustainability and are implementing sustainable practices in the workplace." The next step is a silver certification denoting businesses "who are deepening their commitment to both social and environmental responsibility by spearheading changes in their core business practices, along their supply chain, and in the community." The gold certification is reserved for "green businesses showing leadership in their industry" and companies "that embed social responsibility into the DNA of their company." The Green America standards committee, appointed by and reporting to Green America's board of directors, reviews and evaluates data provided by a prospective candidate to "certify that a business has moved beyond general sustainability actions into a deeper focus on their core business area."[99]

For its part, the Green Clean Institute certifies "green" claims in the janitorial industry. Cleaning-service companies can apply for certification based on the size of a particular firm. A company with 10 or fewer full-time employees can seek a "Small Firms" certification. A company with 50 or fewer full-time employees can apply for a "Silver Firms" certification. A company with 50 or more full-time employees can seek a "Gold Firms" certification.[100]

In addition to certifying companies, the Green Clean Institute certifies products. In 2012, the company's website made it clear that "our review is not a product certification program like Green Seal. Our purpose is to prevent improper Green claims and obvious [sic] misleading statements for your product or service. This is a 'Truth in Labeling' program in support of FTC demands. Therefore, our task is a thorough review of what the public impact of your product might be from a consumer viewpoint." The website indicated that the Institute would not tolerate false or misleading claims. "Greenwashing has been described as over-exaggeration or false claims as to the Green or environmental value of a product or service. To prevent this, an independent third party should review the obvious elements of the product and the claims being made. In other words, we look at the veracity of your product or service. We are not a testing lab. If that is needed, there are firms that will do that. Green Clean Institute relies upon existing information, and we validate the legitimacy of the claims being made." Based upon this existing information, and "if your product or service is legitimately Green," it might be "helpful to have a third-party review of your representation."

The fee for this validation service in 2012 was $1,500. "We only review the product claims for fairness. When the review is done, a full report of our finding will be issued that should dispel Greenwashing claims and instill consumer confidence." In other words, the organization does not substantiate claims based on an independent scientific assessment. The Green Clean Institute earns its fee by serving as a shadow U.S. Federal Trade Commission to ensure that no "improper Green claims and obvious [sic] misleading statements for your product or service are made." The Institute achieves the objective by judging "the veracity of your product or service," validating "the legitimacy of the claims being made," and reviewing "the product claims for fairness," thereby ensuring that the product or service can be proclaimed "legitimately Green." The ultimate value of the Green Clean Institute certification presumably derives from its ability to instill consumer confidence.[101]

Green Globe, according to its website, "is the global travel and tourism industries' certification program for sustainable tourism. Green Globe members save energy and water resources, reduce operational costs, positively contribute to local communities and their environment and meet the high expectations of green leisure and business travelers." To assess members' efforts and successes, Green Globe has developed 337 compliance indicators to apply to 41 individual sustainability criteria. The indicators were designed to be flexible and varied by type of certification as well as geographical area. A business applying for Green Globe certification must demonstrate that it has implemented a sustainability management system. The business must also document compliance with all applicable environmental laws and regulations. Employees must be trained "regarding their role in the management of environmental, sociocultural, health, and safety practices." Customer satisfaction plays an integral role in Green Globe certification. If customers are not factored into a travel agency's operations, other sustainability criteria will not matter. Criteria also exist for a company's promotional materials, communications strategy, health and safety measures, and the design and construction of an agency's building.[102]

The Green Key Eco-labeling Program and Green Key Meetings Program apply to motels, hotels, and resorts. The group's website explains, "Based on the results of a comprehensive environmental self-assessment, hoteliers are awarded a 1–5 Green Key rating and provided with guidance on how to 'unlock' opportunities to become a cutting edge, green lodging property, and reduce environmental impacts and operating costs through reduced utility consumption, employee training, and supply chain management." The program covers operational areas throughout the property, including corporate environmental management, housekeeping, food and beverage operations, conference and meeting facilities, and engineering. Green Key assesses energy conservation, water conservation, solid waste management, hazardous waste management, indoor air quality, community outreach, building infrastructure, land use, and general environmental management. The Green Key website provides information on traveling green, holding green meetings, and calculating a property's carbon footprint.[103]

The National Green Building Program was established in February 2008 to provide educational resources and advocacy tools for builders and remodelers involved with the National Association of Home Builders (NAHB) who sought to "build green." The centerpiece of the effort was a "credible green standard" resulting in a "national green home certification system." As the organization's website explained in 2012,

> While some of the impetus to build green is being dictated by policy makers, increased consumer awareness is also driving the growth in this sector, making green a self-fulfilling prophecy—the more consumers are exposed to the benefits of green homes, the more demand we will see. Accordingly, NAHB Green helps prepare the industry for the increasing interest in sustainable construction by educating and credentialing building professionals, developing measurements for green building, and advocating at the national level for credible, voluntary, and market-sensitive green building initiatives.[104]

The American National Standards Institute (ANSI) has approved the organization's ICC 700 National Green Building Standard™ for building single-family homes, low-, mid-, and high-rise multifamily buildings, residential remodeling projects, and site development projects. Four threshold levels—bronze, silver, gold, and emerald—exist. To meet the standard for even the lowest threshold, a builder or remodeler is required to

> incorporate a minimum number of features in the following areas: lot and site development; energy, water, and resource efficiency; indoor environmental quality; and home owner education. In order to attain a higher level of green certification by the NAHB Research Center, a home must accrue successively higher levels of minimum points in every category—the highest level of certification is dictated by the lowest category score level.

The website provides a handy "green scoring" tool.[105]

The Rainforest Alliance, an NGO founded in 1987, proclaimed in 2012 that its mission was to "conserve biodiversity and ensure sustainable livelihoods by transforming land-use practices, business practices and consumer behavior." One method for achieving the goal is by

> delivering sustainability auditing, verification, validation, and certification services based on the best available global standards; with the highest integrity, transparency, and quality; that generate positive economic, ecological, and social benefits to our clients; and that can achieve globally meaningful scale. RA-Cert, the auditing division of the Rainforest Alliance, conducts all certification, verification and validation activities in the areas of forestry, agriculture and carbon/climate.

As with many certifying organizations, the Alliance has both supporters and detractors. The former argue that Rainforest Alliance certifications ensure at least a minimal level of sustainability while the latter contend that the certification process is a revenue-generating tool for an NGO anxious to "partner" with corporations seeking to cash in on trendy environmentalism.[106]

In light of the large number of organizations offering to certify greenness, the running joke among persons working in the environmental field is that a small group of unemployed hacks could create their own self-proclaimed "green-certifying" non-profit organization. If they developed an aesthetically pleasing, technically sophisticated website and coined a suitably pleasant, official-sounding name, for example, the "Green Justice Institute," they could offer, for a fee, to certify corporations and schools as "green." Even without the imprimatur of a government agency or an official standards organization, the Green Justice Institute could carve out a niche in the already crowded certification/management field. As the organization became better known and its funding increased, the Institute could hire staff, file for nonprofit status as a § 501 (c) (3) organization, beef up its website, improve its marketing strategy, and tout itself as a full-fledged environmental NGO. Perhaps this hypothetical scenario sounds far-fetched, cynical, and implausible. Yet the state of modern American environmentalism is such that discerning the wheat from the chaff and the genuine expert from the pretentious poseur is all but impossible for the average consumer who wants to protect the natural environment without necessarily becoming an expert in environmental science.

Environmentalism for Dummies

The confusion about standards and certifications reveals a major challenge in American environmentalism. With the proliferation of individuals and groups claiming to be experts in protecting the natural environment, the field displays an "Old West" character where almost anything goes. Making sense of the players and the playing field is a daunting enterprise; even the most gifted prose stylist would find the task of defining and explaining environmentalism in general, and sustainability in particular, to be problematic.

A series of popular books ending with the title *For Dummies*, including *Environmental Science for Dummies* and *Green Careers for Dummies*, promises to elucidate conceptually difficult material in a down-to-earth, easy-to-understand format. If the unwritten rules and chaotic "marketplace of ideas" that characterize American environmentalism of the early twenty-first century could be summarized in a book titled *Environmentalism for Dummies*, the text would have to acknowledge the diverse, disorderly, almost dysfunctional nature of the landscape before outlining the choices available to any environmentalist seeking to act in good faith.[107]

Environmentalism for Dummies, if such a text existed, would set forth a series of steps. The first step in forging a path forward in the environmental field is for an activist to join with other like-minded activists to form a group, recognizing the power of numbers. The new group must frame the type of issue it seeks to address. Pursuing a narrow issue, such as banning a specific product or calling attention to the negative effects of a manufacturing process, has the benefit of clarifying an issue easily and establishing a feasible objective. At the same time, addressing narrow objectives limits the universe of would-be contributors and, ironically, creates a dilemma for the successful environmentalist. If the product is banned or the process is discontinued, the *raison d'être* for the group no longer exists. One of the worst things that can happen to any group is to achieve the desired objective; at that point, a new cause must be found or the goals become obsolete and the group is lost in the crowded NGO marketplace.

Pursuing a broad issue such as climate change, deforestation, or the inequity between developed and developing nations expands the universe of potential contributors and provides a large playing field for decades to come. For the dedicated environmentalist who seeks measurable success from incremental progress, however, a broad, amorphous menu of issues can be demoralizing. Operationalizing the variable is difficult when the issues are so large and unfocused, so open to competing interpretations, that it becomes virtually impossible to determine whether genuine progress has been achieved. Recognizing the limitations of the overly narrow and overly broad approaches to environmental issues, the more successful, long-lived organizations pursue a hybrid-blend of the narrow and the broad. They articulate an ambitious, amorphous mission while simultaneously seeking specific victories in smaller venues.

As a second step, the environmental group must determine how it will solicit funding. Often the scope of the issue will determine the means of finance. Direct appeals through e-mail blasts, website messages, and snail-mail marketing can be lucrative if the group is well known and if a highly visible environmental catastrophe recently has captured media attention. Nothing fills the coffers of an environmental group faster than a well-publicized oil spill or a disastrous fire at a chemical plant. For less well-known environmental groups, forging alliances with existing organizations or seeking foundation support can prove to be advantageous.

A third step is to identify the intended audience. An environmental group appealing to ordinary citizens develops a different strategy and tactics than a group that seeks to engage policymakers or members of the scientific community. An environmental NGO reaching out to activists is radically different from an interest group hoping to convince legislators to change the law.

An environmentalist's strategy will necessarily dictate appropriate tactics. An interest group that makes a good-faith effort to understand the latest scientific studies has a different agenda than a group that deliberately relies on "scare tactics" to rustle up dollars and members. Thus, a science-based interest group must be staffed by scientifically literate, presumably credentialed personnel familiar with technical data and processes. A lobbying group will employ professional, experienced lobbyists. A group devoted to litigation will be staffed by attorneys. A group dedicated to aggressive confrontation must find activists willing and able to pay the price for their outrageous tactics, whether the cost is time, money, or a willingness to be arrested for the cause.

The American environmental movement has room for all manner of environmental activist—the well-intentioned, uneducated novice who targets a "pet peeve"; the shrewd, cynical charlatan who seeks an outlet for his outrageous behavior; the dedicated "true believer" with a talent for histrionics who genuinely believes the fate of the earth lies in the balance; the earnest scientist who has read the latest studies and is alarmed that sovereign governments refuse to act on credible scientific data; the "white-knight" attorney who, having been born too late to experience the civil rights movement or the anti-Vietnam War protests of yesteryear, belatedly finds a cause; the idealistic legislator who runs for elective office so he can enact laws protecting the natural environment; the tenacious regulator who sees himself as a champion of the environment against the dark, oppressive forces of powerful business interests; the contemplative industry representative who seeks a middle approach allowing for industrial development along with environmental protection; and combinations thereof. The movement is an eclectic mix of motives, persons, and interests. The American polity can accommodate one and all, exactly as the Federalists and Tocqueville envisioned.

Yet for all of the attention and money afforded to the movement during the twenty-first century, American environmentalism is fractured and frequently self-defeating. The central question now, as ever, is how do we move forward? Where do we go from here?

Notes

1. This phrase is from a famous cartoon produced by illustrator Walt Kelly. It also became a well-known catch phrase during the twentieth-century environmental movement. See, for example, Walt Kelly, *Pogo: We Have Met the Enemy and He Is Us* (New

York: Simon & Schuster, 1987). See also Finis Dunaway, "Gas Masks, Pogo, and the Ecological Indian: Earth Day and the Visual Politics of American Environmentalism," *American Quarterly* 60, 1 (March 2008): 67–99.

2. Giorel Curran, "Whither Environmentalism? Environmental Politics in the 21st Century," *Social Alternatives* 25, 2 (Second Quarter 2006): 48–53; Philip Shabecoff, *Earth Rising: American Environmentalism in the 21st Century* (Washington, D.C. and Covelo, Calif.: Island Press, 2001), 170–194.
3. Senator Al Gore, *Earth in the Balance: Ecology and the Human Spirit* (Boston: Houghton Mifflin, 1992), 305–307, 368.
4. David M. McEvoy and John K. Stranlund, "Self-Enforcing International Environmental Agreements with Costly Monitoring for Compliance," *Environmental & Resource Economics* 42, 4 (April 2009): 491–508; Dritan Osmani and Richard S. J. Tol, "The Case of Two Self-Enforcing International Agreements with Environmental Protection with Asymmetric Countries," *Computational Economics* 36, 2 (August 2010): 93–119; Fang Rong, "Understanding Developing Country Stances on Post-2012 Climate Change Negotiations: Comparative Analysis of Brazil, China, India, Mexico, and South Africa," *Energy Policy* 38, 8 (August 2010): 4582–4591; Andrew K. Rose and Mark M. Spiegel, "Noneconomic Engagement and International Exchange: The Case of Environmental Treaties," *Journal of Money, Credit and Banking* 41, 2–3 (March–April 2009): 337–363.
5. Brian Jack, "Enforcing Member State Compliance with EU Environmental Law: A Critical Evaluation of the Use of Financial Penalties," *Journal of Environmental Law* 23, 1 (March 2011): 73–95; Roger Scruton, *How to Think Seriously about the Planet: The Case for an Environmental Conservatism* (New York and Oxford: Oxford University Press, 2012), 312–317.
6. Charles H. Eccleston and Frederic March, *Global Environmental Policy: Concepts, Principles, and Practice* (Boca Raton, Fla.: CRC Press, 2011), 264–267; Scruton, *How to Think Seriously about the Planet*, 314.
7. Daniel C. Matisoff, "Making Cap-and-Trade Work: Lessons from the European Union Experience," *Environment* 52, 1 (January/February 2010): 10–19; Scruton, *How to Think Seriously about the Planet*, 314–315.
8. See, for example, Scott Barrett, "Rethinking Climate Change Governance and Its Relationship to the World Trading System," *The World Economy* 34, 11 (November 2011): 1863–1882; Ann E. Carlson, "Designing Effective Climate Policy: Cap-and-Trade and Complementary Policies," *Harvard Journal on Legislation* 49, 1 (July 2012): 207–248; W. Bradnee Chambers and Jessica F. Green, editors, *Reforming International Environmental Governance: From Institutional Limits to Innovative Solutions* (Tokyo, New York, and Paris: United Nations University Press, 2005); Matisoff, "Making Cap-and-Trade Work," 10–19; Sevasti-Eleni Vezirgiannidou, "The Climate Change Regime Post-Kyoto: Why Compliance Is Important and How to Achieve It," *Global Environmental Politics* 9, 4 (November 2009): 41–63.
9. No Foam Chicago Website: "Breaking Styrofoam Grip," http://nofoamchicago.org/BreakingStyrofoamGrip.pdf (Accessed November 23, 2012).
10. One major argument was that polystyrene foam takes centuries to biodegrade inside a landfill. Yet researchers for decades have argued that landfills are designed to retard biodegradation; consequently, the group's major contention was not based on scientific data. No Foam Chicago next argued that foam packaging clogs landfills. Yet two major researchers, William L. Rathje and Cullen Murphy, concluded that expanded polystyrene foam accounts for less than 1 percent of the volume of landfill garbage. "It

is worth remembering that if such foam were banned, the relatively small amount of space that it takes up in landfills would not be saved. Eggs, hamburgers, coffee and stereos must still be put in something." In their view, fast-food packaging, polystyrene foam, and disposable baby diapers "have acquired high visibility because they are so noticeable among casual litter, and people think the components of everyday litter are the same as landfilled garbage. As a result, these items have become powerful symbolic targets. But if they disappeared tomorrow, landfill operators would hardly notice." William L. Rathje and Cullen Murphy, "Five Major Myths about Garbage, and Why They're Wrong," *Smithsonian* 23, 4 (July 1992): 114, 115.

11. No Foam Chicago argued that alternative products have fewer negative environmental consequences than polystyrene foam, but this "gut feeling" assumption ignored credible data on life cycle management (LCM). Even if someone were to assess the LCM data and agree with No Foam Chicago's conclusions, the important point in this context is that the environmental group chose to focus on a relatively minor issue and refused to debate the science. As a result, No Foam Chicago represented the worst attributes of American environmentalism. It emphasized minutiae using scare tactics that ignored credible science. The group was highly politicized and impervious to argumentation or reasoned debate. See, for example, Margaret Hardy and John A. Charles, Jr., *Sustainable Failure: Why Portland's Polystyrene Foam Ban Should Be Repealed* (Portland, Oreg.: Cascade Policy Institute, November 2007); Martin B. Hocking, "Is Paper Better Than Plastic?," *Consumer's Research* 74, 10 (October 1991): 28–29; Martin B. Hocking, "Relative Merits of Polystyrene Foam and Paper in Hot Drink Cups: Implications for Packaging," *Environmental Management* 15, 6 (1991): 731–747; Martin B. Hocking, "Reusable and Disposable Cups: An Energy-Based Evaluation," *Environmental Management* 18, 6 (1994): 889–899; James Michael Martinez, "Packaging Legislation and Unintended Consequences: A Case Study on the Necessity of Life Cycle Management," in *Towards Life Cycle Sustainability Management*, Matthias Finkbeiner, editor (London and New York: Springer, 2011), 371–380.
12. No Foam Chicago Website: "Why Now?," http://nofoamchicago.org/WhyNow.html (Accessed November 23, 2012). For scientific literature containing contrary scientific data, see, for example, Gregory L. Durst and Edward A. Laperle, "Styrene Monomer Migration as Monitored by Purge and Trap Gas Chromatography and Sensory Analysis for Polystyrene Containers," *Journal of Food Science* 55, 2 (March 1990): 522–524; Harvard Center for Risk Analysis (Joshua T. Cohen, Gary Carlson, Gail Charnley, David Coggin, Elizabeth Delzell, John D. Graham, Helmut Greina, Daniel Krewski, Michele Medinsky, Richard Monson, Dennis Paustenbach, Barbara Petersen, Stephen Rapport, Lorenz Rhomberg, P. Barry Ryan, and Kimberly Thompson), "A Comprehensive Evaluation of the Potential Health Risks Associated with Occupational and Environmental Exposure to Styrene," *Journal of Toxicology and Environmental Health* 5, 1–2 (January 2002): 1–263; David H. Steele, Michael J. Thornburg, John S. Stanley, Roland R. Miller, Richard Brooke, Janette R. Cushman, and George Cruzan, "Determination of Styrene in Selected Foods," *Journal of Agricultural and Food Chemistry* 42, 8 (August 1994): 1661–1665; Styrene Information & Research Center (SIRC) Website: "Home Page," http://www.styrene.org/ (Accessed November 23, 2012); Styrene Information & Research Center (SIRC) Website: "You Know Styrene," http://youknowstyrene.org/ (Accessed November 23, 2012); United States Department of Health and Human Services, National Toxicology Program, Center for the Evaluation of Risks to Human Reproduction, *NTP-CERHR Monograph on the*

Potential Human Reproductive and Developmental Effects of Styrene (Washington, D.C.: U.S. Government Printing Office, NIH Publication No. 06-4475, February 2006); S. L. Varner and Charles V. Breder, "Headspace Sampling and Gas Chromatographic Determination of Styrene Migration from Food-Contact Polystyrene Cups into Beverages and Food Simulants," *Journal of the Association of Official Analytical Chemists* 64, 5 (September 1981): 1122–1130; Tang Weici, Ingrid Hemm, and Gerhard Eisenbrand, "Estimation of Human Exposure to Styrene and Ethylbenzene," *Toxicology* 144, 1–3 (April 2000), 39–50.

13. See, for example, C. T. McElroy and P. F. Fogal, "Ozone: From Discovery to Protection," *Atmosphere-Ocean* 46, 1 (March 2008): 1–13; P.A. Newman, L. D. Oman, A. R. Douglass, E. L. Fleming, S. M. Frith, M. M. Hurwitz, S. R. Kawa, C. H. Jackman, N. A. Krotkov, E. R. Nash, J. E. Nielsen, S. Pawson, R. S. Stolarski, and G. J. M. Velders, "What Would Have Happened to the Ozone Layer if Chlorofluorocarbons (CFCs) Had Not Been Regulated?" *Atmospheric Chemistry & Physics* 9, 6 (2009): 2113–2128; A. Voulgarakis, D. T. Shindell, and G. Faluvegi, "Linkages between Ozone Depleting Substances, Tropospheric Oxidation and Aerosols," *Atmospheric Chemistry & Physics Discussions* 12, 9 (2012): 25551–25572.
14. United Nations Environment Programme, *Montreal Protocol on Substances that Deplete the Ozone Layer: Report of the UNEP Technology and Economic Assessment Panel Quarantine and Preshipment Taskforce—Final Report* (Nairobi, Kenya: October 2009).
15. See, for example, Kurt J. Beron, James C. Murdoch, and Wim P. M. Vijverberg, "Why Cooperate? Public Goods, Economic Power, and the Montreal Protocol," *The Review of Economics and Statistics* 85, 2 (May 2003): 286–297; "Editorial Comment—The Montreal Protocol," *International Journal of Remote Sensing* 29, 19 (October 10, 2008): 5455–5459; Chris Peloso, "Crafting an International Climate Change Protocol: Applying the Lessons Learned from the Success of the Montréal Protocol and the Ozone Depletion Problem," *Journal of Land Use & Environmental Law* 25, 2 (Spring 2010): 305–329.
16. Matthew J. Kiernan, "The Age of Eco-Strategy: Thoughts for the Earth Summit in Rio de Janeiro, June 1992," *The International Executive* 34, 3 (May/June 1992): 197–214; Chris C. Park, *The Environment: Principles and Applications* (New York and London: Routledge, 2001): 13–14.
17. Quoted in Miranda A. Schreurs, "20th Anniversary of the Rio Summit: Taking a Look Back and at the Road Ahead," *GAIA: Ecological Perspectives for Science & Society* 21, 1 (March 2012): 13. See also Günther Handl, *Declaration of the United Nations Conference on the Human Environment Stockholm, 16 June 1972,* at Audiovisual Library of International Law, http://untreaty.un.org/cod/avl/ha/dunche/dunche.html (Accessed December 3, 2012).
18. Quoted in Roger Martella and Kim Smacziak, "Introduction to Rio + 20: A Reflection on Progress since the First Earth Summit and the Opportunities That Lie Ahead," *Sustainable Development Law & Policy* 12, 3 (Spring 2012): 4. See also Schreurs, "20th Anniversary of the Rio Summit: Taking a Look Back and at the Road Ahead," 13–15.
19. Eccleston and March, *Global Environmental Policy*, 104; Risa Kumazawa and Michael S. Callaghan, "The Effect of the Kyoto Protocol on Carbon Dioxide Emissions," *Journal of Economics & Finance* 36, 1 (January 2012): 201–210; Schreurs, "20th Anniversary of the Rio Summit: Taking a Look Back and at the Road Ahead," 14.

20. Eccleston and March, *Global Environmental Policy*, 104; Kumazawa and Callaghan, "The Effect of the Kyoto Protocol on Carbon Dioxide Emissions," 202–202; Peloso, "Crafting an International Climate Change Protocol," 316–318.
21. Quoted in Peloso, "Crafting an International Climate Change Protocol," 320. See also Lin Feng and Jason Buhi, "The Copenhagen Accord and the Silent Incorporation of the Polluter Pays Principle in International Climate Law: An Analysis of Sino-American Diplomacy at Copenhagen and Beyond," *Buffalo Environmental Law Journal* 18, 1 (2010/2011): 12–17.
22. Damilola S. Olawuyi, "From Kyoto to Copenhagen: Rethinking the Place of Flexible Mechanisms in the Kyoto Protocol's Post 2012 Commitment Period," *LEAD Journal (Law, Environment & Development Journal)* 6, 1 (2010): 23–24; Osamu Yoshida, "Procedural Aspects of the International Legal Regime for Climate Change: Early Operation of the Kyoto Protocol's Compliance System," *Journal of East Asia & International Law* 4, 1 (Spring 2011): 41–61.
23. Andreja Cirman, Polona Domadenik, Matjaz Koman, and Tjasa Redek, "The Kyoto Protocol in a Global Perspective," *Economic and Business Review* 11, 1 (February 2009): 29–54; Kumazawa and Callaghan, "The Effect of the Kyoto Protocol on Carbon Dioxide Emissions," 201–210; Olawuyi, "From Kyoto to Copenhagen," 21–35; Peloso, "Crafting an International Climate Change Protocol," 312–315, 318–321; Yoshida, "Procedural Aspects of the International Legal Regime for Climate Change," 41–61.
24. Feng and Buhi, "The Copenhagen Accord and the Silent Incorporation of the Polluter Pays Principle in International Climate Law," 3–14; Paul J. J. Welfens, Jens K. Perret, and Deniz Erdem, "Global Economic Sustainability Indicator: Analysis and Policy Options for the Copenhagen Process," *International Economics and Economic Policy* 7, 2–3 (August 2010): 153–185.
25. Carl Death, "Summit Theatre: Exemplary Governmentability and Environmental Diplomacy in Johannesburg and Copenhagen," *Environmental Politics* 20, 1 (February 2011): 1–19; Feng and Buhi, "The Copenhagen Accord and the Silent Incorporation of the Polluter Pays Principle in International Climate Law," 3–14; Welfens, Perret, and Erdem, "Global Economic Sustainability Indicator," 153–185.
26. Emaculate Ingwani, Tendayi Gondo, and Trynos Gumbo, "The Polluter Pay [sic] Principle and the Damage Done: Controversies for Sustainable Development," *Economia: Seria Management* 13, 1 (July 2010): 53–60; Barbara Luppi, Francesco Parisi, and Shruti Rajagopalan, "The Rise and Fall of the Polluter-Pays Principle in Developing Countries," *The Economics of Efficiency and the Judicial System, International Review of Law and Economics* 32, 1 (March 2012): 135–144; Martin Stoczkiewicz, "The Polluter Pays Principle and State Aid for Environmental Protection," *Journal for European Environmental & Planning Law* 6, 2 (September 2009): 171–196; Edwin Woerdman, Alessandra Arcuri, and Stefano Clo, "Emissions Trading and the Polluter-Pays Principle: Do Polluters Pay under Grandfathering?" *Review of Law and Economics* 4, 2 (December 2008): 565–590.
27. Peter Christoff, "Cold Climate in Copenhagen: China and the United States at COP15," *Environmental Politics* 19, 4 (July 2010): 637–656; Lee Chung Lau, Keat Teong Lee, and Abdul Rahman Mohamed, "Global Warming Mitigation and Renewable Energy Policy Development from the Kyoto Protocol to the Copenhagen Accord—A Comment," *Renewable and Sustainable Energy Reviews* 16, 7 (September 2012): 5280–5284.

28. Steven Ferrey, *Environmental Law: Examples and Explanations* (New York: Aspen Law & Business, 1997), 14–24; Bruce Yandle, "The Common Law and the Environment in the Courts: Discussion of Code Law and Common Law," *Case Western Reserve Law Review* 58, 3 (Spring 2008): 598–599; *Martin v. Reynolds Metal Company*, 221 Or. 86, 342 P.2d 790 (1959); *cert. denied* 362 U.S. 918 (1960).
29. Hope M. Babcock, "The Public Trust Doctrine: What a Tall Tale They Tell," *South Carolina Law Review* 61, 2 (Winter 2009): 393–414; Michael C. Blum, "The Public Trust Doctrine and Private Property: The Accommodation Principle," *Pace Environmental Law Review* 27, 3 (Summer 2010): 649–667; Henry N. Butler, "A Defense of Common Law Environmentalism: The Discovery of Better Environmental Policy," *Case Western Reserve Law Review* 58, 3 (Spring 2008): 705–752; Robin Kundis Craig, "A Comparative Guide to the Western States' Public Trust Doctrines: Public Values, Private Rights, and the Evolution toward an Ecological Public Trust," *Ecology Law Quarterly* 37, 1 (2010): 53–197; Ferrey, *Environmental Law: Examples and Explanations*, 14–24. The Mono Lake scenario in California demonstrated the importance of the public trust doctrine to environmental law. See, for example, Sherry A. Enzler, "How Law Mattered to the Mono Lake Ecosystem," *William and Mary Environmental Law & Policy Review* 35, 2 (Winter 2011): 413–513.
30. The National Environmental Policy Act (NEPA) of 1969, 42 U.S.C. § 4321 *et seq.* (1969), especially § 102(2) (C), 42 U.S.C. § 4332 (2) (C); § 202, 42 U.S.C. § 4341. See also Susan J. Buck, *Understanding Environmental Administration and Law*, 2nd ed. (Washington, D.C. and Covelo, Calif.: Island Press, 1996), 17–18; Ferrey, *Environmental Law: Examples and Explanations*, 59–76; Olga L. Moya and Andrew L. Fono, *Federal Environmental Law: The User's Guide* (St. Paul, Minn.: West, 1997), 55–58.
31. 40 CFR § 1503.13. See also Charles H. Eccleston, "Does NEPA Suffer from the Pike Syndrome?" *Environmental Practice* 4, 1 (March 2002): 8–9; Bradley C. Karkkainen, "Toward a Smarter NEPA: Monitoring and Managing Government's Environmental Performance," *Columbia Law Review* 102, 4 (May 2002): 903–972; Sonja Klopf, Nada Wolff Culver, and Pete Morton, "A Road Map to a Better NEPA: Why Environmental Risk Assessments Should Be Used to Analyze the Environmental Consequences of Complex Federal Actions," *Sustainable Development Law & Policy* 8, 1 (Fall 2007): 38–43, 84–85; Moya and Fono, *Federal Environmental Law*, 58. For a thorough discussion of NEPA's strengths and weaknesses, see, for example, Lynton K. Caldwell, *Science and the National Environmental Policy Act: Redirecting Policy through Procedural Reform* (Tuscaloosa: University of Alabama Press, 1982).
32. Harvey Black, "Imperfect Protection: NEPA at 35 Years," *Environmental Health Perspectives* 112, 5 (April 2004): A292–A295; Roger C. Crampton and Richard K. Berg, "On Leading a Horse to Water: NEPA and the Federal Bureaucracy," *Michigan Law Review* 71, 3 (January 1973): 511–536; United States House of Representatives, Committee on Resources, One Hundred Ninth Congress, First Session, Oversight Hearing, Thursday, November 17, 2005, *NEPA: Lessons Learned and Next Steps* (Washington, D.C.: U.S. Government Printing Office, 2006).
33. Bruce A. Ackerman and Richard B. Stewart, "Reforming Environmental Law: The Democratic Case for Market Incentives," *Columbia Journal of Environmental Law* 13, 1 (1987): 171–199; Morris Altman, "When Green Isn't Mean: Economic Theory and the Heuristics of the Impact of Environmental Regulations on Competitiveness and Opportunity Cost," *Ecological Economics* 36, 1 (January 2001): 31–44; Michael A.

Berry and Dennis A. Rondinelli, "Proactive Corporate Environmental Management: A New Industrial Revolution," *Academy of Management Executive* 12, 2 (May 1998): 38–50.

34. John Bachmann, "Will the Circle Be Broken? A History of the U.S. National Ambient Air Quality Standards," *Journal of the Air & Waste Management Association* 57, 6 (June 2007): 662–663; Jonathan Davidson and Joseph M. Norbeck, *An Interpretive History of the Clean Air Act: Scientific and Policy Perspectives* (Waltham, Mass.: Elsevier, 2012), 7–10; "Evolution of Present Federal Law," *Congressional Digest* 49, 8/9 (August/September 1970): 196–197.
35. The Clean Air Act (CAA) of 1970, 42 U.S.C. § 7401 *et seq.* (1970), especially § 109, 42 U.S.C. § 7409. See also Bachmann, "Will the Circle Be Broken?" 653–654; Davidson and Norbeck, *An Interpretive History of the Clean Air Act*, 5; Ferrey, *Environmental Law: Examples and Explanations*, 141–144.
36. §§ 111-12, 42 U.S.C. §§ 7411-12. See also Davidson and Norbeck, *An Interpretive History of the Clean Air Act*, 15; Ferrey, *Environmental Law: Examples and Explanations*, 149–150.
37. § 111, 42 U.S.C. § 7411; § 162, 42 U.S.C. § 7472; § 173, 42 U.S.C. § 7503. See also Bachmann, "Will the Circle Be Broken?" 654–660; Ferrey, *Environmental Law: Examples and Explanations*, 155–157; John-Mark Stensvaag, "Preventing Significant Deterioration under the Clean Air Act: Area Classification, Initial Allocation, and Redesignation," *Environmental Law Reporter: News & Analysis* 41, 1 (January 2011): 10008–10023.
38. § 109, 42 U.S.C. § 7409; § 502, *et seq.*, 42 U.S.C. § 7661, *et seq.* See also Jed Anderson, "The Disappearing Distinction," *The Environmental Forum* 29, 6 (December 2012): 30–33; Scott E. Atkinson and T. H. Tietenberg, "Approaches for Reaching Ambient Standards in Non-Attainment Areas: Financial Burden and Efficiency Considerations," *Land Economics* 60, 2 (May 1984): 148–159; Ferrey, *Environmental Law: Examples and Explanations*, 162–171.
39. Gary C. Bryner, "Congress and Clean Air Policy," in *Business and Environmental Policy: Corporate Interests in the American Political System*, Michael E. Kraft and Sheldon Kamieniecki, editors (Cambridge, Mass.: MIT Press, 2007), 139–147; Davidson and Norbeck, *An Interpretive History of the Clean Air Act*, 57–76.
40. The Clean Water Act (CWA) (Federal Water Pollution Control Amendments of 1972), 33 U.S.C. § 1251 *et seq.* (1972). See also Claudia Copeland, *Clean Water Act: A Summary of the Law* (Washington, D.C.: Congressional Research Service, April 23, 2010): 1–2; Jeffrey M. Gaba, "Generally Illegal: NPDES General Permits under the Clean Water Act," *Harvard Environmental Law Review* 31, 2 (2007): 419–420.
41. §§ 101-02, 33 U.S.C. §§ 1251-52. See also Copeland, *Clean Water Act*, 5–7; Gaba, "Generally Illegal," 424–432.
42. § 402, 33 U.S.C. § 1342. See also Gaba, "Generally Illegal," 424–455; Brian P. Gaffney, "A Divided Duty: The EPA's Dilemma under the Endangered Species Act and Clean Water Act Concerning the National Pollutant Discharge Elimination System," *The Review of Litigation* 26, 1 (April 2007): 498–500.
43. See, for example, Daniel E. Burgoyne, "Attorneys' Fees and the Conflict between Rule 68 and the Clean Water Act's Citizen Suit Provisions," *Boston College Environmental Affairs Law Review* 33, 2 (2006): 627–660; Christine L. Rideout, "Where Are All the Citizen Suits?: The Failure of Safe Drinking Water Enforcement in the United States," *Health Matrix: Journal of Law-Medicine* 21, 2 (January 2011): 685–686.

44. The Toxic Substances Control Act (TSCA) of 1976, 15 U.S.C. § 2601 *et seq.* (1976). See also John S. Applegate, "The Perils of Unreasonable Risk: Information, Regulatory Policy, and Toxic Substances Control," *Columbia Law Review* 91, 2 (March 1991): 261–333; Blake A. Biles and Lawrence E. Culleen, "TSCA, Redux," *The Environmental Forum* 27, 1 (January/February 2010): 30; Walter A. Rosenbaum, *Environmental Politics and Policy*, 4th ed. (Washington, D.C.: Congressional Quarterly Press, 1998), 138–142; "Toxic Substances Control Act Overview: Current Laws and Policies," *Congressional Digest* 89, 8 (October 2010): 226–227.
45. Biles and Culleen, "TSCA, Redux," 30–31; Rosenbaum, *Environmental Politics and Policy*, 138–141; "Toxic Substances Control Act Overview," 226; "TSCA Reform: The Standard of Safety," *Environmental Law Reporter: News & Analysis* 41, 12 (December 2011): 11082–11083; William J. Walsh and Michelle M. Skjoldal, "Sustainability Is Driving Toxic Chemicals from Products," *Natural Resources & Environment* 25, 3 (Winter 2011): 16–17.
46. §§ 4–7, 15 U.S.C. §§ 2603–2606. See also Biles and Culleen, "TSCA, Redux," 30–32; "Toxic Substances Control Act Overview," 227–228.
47. § 8, 15 U.S.C. § 2607; § 20, 15 U.S.C. § 2619. See also Biles and Culleen, "TSCA, Redux," 31–32; "Toxic Substances Control Act Overview," 228, 256; "TSCA Reform," 11084–11085.
48. American Chemistry Council Website: "TSCA Modernization," http://www.americanchemistry.com/Policy/Chemical-Safety/TSCA (Accessed November 16, 2012). See also Biles and Culleen, "TSCA, Redux," 30–32; "EPA Announces Efforts to Strengthen Chemical Management under TSCA," *Hazardous Waste Consultant* 27, 6 (2009): 2.16–2.19; Steve Owens, "A Tool in Need of Updating," *The Environmental Forum* 27, 1 (January/February 2010): 33; "Toxic Substances Control Act Reform: Risk Management," *Environmental Law Reporter: News & Analysis* 42, 7 (July 2012): 10591–10605; "TSCA Reform," 11084–11085.
49. Richard A. Denison, "Ten Essential Elements in TSCA Reform," *Environmental Law Reporter* 39, 1 (January 2009): 10020, 10022. The literature on REACH is large and growing. See, for example, Adam D. K. Abelkop, Agnes Botos, Lois R. Wise, and John D. Graham, "Regulating Industrial Chemicals: Lessons for U.S. Lawmakers from the European Union's REACH Program," *Environmental Law Reporter* 42, 11 (November 2012): 11042–11065; Valerie J. Brown, "Why Is It So Difficult to Choose Safer Alternatives for Hazardous Chemicals?" *Environmental Health Perspectives* 120, 7 (July 2012): A282–A283; Lars Koch and Nicholas A. Ashford, "Rethinking the Role of Information in Chemicals Policy: Implications for TSCA and REACH," *Journal of Cleaner Production* 14, 1 (2009): 31–46; Stephen Van Heerden, "Recent Developments in Global Regulatory Framework in the Chemical Industry," *Chemical Business* 26, 7 (July 2012): 8–12; Michael P. Wilson and Megan R. Schwarzman, "Toward a New U.S. Chemicals Policy: Rebuilding the Foundation to Advance New Science, Green Chemistry, and Environmental Health," *Environmental Health Perspectives* 117, 8 (August 2009): 1202–1209.
50. The literature on the precautionary principle is voluminous. See, for example, Marcello Basili, Alain Chateauneuf, and Fulvio Fontini, "Precautionary Principle as a Rule of Choice with Optimism on Windfall Gains and Pessimism on Catastrophic Losses," *Ecological Economics* 67, 3 (October 2008): 485–491; Kenneth R. Foster, Paolo Vecchia, and Michael H. Rapacholi, "Risk Management: Science and the Precautionary Principle," *Science* 288, 5468 (May 12, 2000): 979–981; Sabrina Genter,

"Decision-Making and the Precautionary Principle," *Impact Assessment & Project Appraisal* 24, 2 (June 2006): 160–161; Steve Maguire and Jaye Ellis, "Redistributing the Burden of Scientific Uncertainty: Implications of the Precautionary Principle for State and Nonstate Actors," *Global Governance* 11, 4 (October-December 2005): 505–526; Julian Morris, editor, *Rethinking Risk and the Precautionary Principle* (Oxford, U.K. and Woburn, Mass.: Butterworth-Heinemann, 2000); Timothy O'Riordan and Andrew Jordan, "The Precautionary Principle in Contemporary Environmental Politics," *Environmental Values* 4, 3 (August 1995): 191–212; David Ozonoff, "On Being Careful What We Wish For: Some Difficulties with Operationalizing the Precautionary Principle," *Human & Ecological Risk Assessment* 11, 1 (February 2005): 53–62; Mariachiara Tallacchini, "Before and Beyond the Precautionary Principle: Epistemology of Uncertainty in Science and Law," *Toxicology & Applied Pharmacology* 207, Supplement 1 (September 2005): 645–651.

51. The Resource Conservation and Recovery Act (RCRA) of 1976, 42 U.S.C. § 6901 *et seq*. (1976). See also Ferrey, *Environmental Law: Examples and Explanations*, 251–252; Robin R. Jenkins, Elizabeth Kopits, and David Simpson, "The Evolution of Solid and Hazardous Waste in the United States," *Review of Environmental Economics and Policy* 3, 1 (Winter 2009): 104–120; Moya and Fono, *Federal Environmental Law*, 91–92; Wagner, "Hazardous Waste: Evolution of a National Environmental Problem," *Journal of Policy History* 16, 4 (October 2004): 308–318.
52. § 1004(27), 42 U.S.C. § 6903(27). See also Ferrey, *Environmental Law: Examples and Explanations*, 253–254; Moya and Fono, *Federal Environmental Law*, 95–98; United States Environmental Protection Agency, *RCRA, Superfund & EPCRA Call Center* (Washington, D.C.: U.S. Government Printing Office, Solid Waste and Emergency Response Publication No. EPA530-R-04-010, September 2003), 2.
53. § 1004(5), 42 U.S.C. § 6903(5) and § 3001(a)-(b), 42 U.S.C. § 6921(a)-(b). See also Moya and Fono, *Federal Environmental Law*, 98; United States Environmental Protection Agency, *RCRA, Superfund & EPCRA Call Center*, 2.
54. § 3002, 42 U.S.C. § 6922; 40 C.F.R. Part 263. See also Ferrey, *Environmental Law: Examples and Explanations*, 264–266; Moya and Fono, *Federal Environmental Law*, 112–119; United States Environmental Protection Agency, *RCRA, Superfund & EPCRA Call Center*, 3–5.
55. §§ 3004–3005, 42 U.S.C. §§ 6924–6925. See also Ferrey, *Environmental Law: Examples and Explanations*, 268–272; Moya and Fono, *Federal Environmental Law*, 125–141; "RCRA Questions and Answers Provide Regulatory Insight," *Hazardous Waste Consultant* 30, 4 (2012): 4.1–4.23.
56. § 7002(a), 42 U.S.C. § 6972(a). See also Ferrey, *Environmental Law: Examples and Explanations*, 281–282; Michael F. Hearn, "One Person's Waste Is Another Person's Liability: Closing the Liability Loophole in RCRA's Citizen Enforcement Action," *McGeorge Law Review* 42, 2 (April 2011): 471–473; Jason M. Levy, "Conflicting Enforcement Mechanisms under RCRA: The Abstention Battleground between State Agencies and Citizen Suits," *Ecology Law Quarterly* 39, 2 (2012): 380–382, 403–404; Moya and Fono, *Federal Environmental Law*, 137–141; Christopher Rizzo, "RCRA's 'Imminent and Substantial Endangerment' Citizen Suit Turns 25," *National Resources & Environment* 23, 2 (Fall 2008): 50–51.
57. Robert Frederickson, "A Green Bird in the Hand: An Example of Environmental Regulations Operating to Stifle Environmentally Conscious Industry," *Boston College Environmental Affairs Law Review* 34, 1 (2007): 303–333; Hearn, "One Person's

Waste Is Another Person's Liability," 496; Harold L. Segall, Katherine T. Gates, and Annise K. Maguire, "Defeating RCRA Claims Based on Failure to Establish That an Alleged Harm Is 'Imminent,'" *Environmental Litigation Committee Newsletter* 22, 1 (Fall 2010): 1–16; John L. Wittenborn and Eric Waeckerlin, "The Saga Continues—*Howmet* and the Ongoing Uncertainty of Solid Waste Regulation under RCRA," *National Wetlands Newsletter* 33, 1 (January/February 2011): 10005–10007.

58. The Comprehensive Environmental Response, Compensation and Liability Act (CERCLA) of 1980, 42 U.S.C. § 9601 *et seq.* (1980). See also Martin A. McCrory, "Who's on First: CERCLA Cost Recovery, Contribution, and Protection," *American Business Law Journal* 37, 1 (Fall 1999): 4–7; Emilee Mooney Scott, "Bona Fide Protection: Fulfilling CERCLA's Legislative Purpose by Applying Differing Definitions of 'Disposal,'" *Connecticut Law Review* 42, 3 (February 2010): 957–990; Maryam Tabatabai, "Comparing U.S. and EU Hazardous Waste Liability Frameworks: How the EU Liability Directive Competes with CERCLA," *Houston Journal of International Law* 34, 3 (Summer 2012): 656–657.
59. Phil Brown, PhD, and Richard Clapp, DSc, "Looking Back on Love Canal," *Public Health Reports* 117, 2 (March/April 2002): 95–98; Buck, *Understanding Environmental Administration and Law*, 110–111; Lois Marie Gibbs, "Housewife's Data," *American Journal of Public Health* 101, 9 (September 2011): 1556–1559; Andrew J. Hoffman, "An Uneasy Rebirth at Love Canal," *Environment* 37, 2 (March 1995): 4–9, 25–31; Moya and Fono, *Federal Environmental Law*, 156; Alicia Saunte Phillips, Yung-Tse Hung, and Paul A. Bosela, "Love Canal Tragedy," *Journal of Performance of Constructed Facilities* 21, 4 (July/August 2007): 313–319; Scott, "Bona Fide Protection," 965–966.
60. § 107, 42 U.S.C. § 9607. See also Ronald G. Aronovsky, "Federalism and CERCLA: Rethinking the Role of Federal Law in Private Cleanup Cost Disputes," *Ecology Law Quarterly* 33, 1 (2006): 24–34, 50–54, 104; Stefanie Gitler, "Settling the Tradeoffs between Voluntary Cleanup of Contaminated Sites and Cooperation with the Government under CERCLA," *Ecology Law Quarterly* 35, 1 (2008): 344–350; Robert M. Guo, "Reasonable Bases for Apportioning Harm under CERCLA," *Ecology Law Quarterly* 37, 1 (2010): 320–327; Amy Luria, "CERCLA Contribution: An Inquiry into What Constitutes an Administrative Settlement," *North Dakota Law Review* 84, 1 (2008): 334–337; McCrory, "Who's on First," 5–7; Moya and Fono, *Federal Environmental Law*, 156–158.
61. Ferrey, *Environmental Law: Examples and Explanations*, 298–302; Moya and Fono, *Federal Environmental Law*, 158–160; Rosenbaum, *Environmental Politics and Policy*, 244–245.
62. §§ 103–104, 42 U.S.C. §§ 9603–9604. See also Milos Jekic, "Lowering the Jurisdictional Bar: A Call for an Equitable-Factors Analysis under CERCLA's Timing-of-Review Provision," *Kansas Law Review* 59, 1 (2011): 157–190; Arnold W. Reitze, Jr., "Emergency Response and Planning Requirements Applicable to Unpermitted Air Pollution Releases," *Brigham Young University Law Review* 2005, 5 (2005): 1078–1081; Larry Schnapf, "How the CERCLA Notification Requirements Facilitate the Creation of Brownfields and What EPA Can Do to Address the Problem," *Sustainable Development Law & Policy* 11, 1 (2010): 19–26, 63–65.
63. §§ 104–105, 42 U.S.C. §§ 9604–9605. See also Ferrey, *Environmental Law: Examples and Explanations*, 306–310; Moya and Fono, *Federal Environmental Law*, 163–164; Reitze, Jr., "Emergency Response and Planning Requirements Applicable to Unpermitted Air Pollution Releases," 1075–1193.

64. §§ 104–105, 42 U.S.C. §§ 9604–9605. See also Harold C. Barnett, "The Allocation of Superfund," *Land Economics* 61, 3 (August 1985): 255; Ferrey, *Environmental Law: Examples and Explanations*, 307–309; Amy Luria, "CERCLA Contribution," 333–364; Moya and Fono, *Federal Environmental Law*, 163–167; Maya Waldron, "A Proposal to Balance Polluter and Community Intervention in CERCLA Litigation," *Ecology Law Quarterly* 38, 2 (2011): 402–413.
65. § 104(i), 42 U.S.C. § 9604(i). See also "Communicating Results to Community Residents: Lessons from Recent ATSDR Health Investigations," *Journal of Exposure Analysis and Environmental Epidemiology* 14, 7 (November 2004): 484–491; Christopher J. Portier, PhD, "ATSDR in the 21st Century," *Journal of Environmental Health* 74, 7 (March 2012): 30–31; Yee-Wan Stevens, Mildred M. Williams-Johnson, and William Cibulas, Jr., "Findings and Accomplishments of ATSDR's Superfund-Mandated Substance-Specific Applied Research Program," *International Journal of Hygiene and Environmental Health* 205, 1 (2002): 29–39; United States General Accountability Office, *Agency for Toxic Substances and Disease Registry: Policies and Procedures for Public Health Product Preparation Should Be Strengthened* (Washington, D.C.: U.S. Government Printing Office, GAO Publication No. GAO-10-449, April 2010): 7–8.
66. §§ 111–112, 42 U.S.C. §§ 9611–9612. See also Barnett, "The Allocation of Superfund," 255–262; Steven Ferrey, "Inverting the Law: Superfund Hazardous Substance Liability and Supreme Court Reversal of All Federal Circuits," *William & Mary Environmental Law & Policy Review* 33, 3 (May 2009): 638–649; Brian L. Murphy, "Allocation by Contribution to Cost and Risk at Superfund Sites," *Journal of Environmental Forensics* 1, 3 (September 2000): 117–120.
67. § 107, 42 U.S.C. § 9607. See also § 113, 42 U.S.C. § 9613; Ferrey, "Inverting the Law," 647–662; Guo, "Reasonable Bases for Apportioning Harm under CERCLA," 322–327; McCrory, "Who's on First," 5–7.
68. § 121, 42 U.S.C. § 9621. See also Ferrey, *Environmental Law: Examples and Explanations*, 311–312; Ferrey, "Inverting the Law," 664–665; Moya and Fono, *Federal Environmental Law*, 158; Rosenbaum, *Environmental Politics and Policy*, 245; "The Unrealized Potential of SARA," *Environment* 29, 4 (May 1987): 6–11, 40–44.
69. Ferrey, *Environmental Law: Examples and Explanations*, 311–312; Guo, "Reasonable Bases for Apportioning Harm under CERCLA," 320–326; Murphy, "Allocation by Contribution to Cost and Risk at Superfund Sites," 117–120; "The Unrealized Potential of SARA," 6–10.
70. Ferrey, *Environmental Law: Examples and Explanations*, 311–312; Ferrey, "Inverting the Law," 638–649; Guo, "Reasonable Bases for Apportioning Harm under CERCLA," 322–327; McCrory, "Who's on First," 5–7; Murphy, "Allocation by Contribution to Cost and Risk at Superfund Sites," 117–120.
71. The Federal Insecticide, Fungicide and Rodenticide Act (FIFRA), as amended by the Federal Environmental Pesticide Control Act of 1972, 7 U.S.C. § 136 *et seq.* (1972). See also Joseph Frueh, "Comment: Pesticides, Preemption, and the Return of Tort Protection," *Yale Journal on Regulation* 23, 1 (2006): 299–309; John H. Minan and Tracy M. Frech, "Pesticides as 'Pollutants' under the Clean Water Act," *San Diego Law Review* 47, 1 (Winter 2010): 109–144; Douglas T. Nelson, Joshua Saltzman, William K. Rawson, Claudia O'Brien, and Bart J. Kempf, "Real Environmental Protection: Not a Paper Exercise," *Environmental Law Reporter: News & Analysis* 42, 2 (February 2012): 10166–10171.

72. Minan and Frech, "Pesticides as 'Pollutants' under the Clean Water Act," 116–121; Nelson, et al., "Real Environmental Protection," 10167–10168; Rosenbaum, *Environmental Politics and Policy*, 226, 228.
73. The Endangered Species Act (ESA) of 1973, 16 U.S.C. § 1531, *et seq.* (1973). See also Ferrey, *Environmental Law: Examples and Explanations*, 481–482; J. B. Ruhl, "The Endangered Species Act's Fall from Grace in the Supreme Court," *Harvard Environmental Law Review* 36, 2 (2012): 488–497; Scott Schwartz, "The Hapless Ecosystem: A Federalist Argument in Favor of an Ecosystem Approach to the Endangered Species Act," *Virginia Law Review* 95, 4 (2009): 1325–1360.
74. § 4, 16 U.S.C. § 1533; § 7, 16 U.S.C. § 1536; § 11, 16 U.S.C. § 1540. See also *Bennett v. Spear*, 520 U.S. 154 (1997); Ferrey, *Environmental Law: Examples and Explanations*, 485–488; William H. Rodgers, Jr., *Environmental Law*, 2nd ed. (St. Paul, Minn.: West, 1994), 996–998, 1007–1010, 1019–1020; Ruhl, "The Endangered Species Act's Fall from Grace in the Supreme Court," 494–495.
75. *Tennessee Valley Authority v. Hill*, 437 U.S. 153 (1978). See also Kenneth M. Murchison, *The Snail Darter Case: TVA versus the Endangered Species Act* (Lawrence: University Press of Kansas, 2007), 1–4, 18–19, 57–58, 63–64; Rodgers, *Environmental Law*, 999; Ruhl, "The Endangered Species Act's Fall from Grace in the Supreme Court," 497–498. Aside from the *Hill* case, the ESA led to one of the most famous standing cases in U.S. Supreme Court history, *Luhan v. Defenders of Wildlife*, 504 U.S. 555 (1992), which held that the plaintiffs, members of a private environmental group, did not have standing to sue under the Endangered Species Act because the threat of a species' extinction absent other circumstances did not establish an individual and non-speculative private injury.
76. § 7, 16 U.S.C. § 1536. See also Ferrey, *Environmental Law: Examples and Explanations*, 488–489; Ted Gup, "Down with the God Squad," *Time* 136, 20 (November 5, 1990): 102; Schwartz, "The Hapless Ecosystem," 1328–1329.
77. See, for example, Gup, "Down with the God Squad," 102; Barry R. Noon and Jennifer Blakesley, "Conservation of the Northern Spotted Owl under the Northwest Forest Plan," *Conservation Biology* 20, 2 (April 2006): 288–296; Rodgers, *Environmental Law*, 1008, 1009; Brendon Swedlow, "Reason for Hope? The Spotted Owl Injunctions and Policy Change," *Law & Social Inquiry* 34, 4 (Fall 2009): 825–867; *Robertson v. Seattle Audubon Society*, 503 U.S. 429 (1992).
78. § 404, 33 U.S.C. § 1344. See also Peg Bostwick, "Integrating State and Federal Needs under the Clean Water Act," *National Wetlands Newsletter* 33, 4 (July/August 2011): 5–6; David Evans, "Clean Water Act § 404 Assumption: What Is It, How Does It Work, and What Are the Benefits?" *National Wetlands Newsletter* 31, 3 (May-June 2009): 18–21; Ferrey, *Environmental Law: Examples and Explanations*, 423–427; Palmer Hough and Morgan Robertson, "Mitigation under Section 404 of the Clean Water Act: Where It Comes from, What It Means," *Wetlands Ecology & Management* 17, 1 (January 2009): 15–33; Margaret "Peggy" Strand and Lowell M. Rothschild, "What Wetlands Are Regulated? Jurisdiction of the § 404 Program," *Environmental Law Reporter: News & Analysis* 40, 4 (April 2010): 10372–10393.
79. For a discussion of the value of "green" corporations as well as strategies for developing them, see, for example Jorge Alexis Arevalo, "Critical Reflective Organizations: An Empirical Observation of Global Active Citizenship and Green Politics," *Journal of Business Ethics* 96, 2 (October 2010): 299–316; Donald S. Siegel, "Green Management Matters Only if It Yields More Green: An Economic/Strategic Perspective," *Academy of Management Perspectives* 23, 3 (August 2009): 5–16.

80. International Organization for Standardization (ISO) Website: "About Us," http://www.iso.org/iso/home/about.htm (Accessed November 21, 2012). See also David A. Wirth, "The International Organization for Standardization: Private Voluntary Standards as Swords and Shields," *Boston College Environmental Affairs Law Review* 36, 1 (2009): 80–81.
81. Eccleston and March, *Global Environmental Policy*, 174; International Organization for Standardization (ISO) Website: "ISO 9000—Quality Management," http://www.iso.org/iso/home/standards/management-standards/iso_9000.htm (Accessed November 21, 2012).
82. The literature on ISO 14000 standards is voluminous. See, for example, Niels-Hugo Blunch and Paula Castro, "Enterprise-level Training in Developing Countries: Do International Standards Matter?" *International Journal of Training and Development* 11, 4 (December 2007): 314–324; Pavel Castka and Michaela A. Balzarova, "The Impact of ISO 9000 and ISO 14000 on Standardization of Social Responsibility—An Inside Perspective," *International Journal of Production Economics* 113, 1 (May 2008): 74–87; International Organization for Standardization (ISO) Website: "ISO 14000—Environmental Management," http://www.iso.org/iso/home/standards/management-standards/iso14000.htm (Accessed November 21, 2012); Aseem Prakash and Matthew Potoski, *The Voluntary Environmentalists: Green Clubs, ISO 14001, and Voluntary Environmental Regulations* (New York and Cambridge, U.K.: Cambridge University Press, 2006), 3–15; Gyula Vastag, "Revisiting ISO 14000 Diffusion: A New 'Look' at the Drivers of Certification," *Production and Operations Management* 13, 3 (Fall 2004): 260–267.
83. Eccleston and March, *Global Environmental Policy*, 176–177; Vastag, "Revisiting ISO 14000 Diffusion," 260–261; Wirth, "The International Organization for Standardization," 90–92.
84. See, for example, American Society for Testing and Materials (ASTM) International Website: "Home Page," http://www.astm.org/ (Accessed November 23, 2012).
85. American National Standards Institute (ANSI) Website: "Home Page," http://www.ansi.org/ (Accessed November 23, 2012).
86. Association Française de Normalisation (AFNOR) Website: "Home Page," http://www.afnor.org/en (Accessed November 23, 2012).
87. BSI Group (National Standards Body) Website: "Home Page," http://www.bsigroup.com/en-GB/about-bsi/uk-national-standards-body/ (Accessed November 23, 2012).
88. DIN German Institute for Standardization Website: "Home Page," http://www.sundyne.com/Products/Pumps/Industry+Standards/DIN+German+Institute+for+Standardization (Accessed November 23, 2012).
89. Julie Cidell, "A Political Ecology of the Built Environment: LEED Certification for Green Buildings," *Local Environment* 14, 7 (August 2009): 621–633; Ajay Garde, "Sustainable by Design? Insights from U.S. LEED-ND Pilot Projects," *Journal of the American Planning Association* 75, 4 (Autumn 2009): 424–440; LEED/U.S. Green Building Council Website: "Home Page," https://new.usgbc.org/leed (Accessed November 22, 2012); Society of the Plastics Industry (SPI) Website: "Plastics Industry Win: USGBC to Revisit LEED Standards, http://www.plasticsindustry.org/AboutSPI/NewsItem.cfm?ItemNumber=9347 (Accessed November 22, 2012); Melinda Tuhus, "The Question of LEED: Why the Country's Leading Green Building Certification May Be Inherently Flawed," *E: The Environmental Magazine* 21, 6 (November/December 2010): 15–19.

90. The Biodegradable Products Institute (BPI) Website: "The BPI Approval Process & Test Requirements," http://www.bpiworld.org/BPI-certification-requirements (Accessed November 23, 2012).
91. Sustainable Plastics Website: "Washington: Seattle's Cedar Grove Composting Facility," http://www.sustainableplastics.org/early-adopters/seattles-cedar-grove-composting-facility (Accessed November 23, 2012).
92. Cedar Grove Website: "What We Do," http://www.cgcompost.com/what-we-do (Accessed November 23, 2012).
93. ENERGY STAR Website: "About ENERGY STAR," http://www.energystar.gov/index.cfm?c=about.ab index (Accessed November 23, 2012); ENERGY STAR Website: "History," http://www.energystar.gov/index.cfm?c=about.ab_history (Accessed November 23, 2012).
94. Forest Stewardship Council (FSC) Website: "Mission and Vision: Protecting Forests for Future Generations," http://us.fsc.org/mission-and-vision.187.htm (Accessed November 24, 2012); Forest Stewardship Council (FSC) Website: "Our History," http://us.fsc.org/our-history.180.htm (Accessed November 24, 2012).
95. Forest Stewardship Council (FSC) Website: "Mission and Vision: Protecting Forests for Future Generations," n.p.; Axel Marx and Dieter Cuypers, "Forest Certification as a Global Environmental Governance Tool: What Is the Macro-Effectiveness of the Forest Stewardship Council?" *Regulation & Governance* 4, 4 (December 2010): 408–434.
96. Donald H. Schepers, "Challenges to Legitimacy at the Forest Stewardship Council," *Journal of Business Ethics* 92, 2 (March 2010): 288; Susan E. Moore, Frederick Cubbage, and Celia Eicheldinger, "Impacts of Forest Stewardship Council (FSC) and Sustainable Forestry Initiative (SFI) Forest Certification in North America," *Journal of Forestry* 110, 2 (March 2012): 79.
97. Green Advantage Environmental Certification Website: "About GA Certification," http://www.greenadvantage.org/about-ga-certification (Accessed November 24, 2012); Green Advantage Environmental Certification Website: "Vision, Mission & Deliverables," http://www.greenadvantage.org/vision-mission-deliverables (Accessed November 24, 2012).
98. Green America Website: "About Us," http://www.greenamerica.org/about/ (Accessed November 24, 2012).
99. Green America Website: "Certification for Green Businesses," http://www.greenamerica.org/greenbusiness/certification.cfm (Accessed November 24, 2012).
100. Green Clean Institute Website: "About GCI," http://www.greencleaninstitute.com/about-gci (Accessed November 24, 2012); Green Clean Institute Website: "GCI Firm Certifications,"http://www.greencleaninstitute.com/content/firm-certification (Accessed November 24, 2012).
101. Green Clean Institute Website: "Green Product Certification," http://www.greencleaninstitute.com/green-product-certification (Accessed November 24, 2012).
102. Green Globe Website: "Home Page," http://greenglobe.com/ (Accessed November 24, 2012); Green Globe Website: "Standard Criteria and Indicators," http://greenglobe.com/register/standard/ (Accessed November 24, 2012).
103. Green Key Website: "About the Green Key Eco-Rating Program," http://www.greenkeyglobal.com/about_the_program.asp (Accessed November 24, 2012); Green Key Website: "Home Page," http://www.greenkeyglobal.com/default.asp (Accessed November 24, 2012).

104. National Green Building Website: "About NAHB Green," http://www.nahbgreen.org/AboutNAHBGreen/default.aspx (Accessed November 25, 2012).
105. National Green Building Website: "Green Scoring Tool," http://www.nahbgreen.org/ScoringTool.aspx (Accessed November 25, 2012); National Green Building Website: "ICC 700 National Green Building Standard," http://www.nahbgreen.org/NGBS/default.aspx (Accessed November 25, 2012).
106. "Fairtrade v. Rainforest Alliance," *Marketing Week* 33, 21 (May 20, 2010): 17; Rainforest Alliance Website: "About Us," http://www.rainforest-alliance.org/about (Accessed November 25, 2012); Rainforest Alliance Website: "Certification, Verification, and Validation Services," http://www.rainforest-alliance.org/certification-verification (Accessed November 25, 2012); Tom Wall, "The Rich Aroma of Profits," *New Statesman* 134, 4729 (February 28, 2005): 30–31.
107. Dummies.com Website: "Home Page," http://www.dummies.com/ (Accessed November 29, 2012).

Chapter 10

Conclusion: The Path Forward

Dear earth, I do salute thee with my hand,
Though rebels wound thee with their horses' hoofs:
As a long-parted mother with her child
Plays fondly with her tears and smiles in meeting,
So, weeping, smiling, greet I thee, my earth,
And do thee favours with my royal hands.

—William Shakespeare, *King Richard II*, Act III, Scene ii[1]

American environmentalism has evolved throughout the years, waxing and waning, advancing and retreating, alternately receiving approbation and opprobrium, depending on changing sensibilities and economic fortunes. As discussed in this book, the movement is comprised of a diverse, complex mixture of local, state, regional, national, and international governmental and nongovernmental organizations (NGOs) and policymakers. Its adherents have been, and are, men and women of differing preferences, mixed motives, and varying talents, all employing an array of strategies and tactics to achieve sometimes-eclectic goals. The multifaceted, diverse nature of the issues and participants in this incremental, pluralist landscape has been both a blessing and a curse. Environmentalism has room for everyone, thereby expanding its appeal and efficacy as a democratizing feature of twenty-first-century American life. With the relative convenience and ease of using the Internet to communicate and relying on modern jet aircraft to travel around the world, even average citizens can express an interest in, and concern for, the natural environment by disseminating messages and exploring the terrain without relying on the mediating influence of elites.

Yet the increased accessibility of the environmental movement also threatens to undermine its legitimacy. When debates concerning natural-resource conservation and preservation were confined to scientists and policymakers—mostly subject-matter experts trafficking in obscure, jargon-filled, data-driven arguments about inputs, throughputs, and outputs—the citizenry struggled to comprehend both the source of public policy and the means by which it would be implemented. The obscurity and complexity of environmental science ensured its legitimacy—at least in the eyes of like-minded intelligentsia. The average person was forced to accept command-and-control decisions announced by government officials or, alternatively, seek entry into the environmental field through the professions or by acquiring scientific training.

With the birth of environmental activism during the 1960s and 1970s, specialized training no longer served as a prerequisite for participation in the public-policy arena. Environmentalism became a secular religion for some born-again followers, a quasi-mystical, faith-based subject properly mastered not through the head, but with the heart. A generation of self-proclaimed environmentalists resembled itinerant lay preachers of the nineteenth century traveling through the burned-over district in the aftermath of the Second Great Awakening seeking disciples for the sacred cause through a road-to-Damascus conversion experience. For the committed environmentalist-cum-preacher of the new epoch, facts and data interfered with feeling and emotion. As the breadth of citizen involvement came to augment, and in some instances supplant, scientific depth, the question naturally arose whether environmentalism would be transformed into a trendy, passing fad that, after its New Age appeal wore off and the next big thing captured the public imagination, eventually became an antiquated remnant of a bygone era, wistfully reflected upon as people pulled scrapbooks from storage (or, in keeping with the times, logged onto social-media websites) and smiled while leafing past sepia-toned photographs of love beads, Nehru jackets, and polyester leisure suits. The democratization of environmentalism, therefore, was a development fraught with peril. When the masses weigh in on policy decisions, polemicists sometimes replace adherents of idealized science and supporters of the Method. Whether the resultant shift in sensibilities is a positive or negative feature of the policy process depends on one's understanding of the purposes of environmentalism.[2]

Common Pitfalls

If American environmentalism of the twenty-first century is to retain its legitimacy among substantially all parties and effect public policy changes that measurably improve environmental quality, participants in the debate must avoid at least four common pitfalls. First, they must recognize and come to terms with the postmodern critique of environmentalism. Second, they must distinguish between genuine crises and pseudo-hysteria surrounding some environmental debates. This distinction initially sounds simple and intuitive. Yet one person's crisis, as the old adage suggests, is another person's opportunity. In a similar vein, the urge to politicize science as a means of demonizing the "enemy" must be avoided lest junk science displace generally recognized scientific methods and discourse. Finally, environmentalists of all stripes must recognize panaceas and symbols for what they are: convenient, shorthand descriptions of a situation, perhaps, but most certainly distractions from the long, hard work ahead.

Environmentalism and the Postmodern Conundrum

Several sections of this book, notably Chapter 2, discuss industrialists' and environmentalists' dichotomous perspectives. Chapter 3 devotes many pages to exploring the nuances between the anthropocentric and biocentric schools of thought. Participants on the political right have argued, and continue to argue, that radical environmentalism unduly restricts private property rights and threatens individual freedom owing to a nebulous and probably misguided concern for Mother Nature or Mother Earth, the Gaia goddess. Diehard naturalists deride the industrialists for their shortsighted willingness to foul the earth and thereby kill the goose that laid the golden egg. This fundamental rift between industrialists and environmentalists reflects the freedom-authority divide underlying the American regime. One side heralds the centrality of individual choice, even if that choice ultimately harms the collectivity, while the other emphasizes communal values and a willingness to cede a measure of freedom to government if the outcome improves environmental quality. In recent decades, however, the rift has morphed. Criticism of the status quo also has erupted from the left side of the political spectrum among self-styled "postmodernists."

Postmodernists assail social constructions that pass for "objective reality." Throughout human history, and particularly in the western intellectual tradition since the Enlightenment, humankind has confused "what is" with "what is meant to be." Before the United States eradicated the institution of slavery, for example, apologists for human bondage argued that by virtue of its existence, slavery was God's will. It was the natural condition of darker-hued peoples. By the twenty-first century, society denounced such talk as patently offensive, if not downright racist. Anyone save the most unenlightened Neanderthal now recognizes that American slavery, the so-called "peculiar institution," resulted from decades, indeed centuries, of social construction. Slaveholders justified ongoing brutality as an immutable "truth": Slaves naturally were too infantile and backward to govern themselves as a free people. In retrospect, this was a lame, self-serving, manufactured excuse. Man, not God, imposed conditions of servitude on the bondman. What passed for objective fact in an earlier epoch was nothing more than the institutionalization of now-rejected social mores. Societal institutions always lay their cornerstones on such fictions, albeit generally they are less dramatic, if nonetheless invidious.[3]

Building on this insight, the postmodern perspective on environmentalism indicates that human attitudes about nature are socially constructed. Theoretically-neutral scientific data and supposedly empirical conclusions about natural processes are themselves products of human values and interpretations and are no more or less true than other forms of human interaction with the "outside world." In a final rebuke to Aristotle and his ilk, the postmodernists challenge the ancients' assumptions that *a priori* and *a posteriori* judgments can be bifurcated. Upon reflection, the assumptions themselves are veiled *a posteriori* judgments. Yet to be more precise, even the dual distinctions hold little utility or meaning since both rationalism and empiricism are social constructions. All human experiences are filtered through society's values, existing governmental laws and policies, and the idiosyncrasies of the individual. As one commentator has observed, "privileged discourse" endlessly replicated through mainstream scientific literature does not discover "truth." Truth is unattainable. Postmodernism reveals the flaws in the correspondence theory of truth and its foundational assumptions about epistemology. Nature is not a separate physical entity, a disembodied reality subject to morally neutral physical laws. Man does not discover nature; he creates it.[4]

Consider, as an example, use of the term "animal." Human beings speak of the "other" as a lower order of animal. Although humans are also animals, mankind possesses distinctive abilities to use language, make tools, engage in rational thought, and alter the natural environment in myriad ways, or so the standard narrative posits. Postmodernism rejects the anthropocentric construction behind the concept of lower-order animals. When the "other" is relegated to a category reserved for non-human status, this socially constructed exercise serves as justification for all manner of unexamined dominion over animals. It is little wonder, therefore, that so many apologists for anthropocentrism construe biblical passages to allow for God-sanctioned dominion over the earth and its inhabitants. Postmodernists reject such convenient rationalizations as little more than thinly disguised "might-makes-right" propositions. In its unreflective embrace of long-held presuppositions, mainstream society has substituted proverbs in place of thoughtful, deliberative discourse and decision-making.[5]

Even critics of the postmodern enterprise recognize the value of moving beyond the intellectual cul-de-sac in which the environmental movement found itself early in the new millennium. As long as proponents of anthropocentric and biocentric thinking face off against each other across a seemingly unbridgeable gulf, progress cannot be made. By calling for a systematic reappraisal of the old strictures and methods, postmodernism invites environmentally conscious denizens on both sides of the chasm to rethink their most cherished beliefs and challenge the assumptions behind those beliefs. Environmentalism might become another fad, ridiculed and marginalized by mainstream, anthropocentric policymakers, unless the debate can be reframed, a process that begins with appreciating the social construction of concepts once deemed objectively "true."[6]

Postmodernism legitimizes the fragmented nature of American environmentalism because it recognizes that no single perspective on nature is authoritative. The environmentalist pursuing product bans because he or she nurses a "pet peeve" is no more or less righteous than the credentialed scientist laboriously developing research hypotheses, collecting data, and empirically testing for validity and reliability. If environmentalism has never been a church of one creed, the postmodernist figuratively nails his own *95 Theses* to the door at Wittenberg. The postmodern critique celebrates diverse players in the public policy realm and champions the legitimacy of multiple environmental issues of concern.

Critics charge that postmodernism can degenerate into "anything-goes" relativism. They acknowledge a social dimension to human attitudes on the natural environment but resist the conclusion that because discourse can seldom be stripped of its sometimes-mythic presuppositions and institutionalized prejudices, therefore it must be radically altered or entirely rejected. Flawed discourse can hold value provided its flaws are identified and understood. Direct observation and scientific exploration can yield insights into environmental degradation as long as scientists and policymakers recognize built-in biases and limitations in the data. Postmodernism threatens the school of rationality that has dominated western intellectual life since the Enlightenment. If every endeavor is socially constructed and open to multiple interpretations, then no authoritative structures exist. Nature is whatever we make of her. To continue the "church" metaphor, some committed naturalists have transformed American environmentalism into a secular religion. They tout the new religion based on faith and contend their detractors wallow in sin. In the meantime, industrialists dismiss environmentalists as crackpots who have overplayed their hand, fudging data on climate change and other indicia of environmental despoliation owing to religious zealotry. The competing schools of thought cannot reach an accommodation when they do not speak the same language or possess the same values.[7]

Three possible options exist for persons who seek to practice environmentalism in the face of the postmodern conundrum. A critic can attack postmodernism using classic *reductio ad absurdum* argumentation. Postmodernists contend that all privileged discourse contains putative assumptions that cannot withstand sustained scrutiny; therefore, socially constructed meta-narratives must be cast aside. Yet, ironically, a discourse on the fallacies inherent in discourse replaces one set of foundational assumptions with another. After all, the argument that all foundational constructions are inherently flawed is itself a foundational construction. The tautological beast chases its own tail and eventually consumes itself. Nihilism reigns supreme.

A second option is to accept the postmodern critique and seek solutions through an ongoing dialogue with like-minded postmodernists. The difficulty, of course, is that no authoritative common ground exists. Each participant in the dialogue enjoys a measure of legitimacy in the creative enterprise. In a field where every manner of discourse is afforded equal weight and dignity, the multiplicity of resultant standards is tantamount to no standard at all.

In lieu of absolutely rejecting or righteously embracing postmodernism, a few commentators argue in favor of "accepting the intellectual insights of postmodern critics and, at the same time, providing some guidelines for protecting the nonhuman world." Without tackling the problem of whether postmodernism definitively resolves (or exacerbates) outstanding ontological or epistemological problems, a supporter of the third option recognizes that a pragmatic response requires the development of generally accepted standards and processes, acknowledging the socially constructed character of all human concepts and institutions. Nature and nonhuman animals remain the "other," but the other is afforded great weight in human decision-making. Animals may not possess all the rights of human beings, but their welfare cannot be dismissed without due consideration for their value in the chain of natural processes. Whether this third approach falls squarely within the anthropocentric or biocentric school of thought is less important than whether it provides intellectual support for continued progress in American environmentalism.[8]

"Crisis" Thinking in Non-Crisis Situations

Anyone who delves into the literature on environmentalism and sustainability quickly discovers hyperbolic language galore. Each problem is a looming crisis that must be resolved forthwith if the planet is to survive. The consequences of inaction or time-consuming deliberation potentially are catastrophic. The thoughtful reader can draw contradictory conclusions: On one hand, the world is in such terrible shape that little can be done absent effecting wholesale, systemic change, which is unlikely to occur in the near future. For the genuinely angst-ridden individual seeking to live a contemplative, ethical life, only the solace of suicide can nurse a troubled soul through many a dark night. On the other hand, the opposite conclusion reached by the hard-nosed skeptic is to dismiss such Chicken Little claims that "the sky is falling" as jejune, the histrionic rhetoric of the snake-oil salesman or false prophet. If every problem is a crisis, no one crisis eclipses any other crisis. If the world is going to hell in a hand-basket, we should resign ourselves to our fate and live for today—tomorrow be damned. In both instances, the response in the face of the impending crisis is a shrug of the shoulders and a rapid descent into solipsism. With apologies to future generations, the hour has grown too late to fulfill our obligations to our progeny.

To forestall a philosophical surrender, the antidote is to sift through the data and distinguish the crisis from the non-crisis. The literature on crisis management recognizes three stages: pre-crisis, crisis, and post-crisis. Each stage requires different approaches and tools. Each stage provides a separate set of warning signals, learning opportunities, and consequences for success or failure. Unfortunately, the criteria for deciding when a particular problem falls into one of the three stages are highly contested. The appeal of postmodernism is that each participant in the environmental field can decide for himself which stage is appropriate. For critics who eschew postmodernism, however, environmentalism quickly degenerates into a "battle of the experts," which further reinforces the "non-privileged discourse" appeal of the postmodern enterprise. Consider a few examples.[9]

One of the most contentious topics that arose early in the twenty-first century was the question of whether Americans' bodies had been overly burdened by exposure to toxic chemicals. Emblematic of this debate was concern about a common chemical called Bisphenol A (BPA) and whether it should be banned from use in U.S. consumer products. A component of certain types of plastics, BPA has been used to manufacture diverse products such as automobile dashboards, CDs, food-storage containers, polycarbonate water bottles, and baby bottles. Prior to 2007, BPA's safety seemed assured: It was one of the most widely tested chemicals in common usage. More than 4,500 studies had been conducted across multiple decades. In 1982, the National Cancer Institute and the National Toxicology Program concluded that BPA was not a possible carcinogen. The U.S. Environmental Protection Agency (EPA) reviewed the scientific and technical data in 1988, 2008, and 2010, and found no cause for concern. Similarly, the U.S. Food & Drug Administration (FDA) decided that BPA was safe for use in household products.[10]

In 2007, the Environmental Working Group (EWG), an NGO created in 2002 to promote increased reliance on the precautionary principle in American environmentalism, sponsored a study raising questions about possible health effects associated with BPA usage. With assistance from Fenton Communications, a well-known left-leaning public relations firm, EWG promoted the study as evidence of the dangers of BPA. On its website, the group warned, "Causal relationships between environmental exposures and health effects are often difficult to establish because of many factors, including non-standardized diagnosis of diseases, difficulties ascertaining exposure to ubiquitous environmental contaminants, and inadequate statistical power to address other contributing factors." Evidence of cause-and-effect did not demonstrate a link between a specific chemical exposure and negative human health effects, but no matter. Direct, unimpeachable evidence was unnecessary owing to the precautionary principle. The *possibility* of harm was sufficient. "Endocrine disruptors, such as BPA, may very well play a part in the etiology of classic reproductive disorders and cancers, as well as diseases not often linked to hormonal activity—immune system conditions, learning and behavioral disorders, diabetes, and even obesity." To avoid data-driven arguments about the nature of the scientific evidence, the group couched the issue in ambiguous, suggestive terms: "If BPA does indeed contribute to any of these epidemic disorders, the potential ramifications for public health are far-reaching. Toxicity studies involving rodent exposure to BPA should not be considered in a void, but rather should be framed within the current context of clinical concerns for the U.S. population."

Other environmental groups, notably the Natural Resources Defense Council, also joined the anti-BPA campaign. They were careful to argue that conclusive proof did not exist that BPA was dangerous, but the specter of danger was enough to warrant a change.[11]

By 2008, after the BPA issue had become a cause célèbre following a series of sensationalistic stories in the mainstream news media, retailers and consumer-product manufacturers directed their suppliers to discontinue using products containing the chemical. Plastics industry representatives and professional skeptics such as Jon Entine continued to argue about the lack of a credible scientific foundation for banning BPA, but their appeals generally fell on deaf ears. The presence of BPA in baby bottles was especially upsetting to mothers who feared for the safety of their children. Their emotional response all but quelled further scientific debate. The politics of BPA and the precautionary principle had triumphed.[12]

The BPA imbroglio was part of a larger concern over the fear of a "body burden" owing to the seemingly ubiquitous presence of chemicals and their proximity to consumers. Mainstream journalists and self-styled "environmental reporters" had found a cause—the need to reduce the number of toxic chemicals prevalent in American life. Whether an actual risk existed was immaterial. Here the ugly specter of postmodernism yet again reared its head. Scientists often repeat the truism "the dose makes the poison"—hardly a reassuring credo in the midst of a public policy debate with participants who often lack basic scientific literacy—but the underlying point is important. "Risk" for scientists is determined by examining the exposure to a chemical; that is, they search for data on the concentration of the chemical and the length of time to which the person was found in its proximity. They also desire to know the toxicity of the chemical—a measure of the degree to which the substance is poisonous. A risk assessment requires information on the person's exposure multiplied by the toxicity of the chemical. Virtually any substance—salt or water, for example—can be harmful, even fatal, if someone is exposed to enough of the substance for a long enough period of time. These basic scientific principles seldom become part of postmodern public policy debates.[13]

The "body burden" issue was manifested especially in the debate concerning endocrine-disrupting chemicals (EDCs). The endocrine system regulates hormones in the body. Chemicals that disrupt hormones potentially lead to problems in metabolic rate, sex development, insulin production and use, growth, stress response, gender behavior, and reproduction. As with many scientific issues, pinpointing a cause-and-effect relationship between chemical use and exposure leading to damage to the endocrine system has been problematic. In 1996, Congress directed the U.S. EPA to establish a testing program for pesticides, environmental contaminants, and commercially available chemicals to determine the endocrine-disrupting effects, if any, of those substances. In response, EPA created the Endocrine Disruptors Screening Program (EDSP) to fulfill the requirements of the Food Quality Protection Act of 1996. Observers are divided on whether the EDSP has accomplished its objectives.[14]

The EDC issue probably would have remained important for a small group of environmentalists and scientists, but hardly an issue of public concern, had it not been for the publication of a sensationalistic 1996 book, *Our Stolen Future.* Penned by three environmentally conscious activists—Theo Colborn, a zoologist who founded the Endocrine Disruption Exchange (TEDX); Dianne Dumanoski, a long-time environmental activist, journalist, and author; and John Peterson Myers, an environmental activist with the W. Alton Jones Foundation (now called the Blue Moon Fund)—the book argued that widespread use of chemicals disrupts the normal workings of the endocrine system. Even at low-dose exposures, EDCs can trigger negative health effects. Colborn had been developing the thesis since 1993. With the publication of *Our Stolen Future,* which was written for a popular audience, Colborn felt vindicated as the term "endocrine disruptor" entered the vernacular. Vice President Al Gore assisted in promoting the work by writing the foreword. Referring to Rachel Carson (always desirable when touting the virtues of a work that warns of the proliferation of chemicals), Gore urged readers to explore "a book that is in many ways" a sequel to *Silent Spring.*[15]

The brouhaha over genetically modified organisms (GMO) used in agriculture is another example of a twenty-first-century environmental "crisis" that has divided consumers, environmentalists, biotechnology companies, and government regulators. For decades, at least since Paul Ehrlich published his famous diatribe *The Population Bomb* in the 1960s, demographers and government planners have debated methods for improving crop yields to prevent mass starvation. By the 1990s, scientists had discovered a means of genetically engineering crops. Although some observers welcomed the technological advances, a number of environmental groups, notably Greenpeace and the World Wildlife Fund, cautioned against the unknown health effects of producing and consuming modified foods. Raising the specter of a long latency period to determine negative side effects, the groups referred to GMO crops as "FrankenFoods," a not-so-subtle reference to the promethean calamity explored in a classic work of fiction, *Frankenstein.* Opponents urged lawmakers in many countries to label GMO foods and avoid mixing them with organically produced crops.[16]

Hydraulic fracturing—known colloquially as "fracking" or "fraccing"—also became a controversial issue early in the twenty-first century. Although the technique was first used at the end of the 1940s, it became a high-profile topic 60 years later as policymakers and energy company executives searched for a promising means of decreasing America's dependence on foreign energy sources. Fracking is the process of injecting a large volume of water under high pressure, mixed with chemicals and sand, into underground shale formations to establish pathways for capturing natural gas and oil that otherwise would be unavailable. A new technique, directional or horizontal drilling, allows a downward-plotting drill bit to turn as much as 90 degrees and continue drilling within the layer, opening up thousands of additional feet of sequestered layers of methane-rich shale. Fracking enthusiasts estimate that the process may provide access to 827 trillion cubic feet of heretofore-inaccessible shale gas for commercial exploitation, although some scientists caution that the figures may be inflated. In any case, even conservative estimates suggest that fracking will provide decades of plentiful, inexpensive natural gas and oil. The effects will ripple across energy markets, affecting supply and demand and even American foreign policy. If the United States can increasingly rely on domestic sources of energy, the relative independence this affords policymakers will change the country's relationship with oil-producing nations around the world.[17]

As promising as this new process sounds, environmentalists have been concerned about serious threats to the environment and human health. Several chemicals used to treat the water have been identified as potentially harmful to humans. In 2005, Congress exempted hydraulic fracturing fluids and their additives from Safe Drinking Water Act (SDWA) requirements, a continuing source of anguish and concern for the environmental community. If SDWA provisions can be set aside in a cavalier fashion, what measures prevent Congress from exempting fracking fluids from other applicable federal statutes, especially the Clean Water Act? A variety of federal environmental laws enacted to control polluted water and toxic chemicals might be obviated by widespread fracking. In the rush to exploit a developing technology, the United States may undermine citizens' health and safety with little concern for the consequences. If environmental laws are enacted to protect the health and welfare of the citizenry, they must be obeyed even in instances when enforcement is inconvenient.[18]

Ironically, fracking has been viewed as a promising means of accessing natural gas, an alternative to "dirtier" fossil fuels that generate greenhouse gases, which contribute to global warming. Even as environmentalists have raised concerns about fracking, they have lobbied government policymakers to tackle arguably the most urgent and divisive environmental issue of the twenty-first century: climate change caused, or at least exacerbated, by fossil-fuel usage. The issue was debated seriously in the United States as early as the 1970s, although the origins stretched back to 1896. That year, the Swedish scientist Svante Arrhenius calculated that doubling carbon dioxide emissions would increase global temperatures by 5° or 6°C. Arrhenius did not have access to modern computers nor did he know as much about atmospheric chemistry as scientists of a later age. The scientific community of the time greeted his hypothesis with skepticism that eventually led to outright rejection. For the next seven decades, chemists would occasionally revisit the issue as new studies revealed more secrets of climatology. Nonetheless, consensus was lacking in the scientific community.[19]

For a time, scientists debated the possibility that the earth was cooling. Major publications in 1972 and 1973 warned of another ice age pending on the horizon. A popular 1976 book, *The Cooling*, alerted readers, "The danger is apparent. The cooling has already killed hundreds of thousands of people in poor nations." Skeptics gleefully point to this and other popular works of that era as evidence that fears over a coming apocalypse owing to climate change, at least non-cyclical change triggered by human activity, are overblown and not supported by data.[20]

By the late 1970s, improved atmospheric models had debunked the theory that an ice age was lurking over the next horizon. Global temperatures were fluctuating, but the trend was not toward a colder climate. During the twenty-first century, the scientific consensus was that global temperatures were rising. The question was whether increased temperatures were attributable to natural processes, human industrial activity, or a combination thereof. One of the most ferocious scientific and political debates of recent years intensified over precisely this question.[21]

In 1988, the World Meteorological Organization of the United Nations Environment Programme established the Intergovernmental Panel on Climate Change (IPCC). Since its inception, the IPCC has issued a series of reports on the state of global climate-change science. For the most part, the group's moderate views and cautious conclusions have been well received. The IPCC's legitimacy was further enhanced when the organization shared the Nobel Peace Prize with former U.S. Vice President Al Gore in 2007. Owing to the IPCC's credibility, government policymakers in many countries looked to the organization for factual data on the issue, or at least they did prior to late 2009. In November of that year, however, hackers tapped into the computer server at the University of East Anglia Climatic Research Unit (CRU) a few weeks before the Copenhagen Climate Change Conference was set to convene. The IPCC relied on CRU temperature measurements as a major basis of the panel's reports. Much to the delight of climate-change skeptics, the hackers discovered e-mails and computer files suggesting that IPCC data were deliberately manipulated and falsified. One enterprising blogger labeled the ensuing political controversy "Climategate," an unmistakable reference to the Watergate scandal of the 1970s.[22]

In the wake of the Climategate scandal, the global warming/climate-change debate only grew more politically charged and clamorous. Committed environmentalists argued that the hour grows late: If substantial, systemic policy changes are not implemented immediately, the worldwide consequences will be cataclysmic—seas will rise rapidly, changing weather patterns will trigger more frequent and intense hurricanes and tornadoes, and entire species of plants and animals will face imminent extinction. On the far right side of the political spectrum, critics have been equally as bombastic, denouncing environmentalists as hypocritical doomsayers who have fabricated data to alarm the public as a cynical ploy to fill NGO coffers. Even when middle ground can be discovered, little political will seems to exist in most countries to formulate and implement the drastic policy changes necessary to forestall further global warming. The dispassionate observer is left scratching his head in wonder, perplexed by a continuing debate that creates more heat than light (pun intended).[23]

In each example cited here—BPA, endocrine disruptors, GMO foods, fracking, and climate change—as well as other common environmental concerns (the desire to ban specific products; tales of a garbage patch twice the size of Texas floating in the Pacific Ocean; the Alar health scare of the late 1980s and early 1990s; ongoing worries about electromagnetic fields, cell phones, radon, and so forth) a measure of uncertainty exists. Fear is born from that uncertainty. As the precautionary principle suggests, it is possible that one or more of these products or practices will lead to calamitous consequences at some unknown future time so it is preferable to be safe and cautious rather than move forward based on inconclusive evidence. No one can forget the devastating effects of thalidomide, a drug available in Europe during the 1950s to combat insomnia. The drug was slated to be available in the U.S. market until an FDA pharmacologist, Dr. Frances Oldham Kelsey, called attention to a series of toxic side effects that led to birth defects when pregnant women used the drug.

The problem is that the thalidomide example does not apply to each case of uncertainty. Dr. Kelsey had noticed disturbing trends in the research data. She methodically followed up by sending requests for the pharmaceutical companies seeking FDA approval to provide additional scientific information. The existence of conflicting scientific results drove Dr. Kelsey's concerns—a far cry from other types of environmental fears based on the precautionary principle. Environmental scares driven by the absence of data create fears of a crisis even when a crisis does not exist. If scientific uncertainty becomes the rationale for inaction, inaction will become the ultimate decision. A vigorous, spirited, data-driven debate about environmental issues is an integral step in developing effective public policy but the debate must be reasoned, deliberative, thoughtful, and based on generally recognized standards. It must avoid knee-jerk "crisis" responses in the absence of a compelling reason to treat an issue as an urgent matter in need of immediate concern.

Are any or all of the issues identified in this chapter genuine crises and, if so, what can and should be done about them? Undoubtedly, policymakers face daunting political and scientific challenges in coming years and decades. Yet the hyperbole employed by true believers, rather than advancing the cause, can lead to policymaking paralysis. Not every issue can be a looming crisis. To sound the alarm on every issue is to sound no alarm at all. Human beings seem psychologically predisposed to seek out easily identifiable environmental causes for complex effects that may or may not be partially or wholly attributable to these factors. Whatever the risks and uncertainties associated with these possible causes, they must not lead to the debilitating NIMBY (not in my backyard) syndrome.[24]

NIMBY is an understandable human response. No one gleefully looks forward to the day when a homeless shelter, drug-treatment center, halfway house, prison, hazardous-waste incinerator, or landfill appears in the neighborhood. In addition to lowering property values, an undesirable facility becomes a nuisance. The increased traffic and attendant factors—undesirable population, loud noises, and noxious odors, among other things—make it difficult to enjoy one's life and property. Residents of impoverished communities usually lack the political and economic clout to fight city hall, but affluent citizens can and will mobilize their resources and mount an effective public relations campaign to prevent an undesirable facility from being placed in their community. NIMBY becomes especially problematic because it can lead to another syndrome, BANANA (build absolutely nothing anywhere near anything).[25]

Politicization of Science

The contentious debates over BPA, endocrine disruptors, GMO foods, fracking, and climate change partially reflect a major development in twenty-first-century environmentalism: the politicization of science. With the lost luster of idealized science in the mid-twentieth century and the subsequent ascendancy of postmodernism, the thorny issue of how thoughtful people of good faith should approach the scientific process became a central issue. If "practicing science" could never be a neutral, uncontested process, value-laden choices would have to be made—choices about hypotheses, research design, data collection, interpretation and extrapolation, as well as appropriate methods for communicating the results. Ideally, common ground could be identified for undertaking research projects even if, as the postmodernists suggested, the endeavor was socially constructed and the scientific method was not unassailable. Navigating the treacherous shoals would not be simple or easy, but it would be essential if the gains of the western intellectual tradition since the Enlightenment were to be retained.[26]

A late twentieth century phenomenon—the rise of the 24-hour media news-cycle and the desire for "infotainment"—exacerbated the already-difficult quest for scientific legitimacy. Leading figures in the news and entertainment fields recognized fertile ground; they could fashion dual careers by sensationalizing and popularizing heretofore complex, turgid subject matter. Major news outlets began assigning a "science editor" to explore potentially sensationalistic stories, usually billed as lurking dangers caused by fill in the blank—common household detergents, hidden mold, leaking underground storage tanks, pesticides, malicious practices routinely undertaken by a local manufacturing company, etc. Eventually, an intrepid reporter was bound to stumble upon Love Canal, the sequel, and, oh, how the accolades would proliferate.[27]

Nicholas D. Kristof of the *New York Times* became one of the most prominent mainstream journalists to mix science and politics into a potent blend that deliberately commingled fact and opinion. In a series of clever newspaper editorials, Kristof raised the alarm about the imminent dangers of toxic chemicals in American life. His lack of credible data and his limited scientific background were hardly impediments to writing a good, "juicy" editorial. Every story requires familiar elements—a beginning, a middle, and an end, as well as recognizable characters (heroes and villains), conflict, and resolution. The familiar tale of a big, bad industry (villains) manufacturing dangerous products and forcing them onto an unsuspecting public in the name of corporate greed and profit (conflict) almost writes itself. Serving as the hero in the story, Kristof—the modern-day Alice Hamilton or Rachel Carson—can triumph over personal attacks on his integrity to expose the hidden corruption and save the day (resolution). It is a compelling, dramatic, easy-to-understand fable that reinforces existing stereotypes and requires little research or contemplative thought. In the postmodern landscape (and in the wake of Watergate and Love Canal), every cautionary tale of David versus Goliath is instantly believable. Established facts, while they are pleasant embellishments, are not required. Clichés are welcomed because they underscore the believability of the narrative by cloaking it in familiar garb. Complicated scientific descriptions and nuanced discussions of boring laboratory processes would only destroy the pathos and drama of the original story.[28]

If the goal were to educate and inform the public, Kristof could be castigated as a charlatan of the first order. In the postmodern context, however, entertainment is no more or less valuable than education. Kristof advances the environmental cause by stirring up public debate and presenting a compelling polemic, not to mention stimulating newspaper sales in an era of declining subscriptions and readership. His science may be junk science, lacking anything close to a null hypothesis—a standard feature of a credible research design—but such distinctions have no place or meaning in a postmodern world. If "science" is socially constructed, at least to some extent, why wallow in the data muck to debate the validity of complicated research? Who in this day and age possesses the credibility to determine factual accuracy, anyway?

Ironically, the lack of credible data does not mean a story is factually inaccurate. Lurking environmental dangers, in fact, could contribute to serious human health effects. Because it has become exceedingly difficult to distinguish between manufactured crises based on junk science and environmental dangers that cause health concerns, however, genuine public policy issues are not addressed. Hysterical voices screaming out false alarms drown out well-modulated voices raising legitimate, data-driven concerns.

The problem with politicized, sensationalized science is that the search for factual accuracy and reliable standards is ancillary to the goal of presenting an interesting and compelling tale. Rachel Carson sought to present a fact-based narrative on the chemistry of pesticides when she exposed the dangers of DDT. In the highly politicized, postmodern era, the point is to present the compelling narrative with or without the facts. If the facts happen to be accurate, that is an added bonus, but it is hardly the *sine qua non* of the story. If she were alive today, the methodical, deliberative, patient Rachel Carson no doubt would cringe to learn how her influence had been bastardized in the decades since *Silent Spring* appeared. Carson viewed the narrative as the means to an end. So many communicators of environmental information in the twenty-first century view the narrative as the means *and* the end.[29]

Of Symbols and Panaceas

As citizens join the American environmental movement, they seek assurances that genuine, measurable progress is being achieved. Citizen-environmentalists typically request easily understandable "metrics" such as "green" scorecards showing the amount of waste diverted from the landfill, the number of pounds of recyclable materials collected from drop-off locations, the number of times a specific product can be laid end-to-end before it stretches to California, Shangri-La, or the moon, or the number of homes that can be illuminated from the energy savings gained by turning off unused household appliances. It is no wonder that these measurements hold tremendous symbolic appeal for the budding environmentalist. The measurements are expressed in simple, familiar, easily digestible terms. Any suggestion that such measures of "greenness" fail to address systemic, ongoing environmental problems at a macro level meets with puzzlement, if not outright hostility.

Although they still play a key role in the environmental movement, elites have lost their place at the head of the line. Gone are the days when meaningful, incremental progress in environmental science was judged by how many credentialed researchers sifted through Toxics Release Inventory (TRI) data or struggled to master the nuances of the latest International Agency for Research on Cancer (IARC) monogram. As this book has discussed, American environmentalism has been democratized since the late twentieth century. Partially as a result of the entry of non-scientists into a field historically populated by scientists, new converts sometimes search for "quick fix" solutions—a phenomenon that partially explains why niche, "pet-peeve" issues have become popular among citizen-environmentalists.

A familiar inquiry in the twenty-first century is the question, "How green is your product/service/process?" The interlocutor seeks a response containing specific, measurable, easy-to-understand information on the environmental benefits of the product/service/process. Yet absent a well-known, generally accepted standard, "green" is a meaningless concept. The query is tantamount to asking someone, "Are you a nice person?" It depends on what the questioner means by "nice." If the standard of comparison for "niceness" is Adolf Hitler, the respondent does not hesitate to characterize himself as "nice." If the standard is Mother Teresa, the respondent hesitates to make such a comparison. In short, measurements of environmental quality cannot be made in a vacuum. A context and standards must be provided or the questions and answers become nonsensical. Responding to the question without providing a specific context amounts to a non-response, at best, or an unsubstantiated claim—"greenwashing"—at worst. The rise of "green"-certifying organizations, discussed in Chapter 9, is directly related to this desire to develop a context and standards for judging environmental claims.[30]

The difficulty for American environmentalism is that progress sometimes is measured symbolically. If the local environmental group successfully strong-arms timid legislators into banning plastic bag usage, certain types of fish and poultry that were not treated in the expected way during their lives, and construction of an undesirable facility in the neighborhood, the group publicizes its political prowess and declares victory. Yet the underlying issue has not been resolved. If plastic bags are eliminated, retail items must still be transported through other means, all of which raise environmental issues. If fish and poultry must be "cage free," "cruelty free," or "farm fresh" to enter the marketplace, this requirement may inflate food costs and affect suppliers throughout the supply chain. Perhaps this is a cost that proponents of the ban are willing to pay, but it is incumbent upon supporters to understand and communicate the unintended consequences of their actions. The landfill or hazardous waste facility that is not constructed in one neighborhood must be constructed in another. Symbolic victories do not solve the underlying environmental problems; they shift the problem, and the attendant costs, from one party to another. Under such circumstances, a symbolic triumph becomes a pyrrhic victory.

A committed environmentalist is well advised to avoid embracing most symbols and panaceas. Genuine progress is incremental, iterative, and difficult to assess. It requires multiple steps with assistance from numerous parties across a broad expanse of time and space. If environmental sustainability is to mean anything, it must mean a long-term commitment to difficult, sometimes complex and controversial, concepts, processes, and data. Anything less is chimerical.[31]

Quest for Energy and Its Environmental Consequences

It is axiomatic that the more technologically advanced and sophisticated a civilization becomes, the larger its energy appetite grows. A nomadic society built on a foundation of traditional authority has little need of high-yield energy sources. A highly complex, industrialized civilization based on legal-rational authority with a well-developed institutional structure of government and a capitalist economy necessarily will require plentiful, inexpensive sources of energy to operate factories, automobiles, business offices, residences, and myriad electronic mechanisms and devices.

Throughout human history, civilizations that have developed abundant, efficient, reliable, inexpensive sources of energy have thrived even as their competitors have languished. Therefore, the single greatest challenge for "power-hungry" societies in the twenty-first century is to find and maintain sources of energy while factoring environmental concerns into their calculations. All too often in U.S. history, however, short-term political expediency has won out over long-term environmentalism. The consequences have been severe, and they remain a source of great consternation in the environmental community. Recognizing the spotty history of U.S. natural-resource protection, environmentalists have argued, and continue to argue, that the quest for energy and concern for protecting the natural environment are at odds with each other. In an age of smart technology, this traditional division need not always be the case. If a choice has to be made, however, it is clear that short-term energy needs typically outweigh long-term environmental considerations. One need only look to Brazil, India, and China for salient examples of the negative externalities created when burgeoning, energy-intensive societies fail to engage in long-range environmental planning.[32]

In light of the history of environmental despoliation in the United States, "green" policymakers insist that a national energy plan must include significant investments in renewable resources to mitigate environmental damage associated with meeting national energy needs. "Renewables" include a variety of "clean" sources, most of which are not yet commercially viable or are new industries still in their infancy. Wind energy, for example, is produced by technologically advanced turbines concentrated in sufficient numbers inside "wind farms" designed to generate electricity. Another renewable energy source, sunlight, can be thought of in two ways. "Passive" solar energy is used in "green" buildings, which are designed to take advantage of natural, "unenhanced" sunlight to reduce energy consumption. Sunlight can also be captured by photovoltaic cells (PVs), which store the energy for later use. Biomass is a form of renewable energy that uses dung, decaying wood, and other plant material to generate heat for producing electricity. In developing nations, biomass can be burned to heat homes and cook food. Ethanol, biodiesel, and advanced biofuels comprised of algae, cellulose, and similar feedstocks can serve as substitutes for gasoline in internal combustion engines. Geothermal energy requires hot water or steam to be pumped up from underground areas to propel turbines for producing electricity. Similarly, hydropower relies on falling or pressurized water to drive turbines, although some environmentalists question whether building a dam to supply hydroelectric power is worth the resultant environmental degradation from constructing an edifice that will alter natural river flows.[33]

When environmentalists speak of renewables, they positively gush. No adjective is too fanciful, hyperbolic, or far-fetched. They sometimes call for immediate change—rejecting an incremental transition from fossil fuels to renewables as too time-consuming and laborious for the challenges facing the nation and the world—with renewable energy serving as the vanguard of a green revolution. Their depiction of an inexhaustible supply of inexpensive, clean energy and booming "green energy" jobs sounds too good to be true. In a sense, it is. In their telling, unlike fossil fuels such as coal, gas, oil, and natural gas, which contribute to atmospheric smog and pollution when they are burned, renewable resources are readily available, inexhaustible, and clean.[34]

If renewable energy lived up to its advance billing, virtually everyone save traditional energy company executives would readily embrace such sources. Yet supporters frequently ignore serious challenges facing renewable energy development. Four difficulties immediately spring to mind. First, renewables lack power density (e.g., horsepower or kilowatts), which can be thought of as the amount of power that can be harnessed in a specific geographic area. Wind farms can generate electricity, but they require an enormous amount of vacant land on which to site turbines—to say nothing of the large number of birds that perish when they slam into the turning blades—and they produce relatively little power compared with fossil fuels. Similarly, solar power and biofuels provide "little bang for the buck," which severely limits their practical utility. Second, renewables also contain low energy density (e.g., joules per cubic meter or watt-hours per pound), which means a small amount of energy can be contained and stored in a given unit of mass, area, or volume. Storing energy through batteries for availability on demand from renewables is not feasible for most applications because the number and size of the batteries would have to be enormous. With the increasing urbanization of U.S. households, Americans constantly seek to use an increasing amount of power in a smaller space. Renewables go against the trend.

Because renewables possess low power density and low energy density, they raise problems of cost and scale. They simply do not compare favorably with traditional fossil fuels in meeting energy demand, although they hold environmental advantages. "We use hydrocarbons—coal, oil, and natural gas—not because we like them, but because they produce lots of heat energy, from small spaces, at prices we can afford, and in the quantities that we demand," Robert Bryce wrote in *Power Hungry*, a 2010 book on the energy needs of a technologically advanced society. "And that's the absolutely critical point. The energy business is ruthlessly policed by the Four Imperatives: power density, energy density, cost, and scale." This insight is not to suggest that renewable energy sources should be rejected out of hand. If problems and challenges disqualified an energy source from consideration, no source of power would exist. Bryce's point was that renewables have a place in the energy mix, but they are not a panacea. Accordingly, no one can expect to witness a wholesale transition from established energy sources to renewables in the near future. Dependence on a mixture of fossil fuels and perhaps nuclear power will continue for a long time to come (Figure 10.1).[35]

This sobering assessment means that no immediate relief is available from renewables to combat traditional environmental problems. Air pollution, water pollution, concern about toxic chemicals, and deforestation will remain issues of concern. Americans could resolve these issues more expeditiously if they were willing to forgo energy-on-demand or the lifestyle of convenience to which most citizens have grown accustomed. Deep ecologists and radical environmentalists lobby for a return to the halcyon days of a pre-industrial yesteryear, but no policymaker in a position of authority would seriously propose an energy plan that divested citizens of their technological conveniences. It would be political suicide.

Who Pays and Who Plays

Despite the disdain with which some committed environmentalists view adherents of neoclassical economics, the reality is that every environmental issue eventually becomes a financial issue. Environmental laws and regulations enacted by governments seek to reallocate costs and redistribute benefits; they require private firms to consider negative externalities in their financing decisions. Every dollar spent on pollution control measures and regulatory compliance is a dollar unavailable to reinvest in the business. Measures designed to rectify past inequities and satisfy environmental justice considerations first must determine who will bear the costs.

Figure 10.1 The search for reliable and affordable sources of energy largely determines American environmental policy. Fossil fuels, symbolized here by oil wells on the coast, will continue to be a source of energy and environmental concerns well into the future. (Courtesy of the Library of Congress)

The question of "who pays and who plays" for environmental conservation, preservation, and protection could be (and has been) the subject of many a standalone book. For illustrative purposes here, consider one example of the centrality of neoclassical economics involving a prominent environmental issue, namely solid waste (garbage) financing. Twenty-first-century Americans enjoy a lifestyle of convenience unknown to most generations of the past, but that convenience comes at a cost. A "throwaway society" must devise an effective method of handling discarded materials in a responsible manner.[36]

EPA has promoted the concept of "reduce-reuse-recycle," encouraging consumers to cut back on the amount and number of disposable products they use, reuse products whenever possible, and recycle a product so that it goes on to a second life before it is discarded. Discarded materials can be incinerated or sold to foreign countries for myriad purposes, but often such materials are buried in sanitary landfills or wind up as roadside litter.

Landfills were developed after World War II as an alternative to dumping untreated waste in a vacant lot on the edge of town to fester and rot. In the early 1900s, municipalities routinely dumped discarded materials onto swampland, allowing large piles of refuse to proliferate, especially in poor neighborhoods. Over time, the unsightly, reeking, explosive mountains of garbage attracted rats and assorted scavengers, destroying local property values and breeding disease. The sanitary landfill was engineered to be a marked improvement on this haphazard practice. Rather than dumping unwanted materials onto a large tract of land and allowing them to biodegrade, solid waste planners would excavate a site specifically designed to entomb the contents. After lining the pit with an artificial barrier (typically made of lead or plastic), engineers emplaced waste in layers. In between each layer, they packed soil, gravel, clay, or plastic. They also drilled wells to monitor the groundwater. Venting trapped methane gas reduced the likelihood of an explosion. Removing sunlight, oxygen, and water significantly reduced biodegradation caused by aerobic organisms, although anaerobic organisms thrived in the manufactured environment.[37]

This solid-waste management system required municipalities or private companies to collect discarded materials from commercial properties as well as private residences. In most cases, the parties discarding the waste bore the costs of collection and transportation of materials to the landfill. If a local government supplied the service, the consumer paid through *ad valorem* taxes or a similar taxing mechanism. Private companies supplying the service sent an invoice for payment to the customer. The government or private company hauling discarded material to the landfill remitted a "tipping fee" to back the truck up to the site and dump the load (unless the hauler also operated the site). A government entity or private company managed the landfill operation. In time, some solid-waste operators added a materials recovery facility (MRF) to permit intervention between the point at which the consumer discarded material and the time the garbage truck dropped off the load. If recoverable commodities such as aluminum cans, newsprint, plastic, or glass bottles were mixed with other items, MRF operators could extract the materials for resale, thereby removing items with a secondary value from landfill contents.[38]

As concerns about solid waste increased during the latter half of the twentieth century, sanitary landfills, once heralded as the promise of the future because they prevented harmful biodegradation, became controversial, vilified as impediments to biodegradation. To the uninitiated citizen-environmentalist, biodegradation sounded wonderful, a panacea. Psychologically, a problem that rots away is a problem that has been resolved.

Public interest in solid waste had never been widespread, but it grew tremendously after stories circulated during the 1970s and 1980s that state and local governments were running out of landfill space. Matters came to a head in 1987 when the Mobro 4,000 garbage barge set sail from Long Island, New York, bound for Morehead City, North Carolina, to be incinerated for methane. When rumors circulated that the garbage was contaminated with hazardous material, North Carolina health officials refused to allow the barge to unload the trash. For their part, Long Island officials refused to allow the shipment to return. Americans therefore were treated to the spectacle of the garbage barge traveling up and down the eastern seaboard searching for a home. The shipment journeyed as far south as Belize before being incinerated in Brooklyn in October—seven months after it first set sail. Although the episode was the result of poor decision-making by state and local officials as well as the uncertainty of new environmental regulations, it garnered much press attention, especially when media personalities such as Phil Donahue and Johnny Carson repeatedly mocked the absurd incident on their television broadcasts. By the time it was resolved, the Mobro imbroglio had entered the realm of myth. More than any other single episode, the homeless garbage barge suggested to the public that the American solid-waste management system was broken and something must be done—and done soon—to fix it.[39]

Numerous proposals exist for handling materials that are not reduced, reused, or recycled. Every proposal involves the use of financial incentives to alter human behavior. Some measures, such as the perennially discussed advance disposal fee (ADF), would impose a Pigouvian tax on selected disposable products. Presumably, if consumers were forced to pay a fee and made aware of the costs involved, they would modify their purchasing decisions and insist on "greener" products. Even if the tax did not change consumer behavior, ADF-generated funds could be used to improve the solid-waste management infrastructure.

Free-market economists and proponents of the Coase Theorem vehemently disagree with these premises, expressing little faith in tax-and-spend proposals. In their view, consumers would not necessarily alter their behavior. A proportion of the costs associated with keeping track of the ADF funds and submitting payment to the government entity that imposed the fee would be passed on to the consumer, but a percentage of the costs would be borne by wholesalers or retailers. Inefficiency in administration, coupled with the propensity of legislators to deposit collected monies into the general fund, mean that an ADF could become another tax that does not alter consumer behavior or improve the solid-waste management infrastructure.[40]

Each proposal floated to improve solid waste collection and management—imposing landfill bans on selected materials; mandating specific recycling "rates and dates"; replacing a "one-size-fits-all" fee for trash collection from commercial and residential properties with a rate based on the volume of waste discarded; creating a system of extended producer responsibility requiring manufacturers of a product to "take back" the product after it reaches

Figure 10.2 Solid waste became a major environmental concern late in the twentieth century. This photograph shows the Fresh Kills Landfill in New York in 1999. (Courtesy of the Granger Collection, New York)

the end of its useful life; co-locating a facility to incinerate landfill contents and using the resultant heat as an energy source—involves a reallocation of resources. Money is an effective behavioral incentive, but incentives imposed by one party on other parties without adequate consultation sometimes contain hidden costs and trigger unintended consequences. To avoid unforeseen costs and consequences, the party that pays for environmental cleanup must be a major decision-maker. For that reason, an effective partnership with all units of government, private businesses, environmental groups, and the citizenry will be necessary—not just in the realm of solid waste, but for all environmental problems. If one party is vilified or required to bear a burden disproportionate to its share of the problem, solutions will be difficult, if not impossible, to implement. The days of command-and-control environmentalism must yield to a new era of sustainable practices—by all parties, public and private, corporate and citizen (Figure 10.2).[41]

Tradeoffs and Choices: Life Cycle Management (LCM)

Speaking of sustainable practices, the difficulty with postmodern environmentalism is the tendency among some parties to reach conclusions outside of a specific context. As previously discussed, answers to questions such as "Is this process green?" or "Should this product be banned?" cannot be found without considering how the process or product compares with alternatives. The comparison must consider all relevant variables—raw material supplies and cost, energy usage, transportation effects, pollutants emitted because of the process or product manufacture, end of life for the process or product, and so forth. Comparisons must consider realistic alternatives. After the variables and alternatives are identified, researchers must assign a value to inputs, throughputs, and outputs so that quantitative data can be generated. Banning plastic bags in favor of a hypothetical, non-existent competing product, for example, does not achieve a sustainability goal if other types of disposable bags are worse for the environment based on comparisons among measurable data. The "life cycle" of each process or product must be assessed and compared to generate reliable data that will serve as the basis for thoughtful, deliberative, data-driven decision-making.

In the words of one commentator, "Lip service is not enough to achieve a sustainable development of our societies. If we want to make sustainability happen as concrete reality in both public policy-making and corporate strategies, sustainability cannot please everybody. To make it happen, we have to be able to discern good and evil." Discerning "good and evil" requires a comprehensive assessment of environmental, economic, and social performance indicators. "This is where life cycle assessment (LCA) and life cycle management (LCM) come into play. LCA is an internationally accepted method for measuring environmental performance and LCM is in a nutshell about the application of LCA or rather life cycle thinking (LCT)."[42]

LCM techniques vary according to the process or product under consideration. In most instances, the first step is to identify "green stakeholders." In a corporate setting, internal stakeholders exist—primarily shareholders, corporate officers, and employees—as well as external stakeholders (suppliers, wholesalers, retailers, and customers). Elected and unelected government officials also have a stake, as do media representatives and local communities. Not all parties may agree on the specific value to assign to each variable, but the decision to engage in the LCM exercise likely will be popular, assuming the parties genuinely seek to arrive at decisions grounded in quantitative data.

LCM studies became popular after the World Commission on Environment and Development (WCED) published its report, *Our Common Future: From One Earth to One World*, in 1987. During the ensuing years, LCM techniques and indicators became more sophisticated and standardized. International management conferences devoted exclusively to developing scientific and practical solutions for incorporating life-cycle approaches into strategic and operational decision-making have proliferated. Engineers, natural scientists, social scientists, and policymakers have spent years refining life-cycle management tools. Even more robust and rigorous LCM techniques, especially regarding implementation processes and the incorporation of economic and social-science data within sustainability studies, probably will appear in the future.[43]

Whatever else the future holds, it seems reasonably clear that American environmentalism is alive and well, albeit a bit fractured and occasionally chaotic. Citizens are more aware of, and concerned about, the natural environment than at any time in U.S. history. Whether this awareness will translate into effective, thoughtful, well-informed public policy formulation and implementation remains an open question and a never-ending concern. If the pitfalls discussed in this chapter can be avoided and if the democratization of the environmental movement can be directed toward the suitable ends discussed in this book, the future appears promising. All the features necessary to open the door to an environmentally sustainable future exist, particularly a stable political system constructed on the bedrock of a strong institutional structure that protects individual rights, encourages citizen participation, and promotes incremental pluralism as well as a sound economic system that provides incentives for technological innovation to meet changing consumer demands. All that remains is for each participant in the environmental field, armed with a good-faith desire to find solutions to common problems through sound science and shared sacrifice, to walk through the door leading to the rest of the twenty-first century and beyond.

Notes

1. William Shakespeare, *King Richard II* (New York: Airmont Publishing Company, 1966), Act III, scene ii, 63.
2. See, for example, Thomas R. Dunlap, "Environmentalism, a Secular Faith," *Environmental Values* 15, 3 (August 2006): 321–330; Joel Garreau, "Environmentalism as Religion," *The New Atlantis: A Journal of Technology & Society* 28, 1 (Summer 2010): 61–74; Robert H. Nelson, "Economic Religion vs. Environmental Religion," *The Independent: Newsletter of the Independent Institute* 20, 4 (Winter 2011): 1, 7.
3. On the slavery question, see, for example, Eric Foner, *Forever Free: The Story of Emancipation and Reconstruction* (New York: Knopf, 2005); James Oliver Horton and Lois E. Horton, *Slavery and the Making of America* (New York and Oxford: Oxford University Press, 2005); J. Michael Martinez, *Coming for to Carry Me Home: Race in America from Abolitionism to Jim Crow* (Lanham, Md.: Rowman & Littlefield, 2012).
4. See, for example, Jim Cheney, "Postmodern Environmental Ethics: Ethics as Bioregional Narrative," in *Postmodern Environmental Ethics*, Max Oelschlaeger, editor (Albany, N.Y.: State University of New York Press, 1995), 23–42.
5. Paul Maltby, "Fundamentalist Dominion, Postmodern Ecology," *Ethics & The Environment* 13, 2 (Fall 2006): 128, 131–135.
6. Jon Anderson, "From 'Zombies' to 'Coyotes': Environmentalism Where We Are," *Environmental Politics* 19, 6 (November 2010): 973–991; Maltby, "Fundamentalist Dominion, Postmodern Ecology," 135–136.

7. Robert H. Nelson, *Economics as Religion: From Samuelson to Chicago and Beyond* (University Park, Penn.: Pennsylvania State University Press, 2001), 76; Dan Tarlock, "Environmentalism: Postmodern Evangelism or Unitarianism?" *Case Western Reserve Law Review* 56, 3 (Spring 2006): 644–647; Paul Wapner, "Leftist Criticism of 'Nature': Environmental Protection in a Postmodern Age," *Dissent* 50, 1 (Winter 2003): 71–72.
8. The options are discussed in Wapner, "Leftist Criticism of 'Nature,'" 71–74. The quote appears on page 73 of Wapner. See also Maltby, "Fundamentalist Dominion, Postmodern Ecology," 135; Dan Tarlock, "Environmentalism," 646–648.
9. See, for example, Wallace Kaufman, *No Turning Back: Dismantling the Fantasies of Environmental Thinking* (New York: Basic Books, 1994); M. Mazilu and M. Ciobanu, "Deliberative Democracy Establishment in Environmental Policies: Alternatives for the Environmental Conflict Resolution," *Journal of Environmental Science and Engineering* 4, 4 (April 2010): 77–84; Shari R. Veil, "Mindful Learning in Crisis Management," *Journal of Business Communication* 48, 2 (April 2011): 116–147.
10. Lynn Bergeson, "Facts and Consequences," *The Environmental Forum* 27, 6 (November/December 2010): 27; Jon Entine, "BPA: DOA?" *The Environmental Forum* 27, 6 (November/December 2010): 24–30; Alice Park, "Ban BPA?," *Time* 179, 5 (April 16, 2012): 22.
11. Environmental Working Group (EWG) Website: "About the Environmental Working Group," http://www.ewg.org/about (Accessed December 19, 2012); Environmental Working Group (EWG) Website: "BPA and Human Diseases on the Rise," http://www.ewg.org/node/20937 (Accessed December 19, 2012); John Berlau, "The Smear Campaigns," *Insight on the News* 20, 11 (May 11, 2004): 18–20; Louis Jacobson, "Fenton Helps MoveOn Move On," *National Journal* 36, 4 (January 24, 2004): 253; Rowan Scarborough, "Fenton Communications: The Far-Left's Propaganda Machine," *Human Events* 65, 17 (May 11, 2009): 7. See also Natural Resources Defense Council Website: "Bisphenol A," http://www.nrdc.org/living/chemicalindex/bisphenol-a.asp (Accessed December 19, 2012).
12. Alex B. Berezow and Hank Campbell, *Science Left Behind: Feel-Good Fallacies and the Rise of the Anti-Scientific Left* (New York: Public Affairs Books, 2012), 123–124; Entine, "BPA: DOA?," 24–30; Susan Freinkel, *Plastic: A Toxic Love Story* (Boston: Houghton Mifflin Harcourt, 2011), 93–94, 114, 268–269; The Society of the Plastics Industry Website: "The Safety of Polycarbonate Packaging," http://www.plasticsindustry.org/AboutPlastics/content.cfm?ItemNumber=716&navItemNumber=1126 (Accessed December 19, 2012); Sarah A. Vogel, "The Politics of Plastics: The Making and Unmaking of Bisphenol A 'Safety,'" *American Journal of Public Health* 99, S3 (2009): S559–S566.
13. Ashley Ahearn, "Communicating about Chemical Body Burden, with Tracey Woodruff and Rachel Morello-Frosch," *Environmental Health Perspectives* 119, 5 (May 2011): 1–7; Berezow and Campbell, *Science Left Behind*, 124; Melinda Burns, "Toxics Shock," *Miller-McCune.com* 4, 4 (July-August 2011): 74–76; Charles H. Eccleston and Frederic March, *Global Environmental Policy: Concepts, Principles, and Practice* (Boca Raton, Fla.: CRC Press, 2011), 146–148; Jon Entine, *Scared to Death: How Chemophobia Threatens Public Health: A Position Statement of the American Council on Science and Health* (New York: American Council on Science and Health, 2011); Mark Mattson and Edward Calabrese, "Best in Small Doses," *New Scientist* 198, 2668 (August 9, 2008): 36–39; Cristina L. Quinn and Frank Wania, "Understanding Differences in the Body Burden—Age Relationships of Bioaccumulating Contaminants Based on

Population Cross Sections versus Individuals," *Environmental Health Perspectives* 120, 4 (April 2012): 554–559; Sarah A. Vogel, "From 'The Dose Makes the Poison' to 'The Timing Makes the Poison': Conceptualizing Risk in the Synthetic Age," *Environmental History* 13, 4 (October 2008): 667–673.

14. The Food Quality Protection Act (FQPA) of 1996, 7 U.S.C § 136d; 7 U.S.C. § 136a; 7 U.S.C. § 136w; 7 U.S.C. § 136; 7 U.S.C. § 136a; 7 U.S.C. § 136q; 7 U.S.C. § 136d (1996). See also Robin E. Dodson, Marcia Nishioka, Laurel J. Standley, Laura J. Perovich, Julia Green Brody, and Ruthann A. Rudel, "Endocrine Disruptors and Asthma-Associated Chemicals in Consumer Products," *Environmental Health Perspectives* 120, 7 (July 2012): 935–943; Cristina Preda, Maria Christina Ungureanu, and Carmen Vulpoi, "Endocrine Disruptors in the Environment and Their Impact on Human Health," *Environmental Engineering and Management Journal* 11, 9 (September 2012): 1697–1706; United States Environmental Protection Agency Website: "Endocrine Disruptor Screening Program (EDSP)," http://www.epa.gov/endo/ (Accessed December 20, 2012); United States Environmental Protection Agency Website: "The Food Quality Protection Act of 1996 (FQPA) Background," http://www.epa.gov/opp00001/regulating/laws/fqpa/backgrnd.htm (Accessed December 20, 2012); Alana Van Der Mude, "Endocrine-Disrupting Chemicals: Testing to Protect Future Generations," *Boston College Environmental Affairs Law Review* 38, 2 (2011): 509–535; Jason M. Vogel, "Perils of Paradigm: Complexity, Policy Design, and the Endocrine Disruptor Screening Program," *Environmental Health: A Global Access Science Source* 4, 2 (February 2005): 2–11; Jason M. Vogel, "Tunnel Vision: The Regulation of Endocrine Disruptors," *Policy Sciences* 37, 3–4 (December 2004): 277–303.
15. Theo Colborn, Dianne Dumanoski, and John Peterson Myers, *Our Stolen Future: Are We Threatening Our Fertility, Intelligence, and Survival? A Scientific Detective Story* (New York: Plume, 1997). Gore is quoted on page vii. See also The Endocrine Disruption Exchange Website: "About Us," http://www.endocrinedisruption.com/about.introduction.php (Accessed December 20, 2012); Our Stolen Future Website: "Home Page," http://www.ourstolenfuture.org/ (Accessed December 20, 2012); Blue Moon Fund Website: "Home Page," http://www.bluemoonfund.org/ (Accessed December 20, 2012); Dianne Dumanoski Website: "Home Page," http://www.diannedumanoski.com/about.html (Accessed December 20, 2012).
16. Berezow and Campbell, *Science Left Behind*, 46, 220; Eccleston and March, *Global Environmental Policy*, 28; H. Eggert and M. Greaker, "Trade, GMOs, and Environmental Risk: Are Current Policies Likely to Improve Welfare?" *Environmental and Resource Economics* 48, 4 (April 2011): 587–608; Jon Entine, *Crop Chemophobia: Will Precaution Kill the Green Revolution?* (Washington, D.C.: American Enterprise Institute, 2011); Jon Entine, editor, *Let Them Eat Precaution* (Washington, D.C.: American Enterprise Institute, 2006); FrankenFoods.com Website: "Home Page," http://www.frankenfoods.org/ (Accessed December 20, 2012); Greenpeace Website: "GMO Victories," http://www.greenpeace.org/usa/en/campaigns/victories/gmo-victories/ (Accessed December 21, 2012); Pieter Maeseele, "NGOs and GMOs: A Case Study in Alternative Science Communication," *Javnost—The Public* 16, 4 (2009): 55–72; P. Pandey, B. Kumar, and D. K. Tiwari, "Environmental Considerations Concerning the Release of Genetically Modified Organisms," *ProEnvironment*

Promediu 3, 6 (December 2010): 381–384; Lisa H. Weasel, *Food Fray: Inside the Controversy over Genetically Modified Food* (New York: AMACOM, A division of the American Management Association, 2009), 32.

17. Peter Heywood, "Fracking: Safer and Greener?," *TCE: The Chemical Engineer* 850 (April 2012): 42–45; Chris Mooney, "The Truth about Fracking," *Scientific American* 305, 5 (November 2011): 80–85; Bob Weinhold, "The Future of Fracking: New Rules Target Air Emissions for Cleaner Natural Gas Production," *Environmental Health Perspectives* 120, 7 (July 2012): A272–A279; Daniel Yergin, *The Quest: Energy, Security, and the Remaking of the Modern World* (New York: The Penguin Press, 2011), 327–332.
18. The Safe Drinking Water Act (SDWA) of 1974, 42 U.S.C. § 300f *et seq.* (1974). See also Abby W. Schachter, "Energy Independence and Its Enemies: The Bounty of Shale Oil—And the Environmentalists That Want to Keep It Buried," *Commentary* 133, 6 (June 2012): 24–29; Yergin, *The Quest*, 330–331.
19. Spencer Weart, "Global Warming: How Skepticism Became Denial," *Bulletin of the Atomic Scientists* 67, 1 (January 2011): 41–42; Yergin, *The Quest*, 428–429.
20. Lowell Ponte, *The Cooling: Has the Next Ice Age Already Begun? Can We Survive It?* (Upper Saddle River, N.J.: Prentice Hall, 1976), xv. See also Roger Scruton, *How to Think Seriously about the Planet: The Case for an Environmental Conservatism* (New York and Oxford: Oxford University Press, 2012), 39–41; Weart, "Global Warming," 42.
21. Tom Clynes, "The Battle: Climate Scientists Routinely Face Death Threats, Hate Mail, Nuisance Lawsuits, and Political Attacks. How Much Worse Can It Get?" *Popular Science* 281, 1 (July 2012): 37–43, 80; Kruti Dholakia-Lehenbauer and Euel W. Elliott, "Decisionmaking, Risk, and Uncertainty: An Analysis of Climate Change Policy," *Cato Journal* 32, 3 (Fall 2012): 539–556; Alexander A. Golub, "Climate Policy and Uncertainty: A Precautionary Principle versus Real Options Analysis," *Environmental Law Reporter: News & Analysis* 42, 8 (August 2012): 10733–10736; Naomi Klein, "Capitalism vs. Climate: What the Right Gets—and the Left Doesn't—About the Revolutionary Power of Climate Change," *The Nation* 293, 22 (November 28, 2011): 11–21; Fred Powledge, "Scientists, Policymakers, and a Climate of Uncertainty: Can Research Gain a Foothold in the Politics of Climate Change?" *BioScience* 62, 1 (January 2012): 8–13; Weart, "Global Warming," 42–43.
22. Climate Depot Website: "Home Page," http://climatedepot.com/ (Accessed December 21, 2012); The Intergovernmental Panel on Climate Change (IPCC) Website: "Home Page," http://www.ipcc.ch/index.htm#.UNS7YqxNtqA (Accessed December 21, 2012); Scruton, *How to Think Seriously about the Planet*, 38–51.
23. The leading books in the "a-catastrophe-is-a'comin'" school of thought include James Hansen, *Storms of My Grandchildren: The Truth about the Coming Climate Catastrophe and Our Last Chance to Save Humanity* (New York and London: Bloomsbury, 2011) and Mark Lynas, *Six Degrees: Our Future on a Hotter Planet* (London: The Fourth Estate, an Imprint of HarperCollins, 2007). Leading works expressing skepticism about the apocalyptic consequences of climate change include Patrick J. Michaels and Robert C. Balling, *The Satanic Gases: Clearing the Air about Global Warming* (Washington, D.C.: The Cato Institute, 2000); Richard S. Lindzen, "Global Warming: The Origin and Nature of Alleged Scientific Consensus," *Cato Review of Business and Government* 15, 1 (Spring 1992): 87–98; Roy W. Spencer, *Climate Confusion: How Global Warming Hysteria Leads to Bad Science, Pandering Politicians and Misguided Policies That Hurt the Poor* (New York: Encounter Books, 2008). Other sources attempt a more moderate position, that is, that climate change is possible, but the consequences need

not be disastrous. See, for example, Bjørn Lomborg, *The Skeptical Environmentalist: Measuring the Real State of the World* (New York and Cambridge, U.K.: Cambridge University Press, 2001); Anthony Giddens, *The Politics of Climate Change* (Malden, Mass.: Polity Press, 2009); Wendee Holtcamp, "Flavors of Uncertainty: The Difference between Denial and Debate," *Environmental Health Perspectives* 120, 8 (August 2012): A315–A319.

24. See, for example, Emily Anthes, "The Everyday Fear Zone," *Psychology Today* 45, 6 (November/December 2012): 76–83; Lindsey Blomberg, "By the Numbers: The Great Pacific Garbage Patch," *E: The Environment Magazine* 22, 3 (May/June 2011): 8; Linda Bren, "Frances Oldham Kelsey: FDA Medical Reviewer Leaves Her Mark on History," *FDA Consumer* 35, 2 (March/April 2001): 24–29; Gregg Easterbrook, *A Moment on the Earth: The Coming Age of Environmental Optimism* (New York: Viking, 1995), xiii–xv; Freinkel, *Plastic*, 201–232; Kaufman, *No Turning Back*, 11, 17, 121; Aaron M. McCright and Riley E. Dunlap, "The Politicization of Climate Change and Polarization in the American Public's Views on Global Warming, 2001–2010," *The Sociological Quarterly* 52, 2 (Spring 2011): 155–194; MyPlasticFreeLife.com Website: "Home Page," http://myplasticfreelife.com/ (Accessed December 21, 2012); Jedediah Purdy, "The Politics of Nature: Climate Change, Environmental Law, and Democracy," *The Yale Law Journal* 119, 6 (April 2010): 1122–1209; S. F. Singer, "The Science behind Global Environmental Scares," *Consumers' Research Magazine* 74, 10 (October 1991): 17–21; Mark Stephen, "Sinking the Myth of the 'Garbage Patch,'" *Canadian Plastics* 69, 1 (March 2011): 4; Jacqueline Vaughn Switzer, *Green Backlash: The History and Politics of Environmental Opposition in the U.S.* (Boulder, Colo.: Lynne Rienner Publishers, 1997), 291; Beth Terry, *Plastic Free: How I Kicked the Plastic Habit and How You Can Too* (New York: Skyhorse, 2012).

25. Michael Dear, "Understanding and Overcoming the NIMBY Syndrome," *Journal of the American Planning Association* 58, 3 (Summer 1992): 288–300; Helene Hermansson, "The Ethics of NIMBY Conflicts," *Ethical Theory & Moral Practice* 10, 1 (January 2007): 23–34; Don Markley, "It's Not NIMBY Anymore—It's BANANA," *Broadcastengineering.com* 44, 3 (March 2002): 52–53.

26. Berezow and Campbell, *Science Left Behind*, 257–260; Neil Carter, *The Politics of the Environment: Ideas, Activism, Policy*, 2nd ed. (New York and Cambridge, U.K.: Cambridge University Press, 2007), 5–9; Riley E. Dunlap and Aaron M. McCright, "A Widening Gap: Republican and Democratic Views on Climate Change," *Environment* 50, 5 (September 2008): 26–35; Easterbrook, *A Moment on the Earth*, 224; Jon Entine, *Scared to Death*, 4–7; William Happer, "Harmful Politicization of Science," in *Politicizing Science: The Alchemy of Policymaking*, Michael Gough, editor (Stanford, Calif.: Hoover Institution Press, 2003), 27–48; Kaufman, *No Turning Back*, 84; David A. Moss, "Fixing What's Wrong with U.S. Politics," *Harvard Business Review* 90, 3 (March 2012): 134–139; Shawn Lawrence Otto, "America's Science Problem," *Scientific American* 307, 5 (November 2012): 62–71.

27. See, for example, Bonnie Anderson, *News Flash: Journalism, Infotainment, and the Bottom-Line Business of Broadcast News* (San Francisco, Calif.: Jossey-Bass, 2004); Roger Luckhurst, "The Two Cultures, or the End of the World as We Know It," *Interdisciplinary Science Reviews* 32, 1 (March 2007): 55–64; Michael Shermer, "I Want to Believe," *Scientific American* 301, 1 (July 2009): 33–35.

28. See, for example, Nicholas D. Kristof, "The Cancer Lobby," *The New York Times*, October 7, 2012, SR13; Nicholas D. Kristof, "Chemicals and Our Health," *The New York Times*, July 16, 2009, A27.
29. Berezow and Campbell, *Science Left Behind*, 58–59; Mark B. Brown and David H. Guston, "Science, Democracy, and the Right to Research," *Science & Engineering Ethics* 15, 3 (September 2009): 351–366; Bruce Hass and Michael Kleine, "The Rhetoric of Junk Science," *Technical Communication Quarterly* 12, 3 (Summer 2003): 267–284; Dan Hicks, "On the Ideal of Autonomous Science," *Philosophy of Science* 78, 5 (December 2011): 1235–1248; Michael Shermer, *The Believing Brain: From Ghosts and Gods to Politics and Conspiracies—How We Construct Beliefs and Reinforce Them as Truths* (New York: Times Books, 2011), 334–344.
30. The United States Federal Trade Commission (FTC) has developed standards at 40 C.F.R. Part 260, colloquially known as the "green guides," for making "green" environmental marketing claims. The agency originally issued the guides in 1992 and revised them in 1996, 1998, and 2012. See *Guides for the Use of Environmental Marketing Claims*, available at the FTC Website: "Guides for the Use of Environmental Marketing Claims," http://www.ftc.gov/bcp/grnrule/guides980427.htm (Accessed December 26, 2012). See also Lynn L. Bergeson, "The Explosion of Green Marketing Claims: FTC Jumps into the Fray," *Environmental Quality Management* 18, 2 (Winter 2008): 71–78; Lynn L. Bergeson, "Selling Green: U.S. FTC Releases Proposed Revisions to the 'Green Guides,'" *Environmental Quality Management* 20, 3 (Spring 2011): 77–81; David Gibson, "Awash in Green: A Critical Perspective on Environmental Advertising," *Tulane Environmental Law Journal* 22, 2 (June 2009): 423–440; United States Federal Trade Commission (FTC) Website: "FTC Issues Revised 'Green Guides': Will Help Marketers Avoid Making Misleading Environmental Claims,"http://www.ftc.gov/opa/2012/10/greenguides.shtm (Accessed December 26, 2012); Jean-Cyril Walker, "It's Still Not Easy Being Green: Environmental Advertising in Light of the FTC's Proposed Green Guides," *SMT: Surface Mount Technology Magazine* 26, 7 (July 2011): 50–56.
31. Ingolfur Blühdorn, "Sustaining the Unsustainable: Symbolic Politics and the Politics of Simulation," *Environmental Politics* 16, 2 (April 2007): 251–275; William A. Brock and Stephen R. Carpenter, "Panaceas and Diversification of Environmental Policy," *Proceedings of the National Academy of Sciences of the United States of America* 104, 39 (September 25, 2007): 15206–15211; Purdy, "The Politics of Nature," 1207–1209.
32. Robert Bryce, *Power Hungry: The Myths of "Green" Energy and the Real Fuels of the Future* (New York: Public Affairs Books, 2010), 13–20; Robert P. Taylor, Chandrasekar Govindarajalu, Anke S. Meyer, and William A. Ward, *Financing Energy Efficiency: Lessons from Brazil, China, India, and Beyond* (Washington, D.C.: The International Bank for Reconstruction and Development/The World Bank, 2008), 3–20; Yergin, *The Quest*, 12–13, 17–18.
33. Severin Borenstein, "The Private and Public Economics of Renewable Electricity Generation," *Journal of Economic Perspectives* 26, 1 (Winter 2012): 67–92; Bryce, *Power Hungry*, 39–42; "Publisher Llewellyn King on Energy Policy," *Southern Energy Report* 35, 1 (July 1989): 2–3, 4; Yergin, *The Quest*, 523–526.
34. Bryce, *Power Hungry*, 21–23; Thomas L. Friedman, *Hot, Flat, and Crowded: Why We Need a Green Revolution—And How It Can Renew America* (New York: Farrar, Straus and Giroux, 2008), 228–230; Amory B. Lovins and the Rocky Mountain Institute,

Reinventing Fire: Bold Business Solutions for the New Energy Era (White River Junction, Vt.: Chelsea Green Publishing Company, 2011), 228–251; Can Tansel Tugcu, Iihan Ozturk, and Alper Aslan, "Renewable and Non-Renewable Energy Consumption and Economic Growth Relationship Revisited: Evidence from G7 Countries," *Energy Economics* 34, 6 (November 2012): 1942–1950.

35. Quoted at Bryce, *Power Hungry*, 4. See also 3–6, 285–302; Borenstein, "The Private and Public Economics of Renewable Electricity Generation," 67–92; Yergin, *The Quest*, 13–14, 711–717.
36. The term "throwaway society," much used in the United States during the twenty-first century, dates at least to the 1950s. See, for example, Judd H. Alexander, *In Defense of Garbage* (Westport, Conn. and London: Praeger, 1993), 45–47; Gregory M. Kennedy, "Trash Talk," *America: The National Catholic Review* 206, 15 (May 7, 2012): 12–16; Heather Rogers, *Gone Tomorrow: The Hidden Life of Garbage* (New York: The New Press, 2005), 7, 127, 136–139; Helen Spiegelman and Bill Sheehan, "Getting Rid of the Throwaway Society," *In Business* 28, 2 (March/April 2012): 19–21.
37. See for example, Katie Kelly, *Garbage: The History and Future of Garbage in America* (New York: Saturday Review Press, 1973), 14–27; William Rathje and Cullen Murphy, "Five Major Myths about Garbage, and Why They're Wrong," *Smithsonian* 23, 4 (July 1992): 113–122; William Rathje and Cullen Murphy, *Rubbish! The Archaeology of Garbage* (New York: HarperCollins, 1992), 40–52; Susan Strasser, *Waste and Want: A Social History of Trash* (New York: Henry Holt, 1999), 135–136.
38. Alexander, *In Defense of Garbage*, 147–166; Rogers, *Gone Tomorrow*, 14–19; Elizabeth Royte, *Garbage Land: On the Secret Trail of Trash* (New York and Boston: Little, Brown and Company, 2005), 51–52, 71–72.
39. Keri Jansen, "The Garbage No One Wanted," *Waste & Recycling News* 18, 13 (October 1, 2012): 4; Patricia Poore and David Maisel, "Is Garbage an Environmental Problem?" *Garbage* 5, 5 (December 1993): 40–45; Rogers, *Gone Tomorrow*, 200; Royte, *Garbage Land*, 127–128.
40. See for example, Paul Calcott and Margaret Walls, "The Economics of Waste: Can Downstream Waste Disposal Policies Encourage Upstream 'Design for Environment?" *The American Economic Review* 90, 2 (May 2000): 233–237; Takayoshi Shinkuma, "On the Second-Best Policy of Household's Waste Recycling," *Environmental and Resource Economics* 24, 1 (January 2003): 77–95.
41. Edward Humes, *Garbology: Our Dirty Love Affair with Trash* (New York: Avery, 2012), 256-262; Larry S. Luton, *The Politics of Garbage: A Community Perspective on Solid Waste Policy Making* (Pittsburgh, Penn.: University of Pittsburgh Press, 1996), 34–58; Rathje and Cullen Murphy, *Rubbish!* 234–245; Royte, *Garbage Land*, 284–287.
42. Matthias Finkbeiner, "Preface," in *Towards Life Cycle Sustainability Management*, Matthias Finkbeiner, editor (London and New York: Springer, 2011), xi.
43. World Commission on Environment and Development (WCED), *Our Common Future: From One Earth to One World* (Oxford and New York: Oxford University Press, 1987), 59–65. See also, for example, Muhamed Abdomerovic, "Project Management Life Span and Life Cycle Processes," *PM World Today* 14, 2 (February 2012): 1–15; Pere Fullana i Palmer, Rita Puig, Alba Bala, Grau Baquero, Jordi Riba, and Marco Raugei, "From Life Cycle Assessment to Life Cycle Management: A Case Study on Industrial Waste Management Policy Making," *Journal of Industrial Ecology* 15, 3 (June 2011): 458–475; David Hunkeler and Gerald Rebitzer, "The Future of Life Cycle

Assessment," *International Journal of Life Cycle Assessment* 10, 5 (2005): 305–308; Santosh K. Mahapatra, Ajay Das, and Ram Narasimhan, "A Contingent Theory of Supplier Management Initiatives: Effects of Competitive Intensity and Product Life Cycle," *Journal of Operations Management* 30, 5 (July 2012): 406–422.

References

Copy from one, it's plagiarism; copy from two, it's research.

— Wilson Mizner, playwright and raconteur

40 CFR Part 263.

40 CFR § 1503.13.

Abbey, Edward. *The Monkey Wrench Gang.* New York: Harper Perennial Modern Classics, 2000 [1975].

Abdomerovic, Muhamed. "Project Management Life Span and Life Cycle Processes." *PM World Today* 14, 2 (February 2012): 1–15.

Abelkop, Adam D. K., Agnes Botos, Lois R. Wise, and John D. Graham. "Regulating Industrial Chemicals: Lessons for U.S. Lawmakers from the European Union's REACH Program." *Environmental Law Reporter: News & Analysis* 42, 11 (November 2012): 11042–11065.

Ackerman, Bruce A., and Richard B. Stewart. "Reforming Environmental Law: The Democratic Case for Market Incentives." *Columbia Journal of Environmental Law* 13, 1 (1987): 171–199.

ActivistCash.com Website: "Greenpeace." http://activistcash.com/organization_overview.cfm/o/131-greenpeace. Accessed October 19, 2012.

Adams, Henry. *The United States in 1800.* Ithaca, N.Y.: Cornell University Press, 1974.

Adams, John H., and Patricia Adams with George Black. *A Force for Nature: The Story of NRDC and Its Fight to Save Our Planet.* San Francisco, Calif.: Chronicle Books, 2010.

Addison, Heather. "Capitalizing Their Charms: Cinema Stars and Physical Culture in the 1920s." *The Velvet Tray— A Critical Journal of Film and Television* 50, 1 (Fall 2002): 15–35.

Adkins, Sarah R. "Democracy's Quiet Virtues." In *Ethics and Character: The Pursuit of Democratic Virtues.* William D. Richardson, J. Michael Martinez, and Kerry R. Stewart, Editors, 203–223. Durham, N.C.: Carolina Academic Press, 1998.

Adler, Mortimer J., and Seymour Cain. *Ethics: The Study of Moral Values.* Chicago: Encyclopedia Britannica, Inc., 1962.

Agar, Nicholas. "Biocentrism and the Concept of Life." *Ethics* 108, 1 (October 1997): 147–168.

Ahearn, Ashley. "Communicating about Chemical Body Burden, with Tracey Woodruff and Rachel Morello-Frosch." *Environmental Health Perspectives* 119, 5 (May 2011): 1–7.

Alexander, Judd H. *In Defense of Garbage.* Westport, Conn. and London: Praeger, 1993.

Allen, Gay Wilson, and Roger Asselineau. *An American Farmer: The Life of St. John de Crèvecœur*. New York: Penguin, 1987.

Allen, Paul J. "Institutions: Natural Resources Defense Council." *Environment* 32, 10 (December 1990): 2–4.

Altman, Ida. "The Revolt of Enriquillo and the Historiography of Early Spanish America." *The Americas* 63, 4 (April 2007): 587–614.

Altman, Morris. "When Green Isn't Mean: Economic Theory and the Heuristics of the Impact of Environmental Regulations on Competitiveness and Opportunity Cost." *Ecological Economics* 36, 1 (January 2001): 31–44.

Ambrose, Stephen E. *Undaunted Courage: Meriwether Lewis, Thomas Jefferson, and the Opening of the American West*. New York: Touchstone Books, 1996.

American Chemistry Council Website: "TSCA Modernization." http://www.americanchemistry.com/Policy/Chemical-Safety/TSCA. Accessed November 16, 2012.

American National Standards Institute (ANSI) Website: "Home Page." http://www.ansi.org/. Accessed November 23, 2012.

American Society for Testing and Materials (ASTM) International Website: "Home Page." http://www.astm.org/. Accessed November 23, 2012.

Amstutz, David Lee. "Nebraska's Live Stock Sanitary Commission and the Rise of Progressivism." *Great Plains Quarterly* 28, 4 (Fall 2008): 259–275.

Anand, Sudhir, and Kara Hanson. "DALYs: Efficiency versus Equity." *World Development* 26, 2 (February 1998): 307–310.

Anderson, Bonnie. *News Flash: Journalism, Infotainment, and the Bottom-Line Business of Broadcast News*. San Francisco, Calif.: Jossey-Bass, 2004.

Anderson, Jed. "The Disappearing Distinction." *The Environmental Forum* 29, 6 (December 2012): 30–33.

Anderson, Jon. "From 'Zombies' to 'Coyotes': Environmentalism Where We Are." *Environmental Politics* 19, 6 (November 2010): 973–991.

Andreoni, James. "Warm-Glow Versus Cold-Pickle: The Effects of Positive and Negative Framing on Cooperation in Experiments." *Quarterly Journal of Economics* 110, 1 (February 1995): 1–21.

Andrews, Richard N. L. "The EPA at 40: An Historical Perspective." *Duke Environmental Law & Policy Forum* 21, 2 (Spring 2011): 223–258.

Anselment, Raymond A. "Robert Boyle, Izaak Walton, and the Art of Angling." *Prose Studies* 30, 2 (August 2008): 124–141.

Anthes, Emily. "The Everyday Fear Zone." *Psychology Today* 45, 6 (November/December 2012): 76–83.

Appel, Peter A. "Wilderness, the Courts, and the Effect of Politics on Judicial Decision-Making." *Harvard Environmental Law Review* 35, 2 (2011): 275–312.

Applegate, John S. "The Perils of Unreasonable Risk: Information, Regulatory Policy, and Toxic Substances Control." *Columbia Law Review* 91, 2 (March 1991): 261–333.

Arevalo, Jorge Alexis. "Critical Reflective Organizations: An Empirical Observation of Global Active Citizenship and Green Politics." *Journal of Business Ethics* 96, 2 (October 2010): 299–316.

Aronovsky, Ronald G. "Federalism and CERCLA: Rethinking the Role of Federal Law in Private Cleanup Cost Disputes." *Ecology Law Quarterly* 33, 1 (2006): 1–104.

Arrow, Kenneth, Bert Bolin, Robert Costanza, Partha Dasgupta, Carl Folke, C. S. Holling, Bengt-Owe Jansson, Simon Levin, Karl-Goran Maler, Charles Perrings, and David Pimentel. "Economic Growth, Carrying Capacity, and the Environment." *Ecological Applications* 6, 1 (February 1996): 13–15.

Asquith, Christina. "Dow Chemical and The Nature Conservancy Team Up to Ask, What Is Nature Worth?" *Solutions Journal* 2, 6 (November-December 2011): 20–21.

Association Française de Normalisation (AFNOR) Website: "Home Page." http://www.afnor.org/en. Accessed November 23, 2012.

Atkinson, Scott E., and T. H. Tietenberg. "Approaches for Reaching Ambient Standards in Non-Attainment Areas: Financial Burden and Efficiency Considerations." *Land Economics* 60, 2 (May 1984): 148–159.

Babcock, Hope M. "The Public Trust Doctrine: What a Tall Tale They Tell." *South Carolina Law Review* 61, 2 (Winter 2009): 393–414.

Bachmann, John. "Will the Circle Be Broken? A History of the U.S. National Ambient Air Quality Standards." *Journal of the Air & Waste Management Association* 57, 6 (June 2007): 652–697.

Balogh, Brian. "Scientific Forestry and the Roots of the Modern American State: Gifford Pinchot's Path to Progressive Reform." *Environmental History* 7, 2 (April 2002): 198–225.

Banning, Lance. "1787 and 1776: Patrick Henry, James Madison, the Constitution, and the Revolution." In *Toward a More Perfect Union: Six Essays on the Constitution.* Neil L. York, Editor, 59–89. Provo, Utah: Brigham Young University, 1988.

Barber, Benjamin R. *Jihad vs. McWorld: How Globalism and Tribalism Are Reshaping the World.* New York: Ballantine Books, 1996.

Barger, M. Susan, and William B. White. *The Daguerreotype: Nineteenth-Century Technology and Modern Science.* Washington, D.C.: Smithsonian Institution Press, 1991.

Barnes, Barry. *T. S. Kuhn and Social Sciences.* New York: Columbia University Press, 1982.

Barnett, A. H. "The Pigouvian Tax Rule under Monopoly." *The American Economic Review* 70, 5 (December 1980): 1037–1041.

Barnett, Harold C. "The Allocation of Superfund." *Land Economics* 61, 3 (August 1985): 255–262.

Barrett, Nancy J. "The Struggles of Women Industrial Workers to Improve Work Conditions in the Progressive Era." *OAH Magazine of History* 13, 3 (Spring 1999): 43–48.

Barrett, Scott. "Rethinking Climate Change Governance and Its Relationship to the World Trading System." *The World Economy* 34, 11 (November 2011): 1863–1882.

Barry, Norman P. *On Classical Liberalism and Libertarianism.* New York: Macmillan, 1986.

Basili, Marcello, Alain Chateauneuf, and Fulvio Fontini. "Precautionary Principle as a Rule of Choice with Optimism on Windfall Gains and Pessimism on Catastrophic Losses." *Ecological Economics* 67, 3 (October 2008): 485–491.

Beaton, Kendall. "Dr. Gesner's Kerosene: The Start of American Oil Refining." *Business History Review* 29, 1 (March 1955): 28–53.

Becker, Robert A. "Intergenerational Equity: The Capital-Environment Tradeoff." *Journal of Environmental Economics and Management* 9, 2 (June 1982): 165–185.

Beckerman, Wilfred. "'Sustainable Development': Is It a Useful Concept?" In *The Environmental Ethics & Policy Book.* 2nd ed. Donald VanDeVeer and Christine Pierce, Editors, 462–474. Belmont, Calif.: Wadsworth, 1998.

Beermann, Jack M. "End the Failed *Chevron* Experiment Now: How *Chevron* Has Failed and Why It Can and Should Be Overruled." *Administrative & Regulatory Law News* 35, 2 (Winter 2010): 3–15.

Bennett v. Spear, 520 U.S. 154 (1997).

Bentham, Jeremy. *An Introduction to the Principles of Morals and Legislation.* New York: Hafner, 1948.

Berezow, Alex B., and Hank Campbell. *Science Left Behind: Feel-Good Fallacies and the Rise of the Anti-Scientific Left*. New York: Public Affairs Books, 2012.

Berger, Helge, and Mark Spoerer. "Economic Crises and the European Revolutions of 1848." *The Journal of Economic History* 61, 2 (June 2001): 293–306.

Bergeson, Lynn. "Facts and Consequences," *The Environmental Forum* 27, 6 (November/December 2010): 27.

Bergeson, Lynn L. "The Explosion of Green Marketing Claims: FTC Jumps into the Fray." *Environmental Quality Management* 18, 2 (Winter 2008): 71–78.

Bergeson, Lynn L. "Selling Green: U.S. FTC Releases Proposed Revisions to the 'Green Guides.'" *Environmental Quality Management* 20, 3 (Spring 2011): 77–81.

Berlau, John. "The Smear Campaigns." *Insight on the News* 20, 11 (May 11, 2004): 18–20.

Beron, Kurt J., James C. Murdoch, and Wim P. M. Vijverberg. "Why Cooperate? Public Goods, Economic Power, and the Montreal Protocol." *The Review of Economics and Statistics* 85, 2 (May 2003): 286–297.

Berry, Chuck. "National Wildlife Federation Is 75 Years Old." *South Dakota Wildlife Out of Doors* 51, 2 (February 2011): 2–3.

Berry, Michael A., and Dennis A. Rondinelli, "Proactive Corporate Environmental Management: A New Industrial Revolution." *Academy of Management Executive* 12, 2 (May 1998): 38–50.

Bewell, Alan. "Romanticism and Colonial Natural History." *Studies in Romanticism* 43, 1 (Spring 2004): 5–34.

Bierce, Ambrose. *The Devil's Dictionary of Ambrose Bierce: Complete and Unabridged, Special Edition*. El Paso, Tex.: El Paso Norte Press, 2010.

Biermann, Frank, Olwen Davies, and Nicolien van der Grijp, "Environmental Policy Integration and the Architecture of Global Environmental Governance." *International Environmental Agreements* 9, 4 (November 2009): 351–369.

Biles, Blake A., and Lawrence E. Culleen. "TSCA, Redux." *The Environmental Forum* 27, 1 (January/February 2010): 30–34.

Bimes, Terri, and Quinn Mulroy. "The Rise and Decline of Presidential Populism." *Studies in American Political Development* 18, 2 (Fall 2004): 136–159.

The Biodegradable Products Institute (BPI) Website: "The BPI Approval Process & Test Requirements." http://www.bpiworld.org/BPI-certification-requirements. Accessed November 23, 2012.

Birchard, Bill. *Nature's Keepers: The Remarkable Story of How the Nature Conservancy Became the Largest Environmental Organization in the World*. San Francisco, Calif.: Jossey-Bass, 2005.

Bird, Alexander. "Kuhn, Naturalism, and the Positivist Legacy." *Studies in History & Philosophy of Science Part A* 35, 2 (June 2004): 337–356.

BirdLife International Website: About BirdLife." http://www.birdlife.org/worldwide/index.html. Accessed October 23, 2012.

Black, Harvey. "Imperfect Protection: NEPA at 35 Years." *Environmental Health Perspectives* 112, 5 (April 2004): A292–A295.

Block, Robert H. "Frederick Jackson Turner and American Geography." *Annals of the Association of American Geographers* 70, 1 (March 1980): 31–42.

Blomberg, Lindsey. "By the Numbers: The Great Pacific Garbage Patch." *E: The Environment Magazine* 22, 3 (May/June 2011): 8.
Blue Moon Fund Website: "Home Page." http://www.bluemoonfund.org/. Accessed December 20, 2012.
Blühdorn, Ingolfur. "Sustaining the Unsustainable: Symbolic Politics and the Politics of Simulation." *Environmental Politics* 16, 2 (April 2007): 251–275.
Blum, Michael C. "The Public Trust Doctrine and Private Property: The Accommodation Principle." *Pace Environmental Law Review* 27, 3 (Summer 2010): 649–667.
Blumin, Stuart M. "Driven to the City: Urbanization and Industrialization in the Nineteenth Century." *Magazine of History* 20, 3 (May 2006): 47–53.
Blunch, Niels-Hugo, and Paula Castro. "Enterprise-level Training in Developing Countries: Do International Standards Matter?" *International Journal of Training and Development* 11, 4 (December 2007): 314–324.
Boime, Albert. "John James Audubon: A Birdwatcher's Fanciful Flights." *Art History* 22, 5 (December 1999): 728–755.
Bokulich, Alisa. "Heisenberg Meets Kuhn: Closed Theories and Paradigms." *Philosophy of Science* 73, 1 (January 2006): 90–107.
Bookchin, Murray. "Will Ecology Become 'The Dismal Science'?" In *The Environmental Ethics & Policy Book*. 2nd ed. Donald VanDeVeer and Christine Pierce, Editors, 230–235. Belmont, Calif.: Wadsworth, 1998.
Booth, Bibi. "Robert Marshall (1901–1939)." In *Environmental Activists*. John Mongillo and Bibi Booth, Editors, 173–178. Westport, Conn.: Greenwood Press, 2001.
Borenstein, Severin. "The Private and Public Economics of Renewable Electricity Generation." *Journal of Economic Perspectives* 26, 1 (Winter 2012): 67–92.
Bostwick, Peg. "Integrating State and Federal Needs under the Clean Water Act." *National Wetlands Newsletter* 33, 4 (July/August 2011): 5–6.
Botkin, Daniel. *No Man's Garden: Thoreau and a New Vision of Civilization and Nature*. Washington, D.C. and Covelo, Calif.: Island Press/Shearwater Books, 2001.
Bouma-Prediger, Steven. *For the Beauty of the Earth: A Christian Vision for Creation Care*. 2nd ed. Grand Rapids, Mich.: Baker Academic, 2010.
Bowker, Geoffrey C., and Susan Leigh Star. *Sorting Things Out: Classification and Its Consequences*. Cambridge, Mass.: MIT Press, 1999.
Brady, Diane. "Fred Krupp." *Bloomberg Business Week* 4284, 1 (June 18-June 24, 2012): 76.
Bratton, Susan Power. "Thinking Like a Mackerel: Rachel Carson's *Under the Sea-Wind* as a Source for a Trans-Ecotonal Sea Ethic." *Ethics and the Environment* 9, 1 (Spring 2004): 1–22.
Braudel, Ferdnand. *The Wheels of Commerce*. Berkeley and Los Angeles: University of California Press, 1992.
Braun, Eric M. "Coring the Seedless Grape: A Reinterpretation of *Chevron U.S.A. v. NRDC*." *Columbia Law Review* 87, 5 (June 1987): 986–1008.
Brecht, Martin. *Martin Luther: His Road to Reformation, 1483–1521*. Kitchener, Ontario, Canada: Fortress Press, 1985.
Bren, Linda. "Frances Oldham Kelsey: FDA Medical Reviewer Leaves Her Mark on History." *FDA Consumer* 35, 2 (March/April 2001): 24–29.
Brinkley, Douglas. *Cronkite*. New York: Harper, 2012.
Brinkley, Douglas. *The Wilderness Warrior: Theodore Roosevelt and the Crusade for America*. New York: HarperCollins, 2009.

Brock, William A., and Stephen R. Carpenter. "Panaceas and Diversification of Environmental Policy." *Proceedings of the National Academy of Sciences of the United States of America* 104, 39 (September 25, 2007): 15206–15211.

Brockington, Dan, Rosaleen Duffy, and Jim Igoe. *Nature Unbound: Conservation, Capitalism, and the Future of Protected Areas.* New York and Oxford: Routledge, 2008.

Brodie, Fawn. *Thomas Jefferson: An Intimate History.* New York: W.W. Norton, 1974.

Bromley, Daniel W. "Environmental Regulations and the Problem of Sustainability: Moving Beyond 'Market Failure.'" *Ecological Economics* 63, 4 (September 15, 2007): 676–683.

Brookhiser, Richard. *Alexander Hamilton: American.* New York: The Free Press, 1999.

Brown, Mark B., and David H. Guston. "Science, Democracy, and the Right to Research." *Science & Engineering Ethics* 15, 3 (September 2009): 351–366.

Brown, Peter. *The Rise of Western Christendom: Triumph and Diversity AD 200–1000.* Oxford, U.K.: B. H. Blackwell, 1996.

Brown, Phil. "Wilderness Advocate." *New York Conservationist,* 62, 1 (August 2007), 2–6.

Brown, Phil, PhD, and Richard Clapp, DSc. "Looking Back on Love Canal." *Public Health Reports* 117, 2 (March/April 2002): 95–98.

Brown, Valerie J. "Why Is It So Difficult to Choose Safer Alternatives for Hazardous Chemicals?" *Environmental Health Perspectives* 120, 7 (July 2012): A280–A283.

Bryce, Robert. *Power Hungry: The Myths of "Green" Energy and the Real Fuels of the Future.* New York: Public Affairs Books, 2010.

Bryner, Gary C. "Congress and Clean Air Policy." In *Business and Environmental Policy: Corporate Interests in the American Political System.* Michael E. Kraft and Sheldon Kamieniecki, Editors, 127–151. Cambridge, Mass.: MIT Press, 2007.

Buck, Susan J. *Understanding Environmental Administration and Law.* 2nd ed. Washington, D.C. and Covelo, Calif.: Island Press, 1996.

Bunge, Mario. *Matter and Mind: A Philosophical Inquiry.* London and New York: Springer, 2010.

Burgoyne, Daniel E. "Attorneys' Fees and the Conflict between Rule 68 and the Clean Water Act's Citizen Suit Provisions." *Boston College Environmental Affairs Law Review* 33, 2 (2006): 627–660.

Burkett, Paul. "Nature in Marx Reconsidered." *Organization & Environment* 10, 2 (June 1997): 164–183.

Burns, Melinda. "Toxics Shock." *Miller-McCune.com* 4, 4 (July-August 2011): 74–76.

Burstein, Andrew, and Nancy Isenberg. *Madison and Jefferson.* New York: Random House, 2010.

Burt, Elizabeth V. "Working Women and the Triangle Fire: Press Coverage of a Tragedy." *Journalism History* 30, 4 (Winter 2005): 189–199.

Butler, Henry N. "A Defense of Common Law Environmentalism: The Discovery of Better Environmental Policy." *Case Western Reserve Law Review* 58, 3 (Spring 2008): 705–752.

Cagle, M. Christine, and J. Michael Martinez. "Have Gun, Will Travel: The Dispute Between the CDC and the NRA on Firearm Violence as a Public Health Problem." *Politics & Policy* 32, 2 (June 2004): 278–310.

Cagle, M. Christine, J. Michael Martinez, and William D. Richardson. "Privatizing Professional Licensing Boards: Self-Governance or Self-Interest?" *Administration & Society* 30, 6 (January 1999): 734–770.

Caitlin, George. *Letters and Notes on the Manners, Customs, and Condition of the North American Indians.* Vol. I. London: Tosswell and Myers, 1841.

Calcott Paul, and Margaret Walls. "The Economics of Waste: Can Downstream Waste Disposal Policies Encourage Upstream 'Design for Environment?" *The American Economic Review* 90, 2 (May 2000): 233–237.

Caldwell, Lynton K. *Science and the National Environmental Policy Act: Redirecting Policy through Procedural Reform*. Tuscaloosa: University of Alabama Press, 1982.

Calhoun, Charles W. *From Bloody Shirt to Full Dinner Pail: The Transformation of Politics and Governance during the Gilded Age*. New York: Hill & Wang, 2010.

Callicott, J. Baird. *In Defense of the Land Ethic: Essays in Environmental Philosophy*. Albany, N.Y.: State University of New York Press, 1989.

Callicott, J. Baird. "Non-Anthropocentric Value Theory and Environmental Ethics." *American Philosophical Quarterly* 21, 4 (October 1984): 299–309.

Callicott, J. Baird, and Roger T. Ames, Editors. *Nature in Asian Traditions of Thought: Essays in Environmental Philosophy*. Albany, N.Y.: State University of New York Press, 1989.

Callicott, J. Baird, and Karen Mumford. "Ecological Sustainability as a Conservation Concept." *Conservation Biology* 11, 1 (February 1997): 32–40.

Campbell, Ballard C. "Understanding Economic Change during the Gilded Age." *Magazine of History* 13, 4 (Summer 1999): 16–20.

Campbell, James. "The Pragmatism of Benjamin Franklin." *Transactions of the Charles S. Peirce Society* 31, 4 (Fall 1995): 745–792.

Carlson, Ann E. "Designing Effective Climate Policy: Cap-and-Trade and Complementary Policies." *Harvard Journal on Legislation* 49, 1 (July 2012): 207–248.

Carone, Gabriela Roxana. "Plato and the Environment." *Environmental Ethics* 20, 2 (Summer 1998): 115–133.

Carpenter, Ronald H. "Frederick Jackson Turner and the Rhetorical Impact of the Frontier Thesis." *Quarterly Journal of Speech* 63, 1 (April 1977): 117–129.

Carson, Rachel. *Silent Spring*. 40th Anniversary Edition. Boston: Houghton Mifflin Harcourt, 2002 [1962].

Carter, Neil. *The Politics of the Environment: Ideas, Activism, Policy*. 2nd ed. New York and Cambridge, U.K.: Cambridge University Press, 2007.

Cartwright, Helen Morris. "Heraclitus and the Bath Water." *The Philosophical Review* 74, 4 (October 1965): 466–485.

Cassuto, David N. "The Law of Words: Standing, Environment, and Other Contested Terms." *Harvard Environmental Law Review* 28, 1 (2004): 79–128.

Castka, Pavel, and Michaela A. Balzarova. "The Impact of ISO 9000 and ISO 14000 on Standardization of Social Responsibility— An Inside Perspective." *International Journal of Production Economics* 113, 1 (May 2008): 74–87.

Cavalieri, Paola. *The Animal Question: Why Nonhuman Animals Deserve Human Rights*. Catherine Woollard, Translator. Oxford and New York: Oxford University Press, 2001.

Cedar Grove Website: "What We Do." http://www.cgcompost.com/what-we-do. Accessed November 23, 2012.

"A Centennial Celebration, 1892–1992." *Sierra* 77, 3 (May/June 1992): 52–73.

Chambers, W. Bradnee, and Jessica F. Green, Editors. *Reforming International Environmental Governance: From Institutional Limits to Innovative Solutions*. Tokyo, New York, and Paris: United Nations University Press, 2005.

Charmantier, Isabelle. "Carl Linnaeus and the Visual Representation of Nature." *Historical Studies in the Natural Sciences* 41, 4 (October 2011): 365–404.

Cheney, Jim. "Postmodern Environmental Ethics: Ethics as Bioregional Narrative." In *Postmodern Environmental Ethics*. Max Oelschlaeger, Editor, 23–42. Albany, N.Y.: State University of New York Press, 1995.

Chernow, Ron. *Alexander Hamilton*. New York: The Penguin Press, 2004.

Chevron U.S.A. Inc. v. Natural Resources Defense Council, Inc., 467 U.S. 837 (1984).

Christoff, Peter. "Cold Climate in Copenhagen: China and the United States at COP15." *Environmental Politics* 19, 4 (July 2010): 637–656.

Christofferson, Bill. *The Man from Clear Lake: Earth Day Founder Senator Gaylord Nelson*. Madison: The University of Wisconsin Press, 2004.

Churchill, Winston. *Quotations*. Norwich, England: Jarrold Publishing, 1997.

Cialdini, Robert B. "Commitment and Consistency: Hobgoblins of the Mind." In *Readings in Managerial Psychology*. 4th ed. Harold J. Leavitt, Louis R. Pondy, and David M. Boje, Editors, 145–182. Chicago and London: University of Chicago Press, 1989.

Cidell, Julie. "A Political Ecology of the Built Environment: LEED Certification for Green Buildings." *Local Environment* 14, 7 (August 2009): 621–633.

Cirman, Andreja, Polona Domadenik, Matjaz Koman, and Tjasa Redek. "The Kyoto Protocol in a Global Perspective." *Economic and Business Review* 11, 1 (February 2009): 29–54.

Clark, Brett, and John Bellamy Foster. "The Environmental Conditions of the Working Class: An Introduction to Selections from Frederick Engels's *The Condition of the Working Class in England in 1844*." *Organization & Environment* 19, 3 (September 2006): 375–388.

Clark, Kenneth. *Civilization*. New York: Harper & Row, 1969.

Claussen, C. Joachim. "Anaximander and Anaximenes: The Earliest Greek Theories of Change?" *Phronesis* 22, 2 (1977): 89–102.

Clay, Karen, and Randall Jones. "Migrating to Riches? Evidence from the California Gold Rush." *The Journal of Economic History* 68, 4 (December 2008): 997–1027.

The Clean Air Act (CAA) of 1970. 42 U.S.C. § 7401 *et seq*. (1970).

The Clean Water Act (CWA) (Federal Water Pollution Control Amendments of 1972). 33 U.S.C. § 1251 *et seq*. (1972).

Clements, Kendrick A. *Hoover, Conservation, and Consumerism: Engineering the Good Life*. Lawrence: University Press of Kansas, 2000.

Climate Depot Website: "Home Page." http://climatedepot.com/. Accessed December 21, 2012.

Clynes, Tom. "The Battle: Climate Scientists Routinely Face Death Threats, Hate Mail, Nuisance Lawsuits, and Political Attacks. How Much Worse Can It Get?" *Popular Science* 281, 1 (July 2012): 37–43, 80.

Coase, Ronald H. "The Institutional Structure of Production." *The American Economic Review* 82, 4 (September 1992): 713–719.

Cobb, John B. Jr. *Economics, Ecology and Justice*. Maryknoll, N.Y.: Orbis, 1992.

Cohen, Bonnie R. "Natural Resources Defense Council— Weapons in the Environmentalist Arsenal: Lawsuits, Blacklists and Publicity." *Organization Trends* (Washington, D.C.: Capital Research Center, August 2003): 1–5.

Cohen, I. Bernard. *Science and the Founding Fathers: Science in the Political Thought of Thomas Jefferson, Benjamin Franklin, John Adams & James Madison*. New York: W. W. Norton, 1995.

Colborn, Theo, Dianne Dumanoski, and John Peterson Myers. *Our Stolen Future: Are We Threatening Our Fertility, Intelligence, and Survival? A Scientific Detective Story*. New York: Plume, 1997.

Cole, G. D. H. *Fabian Socialism*. Oxford, U.K.: Routledge, 1971.

Coleman, Daniel A. *Ecopolitics: Building a Green Society*. New Brunswick, N.J.: Rutgers University Press, 1993.

Collett, Stacey. "World Wildlife Fund." *Computerworld* 42, 8 (February 2008): 40.

Commoner, Barry. *The Closing Circle: Nature, Man, and Technology*. New York: Bantam, 1980 [1971].

Commoner, Barry. *Making Peace with the Planet*. New York: Pantheon, 1990.

Commoner, Barry. *Science and Survival.* New York: Ballantine Books, 1971 [1966].

"Communicating Results to Community Residents: Lessons from Recent ATSDR Health Investigations." *Journal of Exposure Analysis and Environmental Epidemiology* 14, 7 (November 2004): 484–491.

The Comprehensive Environmental Response, Compensation and Liability Act (CERCLA) of 1980. 42 U.S.C. § 9601 *et seq.* (1980).

Conklin, Paul K. *The State of the Earth: Environmental Challenges on the Road to 2100.* Lexington: University of Kentucky Press, 2007.

Connelly, Owen. *The Wars of the French Revolution and Napoleon, 1792–1815.* New York and Oxford: Routledge, 2006.

Constant, Benjamin. "On a Supposed Right to Lie Because of Philanthropic Concerns." In Immanuel Kant. *Fundamental Principles of the Metaphysics of Morals.* 3rd ed. James W. Ellington, Translator, 63–68. Indianapolis: Hackett Publishing, 1993.

Cooper, David E. *World Philosophies: An Historical Introduction.* Cambridge, Mass.: Blackwell, 1996.

Cooper, Laurence D. *Rousseau, Nature, and the Problem of the Good Life.* University Park, Penn.: Pennsylvania State University Press, 1999.

Copeland, Claudia. *Clean Water Act: A Summary of the Law.* Washington, D.C.: Congressional Research Service, April 23, 2010.

Corcoran, Peter B., and Eric Sievers. "Reconceptualizing Environmental Education: Five Possibilities." *Journal of Environmental Education* 25, 4 (Summer 1994): 4–8.

Cornford, Francis M. *Plato's Cosmology: The Timaeus of Plato.* Indianapolis: Hackett Publishing, 1997 [1935].

Corwin, Edward S. "The 'Higher Law' Background of American Constitutional Law." *Harvard Law Review* 42, 2 (December 1928): 149–185; continued in *Harvard Law Review* 42, 3 (January 1929): 365–409.

Costanza, Robert. *Ecological Economics: The Science and Management of Sustainability.* New York: Columbia University Press, 1991.

Costanza, Robert, Cutler Cleveland, and Charles Perrings. *The Development of Ecological Economics.* Cheltenham, U.K.: Edward Elgar Publishing, 1997.

Coward, Harold. *The Perfectibility of Human Nature in Eastern and Western Thought.* Albany, N.Y.: State University of New York Press, 2008.

Craig, Robin Kundis. "A Comparative Guide to the Western States' Public Trust Doctrines: Public Values, Private Rights, and the Evolution toward an Ecological Public Trust." *Ecology Law Quarterly* 37, 1 (2010): 53–197.

Crampton, Roger C., and Richard K. Berg. "On Leading a Horse to Water: NEPA and the Federal Bureaucracy." *Michigan Law Review* 71, 3 (January 1973): 511–536.

Crèvecœur, John Hector St. John de. *Letters from an American Farmer.* Carlisle, Mass.: Applewood Books, 2006 [1782].

Crowe, Beryl L. "The Tragedy of the Commons Revisited." *Science*, New Series, 166, 3909 (November 28, 1969): 1103–1107.

Curd, Patricia. *The Legacy of Parmenides: Eleatic Monism and Later Presocratic Thought.* Princeton, N.J.: Princeton University Press, 1998.

Curran, Giorel. "Whither Environmentalism? Environmental Politics in the 21st Century." *Social Alternatives* 25, 2 (Second Quarter 2006): 48–53.

Currell, Susan. *American Culture in the 1920s.* Edinburgh, Scotland: Edinburgh University Press, 2009.

Daly, Herman E. "Allocation, Distribution, and Scale: Towards an Economics That is Efficient, Just, and Sustainable." *Ecological Economics* 6, 3 (December 1992): 185–193.

Daly, Herman E."Economics, Environment, and Community." *Earth Ethics* 20, 1 (Fall 1989): 9–11.

Daly, Herman E., and Kenneth N. Townsend. *Valuing the Earth: Economics, Ecology, Ethics.* Cambridge, Mass.: MIT Press, 1993.

Danziger, Sheldon. "Do Rising Tides Lift All Boats? The Impact of Secular and Cyclical Changes on Poverty." *The American Economic Review* 76, 2 (May 1986): 405–410.

"David R. Brower." *Earth Island Journal* 16, 1 (Spring 2001): 21–24.

Davidson, Jonathan, and Joseph M. Norbeck. *An Interpretive History of the Clean Air Act: Scientific and Policy Perspectives.* Waltham, Mass.: Elsevier, 2012.

Davis, Chester C. "Is the A.A.A. Experiment Proving a Success?" *Congressional Digest* 13, 12 (December 1934): 306–312.

Deane, Herbert A. *The Political and Social Ideas of St. Augustine.* New York: Columbia University Press, 1963.

Dear, Michael. "Understanding and Overcoming the NIMBY Syndrome." *Journal of the American Planning Association* 58, 3 (Summer 1992): 288–300.

Death, Carl. "Summit Theatre: Exemplary Governmentability and Environmental Diplomacy in Johannesburg and Copenhagen." *Environmental Politics* 20, 1 (February 2011): 1–19.

Defenders of Wildlife Website: "Home Page." http://www.defenders.org/. Accessed November 1, 2012.

Delcourt, Paul A., and Hazel R. Delcourt. *Prehistoric Native Americans and Ecological Change: Human Ecosystems in Eastern North America since the Pleistocene.* New York and Cambridge, U.K.: Cambridge University Press, 2004.

Demir, Ipek. "Incommensurabilities in the Work of Thomas Kuhn." *Studies in History & Philosophy of Science Part A* 39, 1 (March 2008): 133–142.

Denevan, William M. "The Pristine Myth: The Landscape of the Americas in 1492." In *American Environmental History.* Louis S. Warren, Editor, 5–26. Oxford, U.K.: B. H. Blackwell, 2004.

Denison, Richard A. "Ten Essential Elements in TSCA Reform." *Environmental Law Reporter: News & Analysis* 39, 1 (January 2009): 10020–10028.

Desrochers, Pierre. "How Did the Invisible Hand Handle Industrial Waste? By-product Development before the Modern Environmental Era." *Enterprise & Society* 8, 2 (June 2007): 348–374.

Dettelbach, Michael. "Alexander von Humboldt between Enlightenment and Romanticism." *Northeastern Naturalist* 8, Supplement (January 2001): 9–20.

Devall, Bill. "The End of American Environmentalism?" *Nature and Culture* 1, 2 (Autumn 2006): 157–180.

Devall, Bill, and George Sessions. "Deep Ecology." In *The Environmental Ethics & Policy Book.* 2nd ed. Donald VanDeVeer and Christine Pierce, Editors, 221–226. Belmont, Calif.: Wadsworth, 1998.

DeWitt, Calvin. *The Environment and the Christian: What Does the New Testament Say about the Environment?* Grand Rapids, Mich.: Baker Academic, 1991.

Dholakia-Lehenbauer, Kruti, and Euel W. Elliott. "Decisionmaking, Risk, and Uncertainty: An Analysis of Climate Change Policy." *Cato Journal* 32, 3 (Fall 2012): 539–556.

Diamond, Martin. "Democracy and *The Federalist*: A Reconsideration of the Framers' Intent." *American Political Science Review* 53, 1 (March 1959): 52–68.

Dianne Dumanoski Website: "Home Page." http://www.diannedumanoski.com/about.html. Accessed December 20, 2012.

Dietz, Bettina. "Making Natural History: Doing the Enlightenment." *Central European History* 43, 1 (March 2010): 25–46.

Dietze, Gottfried. *The Federalist: A Classic on Federalism and Free Government.* Westport, Conn.: Greenwood Press, 1960.

Diggins, John Patrick. *The Lost Soul of American Politics: Virtue, Self-Interest, and the Foundations of Liberalism.* Chicago and London: University of Chicago Press, 1986.

Diggins, John Patrick. "Recovering 'First Principles': Critical Perspectives on the Constitution and the Fate of Classical Republicanism." In *Toward a More Perfect Union: Six Essays on the Constitution.* Neil L. York, Editor, 119–143. Provo, Utah: Brigham Young University, 1988.

Diggins, John Patrick. *Thorstein Veblen: Theorist of the Leisure Class.* Princeton, N.J.: Princeton University Press, 1999.

Dimock, Marshall E. *Free Enterprise and the Administrative State.* Westport, Conn.: Greenwood Press, 1972.

DIN German Institute for Standardization Website: "Home Page." http://www.sundyne.com/Products/Pumps/Industry+Standards/DIN+German+Institute+for+Standardization. Accessed November 23, 2012.

Dobel, J. Patrick. "Stewards of the Earth's Resources: A Christian Response to Ecology." *Christian Century* 94, 2 (October 12, 1977): 906–909.

Dodson, Robin E., Marcia Nishioka, Laurel J. Standley, Laura J. Perovich, Julia Green Brody, and Ruthann A. Rudel. "Endocrine Disruptors and Asthma-Associated Chemicals in Consumer Products." *Environmental Health Perspectives* 120, 7 (July 2012): 935–943.

Doherty, Brian. "Friends of the Earth International: Negotiating a Transnational Identity." *Environmental Politics* 15, 5 (November 2006): 860–880.

Dolbeare, Kenneth M. *American Political Thought.* Chatham, N.J.: Chatham House, 1984.

Donovan, Frank. *Mr. Madison's Constitution: The Story Behind the Constitutional Convention.* New York: Dodd Mead & Company, 1965.

The Doors. "When the Music's Over." Words and music by the Doors. Copyright © 1967 Doors Music Co. Copyright renewed. All rights reserved. Used by permission. Reprinted by permission of Hal Leonard Corporation.

Downs, Anthony. *An Economic Theory of Democracy.* New York: Harper & Row, 1957.

Dubos, René Jules. *So Human an Animal: How We Are Shaped by Surroundings and Events.* New Brunswick, N.J.: Transaction Publishers, 1998 [1968].

Dugatkin, Lee Alan. *Mr. Jefferson and the Giant Moose: Natural History in Early America.* Chicago: The University of Chicago Press, 2009.

Dummies.com Website: "Home Page." http://www.dummies.com/. Accessed November 29, 2012.

Dunaway, Finis. "Gas Masks, Pogo, and the Ecological Indian: Earth Day and the Visual Politics of American Environmentalism," *American Quarterly* 60, 1 (March 2008): 67–99.

Dunlap, Riley E., and Aaron M. McCright. "A Widening Gap: Republican and Democratic Views on Climate Change." *Environment* 50, 5 (September 2008): 26–35.

Dunlap, Riley E., and Angela G. Mertig. "The Evolution of the U.S. Environmental Movement from 1970 to 1990: An Overview." *Society & Natural Resources: An International Journal* 4, 3 (1991): 209–218.

Dunlap, Thomas R. "Environmentalism, a Secular Faith." *Environmental Values* 15, 3 (August 2006): 321–330.

Dunning, Mike. "Manifest Destiny and the Trans-Mississippi South: Natural Laws and the Extension of Slavery into Mexico." *Journal of Popular Culture* 35, 2 (Fall 2001): 111–127.

Durlauf, Steven N. "*Bowling Alone*: A Review Essay." *Journal of Economic Behavior and Organization* 47, 3 (March 2002): 259–273.

Durst, Gregory L., and Edward A. Laperle. "Styrene Monomer Migration as Monitored by Purge and Trap Gas Chromatography and Sensory Analysis for Polystyrene Containers." *Journal of Food Science* 55, 2 (March 1990): 522–524.

Earth First! Website: "About Earth First!" http://www.earthfirst.org/about.htm. Accessed October 15, 2012.

Earth Island Institute Website: "Home Page." http://www.earthisland.org/. Accessed November 1, 2012.

Easterbrook, Gregg. *A Moment on the Earth: The Coming Age of Environmental Optimism*. New York: Viking, 1995.

Eccleston, Charles H. "Does NEPA Suffer from the Pike Syndrome?" *Environmental Practice* 4, 1 (March 2002): 8–9.

Eccleston, Charles H., and Frederic March. *Global Environmental Policy: Concepts, Principles, and Practice*. Boca Raton, Fla.: CRC Press, 2011.

"Editorial Comment— The Montreal Protocol." *International Journal of Remote Sensing* 29, 19 (October 10, 2008): 5455–5459.

Egerton, Frank. "A History of the Ecological Sciences: Early Greek Origins." *Bulletin of the Ecological Society of America* 82, 1 (January 2001): 93–97.

Eggert, H., and M. Greaker. "Trade, GMOs, and Environmental Risk: Are Current Policies Likely to Improve Welfare?" *Environmental and Resource Economics* 48, 4 (April 2011): 587–608.

Egnal, Marc. "The Beards Were Right: Parties in the North." *Civil War History* 47, 1 (March 2001): 30–56.

Ehrenfeld, David. *The Arrogance of Humanism*. Oxford and New York: Oxford University Press, 1978.

Ehrlich, Paul R. *The Population Bomb*. Cutchogue, N.Y.: Buccaneer Books, 1995 [1968].

Eisenach, Eldon J. "Progressivism as a National Narrative in Biblical-Hegelian Times." *Social Philosophy & Policy* 24, 1 (January 2007): 59–83.

Ellefson, Paul V. "The Safety of Our Forests and the Prosperity of Our People." *Journal of Forestry* 98, 5 (May 2000): 14–22.

Ellis, Joseph J. *American Sphinx: The Character of Thomas Jefferson*. New York: Vintage, 1998.

Elshtain, Jean Bethke. "A Return to Hull House: Reflections on Jane Addams." *Feminist Issues* 15, 1-2 (1997): 105–113.

Ely, James W., Jr., *Railroads and American Law*. Lawrence: University Press of Kansas, 2001.

The Endangered Species Act (ESA) of 1973. 16 U.S.C. § 1531, *et seq*. (1973).

The Endocrine Disruption Exchange Website: "About Us." http://www.endocrinedisruption.com/about.introduction.php. Accessed December 20, 2012.

ENERGY STAR Website: "About ENERGY STAR." http://www.energystar.gov/index.cfm?c=about.ab_index. Accessed November 23, 2012.

ENERGY STAR Website: "History." http://www.energystar.gov/index.cfm?c=about.ab_history. Accessed November 23, 2012.

Entine, Jon. "BPA: DOA?" *The Environmental Forum* 27, 6 (November/December 2010): 24–30.

Entine, Jon. *Crop Chemophobia: Will Precaution Kill the Green Revolution?* Washington, D.C.: American Enterprise Institute, 2011.

Entine, Jon, Editor. *Let Them Eat Precaution*. Washington, D.C.: American Enterprise Institute, 2006.

Entine, Jon. "NGO Openness Is a Two-Way Street." *The Central New York Business Journal* 21, 52 (December 28, 2007): 23.

Entine, Jon. *Scared to Death: How Chemophobia Threatens Public Health: A Position Statement of the American Council on Science and Health*. New York: American Council on Science and Health, 2011.

Environmental Action Website. "About EA." http://www.environmental-action.org/content/about-ea. Accessed September 13, 2012.

"The Environmental Defense Fund, Conclusion." *Business & The Environment* 22, 6 (June 2011): 3–5.

"The Environmental Defense Fund, Part 1." *Business & The Environment* 22, 5 (May 2011): 4–6.

Environmental Defense Fund Website: "Home Page." http://www.edf.org/. Accessed October 29, 2012.

Environmental Defense Fund Website: "Our Mission and History." http://www.edf.org/about/our-mission-and-history. Accessed October 29, 2012.

Environmental Working Group (EWG) Website: "About the Environmental Working Group." http://www.ewg.org/about. Accessed December 19, 2012.

Environmental Working Group (EWG) Website: "BPA and Human Diseases on the Rise." http://www.ewg.org/node/20937. Accessed December 19, 2012.

Environmental Working Group (EWG) Website: "Home Page." http://www.ewg.org/. Accessed November 1, 2012.

Enzler, Sherry A. "How Law Mattered to the Mono Lake Ecosystem." *William and Mary Environmental Law & Policy Review* 35, 2 (Winter 2011): 413–513.

"EPA Announces Efforts to Strengthen Chemical Management under TSCA." *Hazardous Waste Consultant* 27, 6 (2009): 2.16–2.19.

Epstein, David F. *The Political Theory of The Federalist*. Chicago: The University of Chicago Press, 1984.

Evans, David. "Clean Water Act § 404 Assumption: What Is It, How Does It Work, and What Are the Benefits?" *National Wetlands Newsletter* 31, 3 (May-June 2009): 18–21.

"Evolution of Present Federal Law." *Congressional Digest* 49, 8/9 (August/September 1970): 194–197.

"Fairtrade v. Rainforest Alliance." *Marketing Week* 33, 21 (May 20, 2010): 17.

Fara, Patricia. *Newton: The Making of a Genius*. New York: Macmillan/Columbia University Press, 2002.

Farrell, Joseph. "Information and the Coase Theorem." *The Journal of Economic Perspectives* 1, 2 (Autumn 1987): 113–129.

The Federal Insecticide, Fungicide and Rodenticide Act (FIFRA). As amended by the Federal Environmental Pesticide Control Act of 1972. 7 U.S.C. § 136 *et seq.* (1972).

Fee, Elizabeth, and Theodore M. Brown. "Alice Hamilton: Settlement Physician, Occupational Health Pioneer." *American Journal of Public Health* 91, 11 (November 2001): 1767.

Fehr, Ernst, and Simon Gachter. "Reciprocity and Economics: The Economic Implications of *Homo Reciprocans*." *European Economic Review* 42, 3 (May 1998): 845–859.

Feng, Lin, and Jason Buhi. "The Copenhagen Accord and the Silent Incorporation of the Polluter Pays Principle in International Climate Law: An Analysis of Sino-American Diplomacy at Copenhagen and Beyond." *Buffalo Environmental Law Journal* 18, 1 (2010/2011): 1–74.

Fensterwald, Jr., Bernard. "The Anatomy of American 'Isolationism' and Expansionism, Part I." *The Journal of Conflict Resolution* 2, 2 (June 1958): 111–140.

Ferrey, Steven. *Environmental Law: Examples and Explanations.* New York: Aspen Law & Business, 1997.

Ferrey, Steven. "Inverting the Law: Superfund Hazardous Substance Liability and Supreme Court Reversal of All Federal Circuits." *William & Mary Environmental Law & Policy Review* 33, 3 (May 2009): 633–722.

Final Peer-Reviewed Report: Life Cycle Inventory of Polystyrene Foam, Bleached Paperboard, and Corrugated Paperboard Foodservice Products. Prairie Village, Kans.: Franklin Associates, Ltd., March 2006.

Fine, Gail. *On Ideas: Aristotle's Criticism of Plato's Theory of Forms.* 2nd ed. Oxford and New York: Oxford University Press, 2004.

Finkbeiner, Matthias. "Preface." In *Towards Life Cycle Sustainability Management.* Matthias Finkbeiner, Editor, xi–xiii. London and New York: Springer, 2011.

Fiorino, Daniel J. "Rethinking Environmental Regulation: Perspectives on Law and Governance." *Harvard Environmental Law Review* 23, 2 (1999): 441–468.

Fishman, Ethan. "The Quality of Theodore Roosevelt's Environmentalism." In *A Companion to Theodore Roosevelt.* Serge Ricard, Editor, 173–185. West Sussex, England: Wiley-Blackwell, 2011.

Fleming, Thomas. *Duel: Alexander Hamilton, Aaron Burr and the Future of America.* New York: Basic Books, 1999.

Flint, Richard. *No Settlement, No Conquest: A History of the Coronado Entrada.* Albuquerque: University of New Mexico Press, 2008.

Flores, Dan. "Essay: The Great Plains 'Wilderness' as a Human-Shaped Environment." *Great Plains Research* 9, 2 (Fall 1999): 343–355.

Folke, Carl. "Resilience: The Emergence of a Perspective for Social-Ecological Systems Analyses." *Global Environmental Change* 16, 3 (August 2006): 253–267.

Foner, Eric. *Forever Free: The Story of Emancipation and Reconstruction.* New York: Knopf, 2005.

The Food Quality Protection Act (FQPA) of 1996. 7 U.S.C § 136d. 7 U.S.C. § 136a. 7 U.S.C. § 136w. 7 U.S.C. § 136. 7 U.S.C. § 136a. 7 U.S.C. § 136q. 7 U.S.C. § 136d (1996).

Foote, Christopher L., Kristopher Gerardi, Lorenz Goette, and Paul S. Willen. "Just the Facts: An Initial Analysis of Subprime's Role in the Housing Crisis." *Journal of Housing Economics* 17, 4 (December 2008): 291–305.

Foreman, Dave. *Confessions of an Eco-Warrior.* New York: Crown Books, 1993.

Forest Stewardship Council (FSC) Website: "Mission and Vision: Protecting Forests for Future Generations." http://us.fsc.org/mission-and-vision.187.htm. Accessed November 24, 2012.

Forest Stewardship Council (FSC) Website: "Our History." http://us.fsc.org/our-history.180.htm. Accessed November 24, 2012.

Foster, Charles. *The Story of the Bible.* Philadelphia: A. J. Holman Company, 1911.

Foster, John Bellamy. "Ecology and the Transition from Capitalism to Socialism." *Monthly Review* 60, 6 (November 2008): 1–12.

Foster, Kenneth R., Paolo Vecchia, and Michael H. Rapacholi. "Risk Management: Science and the Precautionary Principle." *Science* 288, 5468 (May 12, 2000): 979–981.

Fox, Warwick. "Deep Ecology and Virtue Ethics." *Philosophy Now* 26, 1 (April/May 2000): 21–23.

Frahm, Sally. "The Cross and the Compass: Manifest Destiny, Religious Aspects of the Mexican-American War." *Journal of Popular Culture* 35, 2 (Fall 2001): 83–99.

FrankenFoods.com Website: "Home Page." http://www.frankenfoods.org/. Accessed December 20, 2012.

Frederickson, Robert. "A Green Bird in the Hand: An Example of Environmental Regulations Operating to Stifle Environmentally Conscious Industry." *Boston College Environmental Affairs Law Review* 34, 1 (2007): 303–333.

Free, Andrew, and Nicholas H. Barton. "Do Evolution and Ecology Need the Gaia Hypothesis?" *Trends in Ecology and Evolution* 22, 11 (November 2007): 611–619.

Freinkel, Susan. *Plastic: A Toxic Love Story*. Boston: Houghton Mifflin Harcourt, 2011.

French, Bryant Morey. *Mark Twain and "The Gilded Age": The Book That Named an Era.* Dallas: Southern Methodist University Press, 1965.

Frese, Stephen J. "Aldo Leopold: An American Prophet." *History Teacher* 37, 1 (November 2003): 99–118.

Friedman, Michael. "Ernst Cassirer and Thomas Kuhn: The Neo-Kantian Tradition in History and Philosophy of Science." *Philosophical Forum* 39, 2 (Summer 2008): 239–252.

Friedman, Milton. *Capitalism and Freedom*. Chicago: University of Chicago Press, 1962.

Friedman, Thomas L. *Hot, Flat, and Crowded: Why We Need a Green Revolution— And How It Can Renew America.* New York: Farrar, Straus and Giroux, 2008.

Friends of the Earth Website: "Achievements." http://www.foe.org/about-us/achievements. Accessed November 6, 2012.

Friends of the Earth Website: "Who We Are." http://www.foe.org/about-us. Accessed November 6, 2012.

Friends of the Earth, Inc. v. Laidlaw Environmental Services (TOC), Inc., 528 U.S. 167 (2000).

Fromme, Michael. *Battle for the Wilderness.* Salt Lake City: University of Utah Press, 1997.

Frueh, Joseph. "Comment: Pesticides, Preemption, and the Return of Tort Protection." *Yale Journal on Regulation* 23, 1 (2006): 299–309.

Gaba, Jeffrey M. "Generally Illegal: NPDES General Permits under the Clean Water Act." *Harvard Environmental Law Review* 31, 2 (2007): 409–473.

Gaffney, Brian P. "A Divided Duty: The EPA's Dilemma under the Endangered Species Act and Clean Water Act Concerning the National Pollutant Discharge Elimination System." *The Review of Litigation* 26, 1 (April 2007): 487–524.

Gaido, Daniel. "The Populist Interpretation of American History: A Materialist Revision." *Science & Society* 65, 3 (Fall 2001): 350–375.

Gambrel, Patrick A., and Rebecca Cianci. "Maslow's Hierarchy of Needs: Does It Apply in a Collectivist Culture?" *Journal of Applied Management and Entrepreneurship* 8, 2 (April 2003): 143–161.

Garde, Ajay. "Sustainable by Design? Insights from U.S. LEED-ND Pilot Projects." *Journal of the American Planning Association* 75, 4 (Autumn 2009): 424–440.

Garrard, Graeme. "The Enlightenment and Its Enemies." *The American Behavioral Scientist* 49, 5 (January 2006): 664–680.

Garreau, Joel. "Environmentalism as Religion." *The New Atlantis: A Journal of Technology & Society* 28, 1 (Summer 2010): 61–74.

Garry, Patrick M. *Liberalism and American Identity*. Kent, Ohio: Kent State University Press, 1992.

Gates, Paul. *The Jeffersonian Dream: Studies in the History of American Land Policy and Development*. Albuquerque: University of New Mexico Press, 1996.

Gay, Kathlyn. *American Dissidents: An Encyclopedia of Troublemakers, Subversives, and Prisoners of Conscience*. Santa Barbara, Calif.: ABC-CLIO, 2012.

Gaylord Nelson and Earth Day Website: "April 22, 1970." http://www.nelsonearthday.net/earth-day/4-22-1970.htm. Accessed October 10, 2012.

Gaylord Nelson and Earth Day Website: "Choosing Not to Be in Charge." http://www.nelsonearthday.net/earth-day/not-incharge.htm. Accessed October 10, 2012.

Gaylord Nelson and Earth Day Website: "Meet Gaylord Nelson, Founder of Earth Day." http://www.nelsonearthday.net/nelson/index.htm. Accessed October 10, 2012.

Gaylord Nelson and Earth Day Website: "Modest Beginnings: Nelson's Idea on a Plane." http://www.nelsonearthday.net/earth-day/beginnings.htm. Accessed October 10, 2012.

Gaylord Nelson and Earth Day Website: "'The Wrong Kind of Pollution': Critical Responses to Earth Day." http://www.nelsonearthday.net/earth-day/critics.htm. Accessed October 10, 2012.

Genter, Sabrina. "Decision-making and the Precautionary Principle." *Impact Assessment & Project Appraisal* 24, 2 (June 2006): 160–161.

Gherasim, Gabriel. "Introductory Sketch to the Analytics and Pragmatism of American Progressivism." *Studia Universitatis Babes-Bolyai. Studia Europaea* 56, 3 (September 2011): 119–142.

Ghins, Michel. "Thomas Kuhn on the Existence of the World." *International Studies in the Philosophy of Science* 17, 3 (October 2003): 265–279.

Gibbs, Lois Marie. "Housewife's Data." *American Journal of Public Health* 101, 9 (September 2011): 1556–1559.

Gibson, David. "Awash in Green: A Critical Perspective on Environmental Advertising." *Tulane Environmental Law Journal* 22, 2 (June 2009): 423–440.

Giddens, Anthony. *The Politics of Climate Change*. Malden, Mass.: Polity Press, 2009.

Giere, Ronald N. *Explaining Science: A Cognitive Approach*. Chicago: The University of Chicago Press, 1988.

Gitler, Stefanie. "Settling the Tradeoffs between Voluntary Cleanup of Contaminated Sites and Cooperation with the Government under CERCLA." *Ecology Law Quarterly* 35, 1 (2008): 337–361.

Glazebrook, Trish. "Art or Nature? Aristotle, Restoration Ecology, and Flowforms." *Ethics and the Environment* 8, 1 (Spring 2003): 22–36.

Glover, James A. *A Wilderness Original: The Life of Bob Marshall*. Seattle, Wash.: The Mountaineers Books, 1986.

Godfray, H. C. J. "Linnaeus in the Information Age." *Nature* 446, 7133 (March 15, 2007): 259–260.

Goldhammer, Arthur. "Tocqueville, Associations, and the Law of 1834." *Historical Reflections* 35, 3 (Winter 2009): 74–84.

Goldwin, Robert A. "John Locke/1632-1704." In *History of Political Philosophy*. 2nd ed. Leo Strauss and Joseph S. Cropsey, Editors, 451–486. Chicago: University of Chicago Press, 1972.

Golub, Alexander A. "Climate Policy and Uncertainty: A Precautionary Principle versus Real Options Analysis." *Environmental Law Reporter: News & Analysis* 42, 8 (August 2012): 10733–10736.

Gordon-Reed, Annette. *The Hemingses of Monticello*. New York: W. W. Norton, 2008.

Gore, Senator Al. *Earth in the Balance: Ecology and the Human Spirit.* Boston: Houghton Mifflin, 1992.

Gorman, Hugh S. "Efficiency, Environmental Quality, and Oil Field Brines: The Success and Failure of Pollution Control by Self-Regulation." *Business History Review* 73, 4 (Winter 1999): 601–640.

Gorski, Philip S. *The Disciplinary Revolution: Calvinism and the Rise of the State in Early Modern Europe.* Chicago and London: University of Chicago Press, 2003.

Gottlieb, Robert. "Reconstructing Environmentalism: Complex Movements, Diverse Roots." In *American Environmental History.* Louis S. Warren, Editor, 245–256. Oxford, U.K.: B. H. Blackwell, 2004.

Gowdy, John, and Sabine O'Hara. "Weak Sustainability and Viable Technologies." *Ecological Economics* 22, 3 (September 1997): 239–247.

Graham, Jr., Frank. *The Audubon Ark: A History of the Audubon Society.* Austin: University of Texas Press, 1992.

Green Advantage Environmental Certification Website: "About GA Certification." http://www.greenadvantage.org/about-ga-certification. Accessed November 24, 2012.

Green Advantage Environmental Certification Website: "Vision, Mission & Deliverables." http://www.greenadvantage.org/vision-mission-deliverables. Accessed November 24, 2012.

Green America Website: "About Us." http://www.greenamerica.org/about/. Accessed November 24, 2012.

Green America Website: "Certification for Green Businesses." http://www.greenamerica.org/greenbusiness/certification.cfm. Accessed November 24, 2012.

Green Clean Institute Website: "About GCI." http://www.greencleaninstitute.com/about-gci. Accessed November 24, 2012.

Green Clean Institute Website: "GCI Firm Certifications." http://www.greencleaninstitute.com/content/firm-certification. Accessed November 24, 2012.

Green Clean Institute Website: "Green Product Certification." http://www.greencleaninstitute.com/green-product-certification. Accessed November 24, 2012.

Green Globe Website: "Home Page." http://greenglobe.com/. Accessed November 24, 2012.

Green Globe Website: "Standard Criteria and Indicators." http://greenglobe.com/register/standard/. Accessed November 24, 2012.

Green Key Website: "About the Green Key Eco-Rating Program." http://www.greenkeyglobal.com/about_the_program.asp. Accessed November 24, 2012.

Green Key Website: "Home Page." http://www.greenkeyglobal.com/default.asp. Accessed November 24, 2012.

Greenpeace Website: "The Bombing of the *Rainbow Warrior*." http://www.greenpeace.org/usa/en/campaigns/history/The-Bombing-of-the-Rainbow-Warrior/. Accessed October 19, 2012.

Greenpeace Website: "The Founders." http://www.greenpeace.org/usa/en/campaigns/history/the-founders/. Accessed October 18, 2012.

Greenpeace Website: "GMO Victories." http://www.greenpeace.org/usa/en/campaigns/victories/gmo-victories/. Accessed December 21, 2012.

Greenpeace Website: "Greenpeace International." http://www.greenpeace.org/international/en/. Accessed October 19, 2012.

Greenpeace Website: "Greenpeace USA." http://www.greenpeace.org/usa/en/about/. Accessed October 19, 2012.

Greenpeace Website: "Our History." http://www.greenpeace.org/usa/en/campaigns/history/. Accessed October 18, 2012.

Greenpeace Website: "Questions about Greenpeace in General." http://www.greenpeace.org/international/en/about/faq_old/questions-about-greenpeace-in/. Accessed October 19, 2012.

Greenstein, Fred I. "The Policy-Driven Leadership of James K. Polk: Making the Most of a Weak Presidency." *Presidential Studies Quarterly* 40, 4 (December 2010): 725–733.

Greve, Michael S. "Friends of the Earth, Foes of Federalism." *Duke Environmental Law & Policy Forum* 12, 1 (Fall 2001): 167–182.

Griffith, James J., and Charles R. Knoeber. "Why Do Corporations Contribute to the Nature Conservancy?" *Public Choice* 49, 1 (1986): 69–77.

Griffith, Mark F. "John Locke's Influence on American Government and Public Administration." *Journal of Management History* 3, 3 (1997): 224–237.

Grizzle, Raymond E., Paul E. Rothrock, and Christopher B. Barrett. "Evangelicals and Environmentalism: Past, Present, and Future." *Trinity Journal* 19, 1 (Spring 1998): 3–27.

Grossmann, Matt. "Environmental Advocacy in Washington: A Comparison with Other Interest Groups." *Environmental Policy* 15, 4 (August 2006): 628–638.

Gruner, Hans Peter. "Redistributive Policy, Inequality, and Growth." *Journal of Economics* 62, 1 (February 1995): 1–23.

Guha, Ramachandra. *Environmentalism: A Global History*. New York: Longman, 2000.

Guha, Ramachandra. "Radical American Environmentalism and Wilderness Preservation: A Third World Critique." In *The Environmental Ethics & Policy Book*. 2nd ed. Donald VanDeVeer and Christine Pierce, Editors, 515–522. Belmont, Calif.: Wadsworth, 1998.

Guo, Robert M. "Reasonable Bases for Apportioning Harm under CERCLA." *Ecology Law Quarterly* 37, 1 (2010): 317–352.

Gup, Ted. "Down with the God Squad." *Time* 136, 20 (November 5, 1990): 102.

Hamer, John H. "Money and the Moral Order in Late Nineteenth Century and Early Twentieth-Century American Capitalism." *Anthropological Quarterly* 71, 3 (July 1998): 139–149.

Hamilton, Alexander. "Report on Manufactures (1791)." In *American Political Thought*. Kenneth M. Dolbeare, Editor, 166–170. Chatham, N.J.: Chatham House Publishers, 1984.

Hamilton, Alexander, James Madison, and John Jay. *The Federalist Papers*. Clinton Rossiter, Editor. New York: New American Library, 1961.

Hamilton, Alice. *Industrial Poisons in the United States*. New York: Macmillan, 1929.

Hammond, Kim. "Monsters of Modernity: Frankenstein and Modern Environmentalism." *Cultural Geographies* 11, 2 (April 2004): 181–198.

Handl, Günther. *Declaration of the United Nations Conference on the Human Environment Stockholm, 16 June 1972*. At Audiovisual Library of International Law. http://untreaty.un.org/cod/avl/ha/dunche/dunche.html. Accessed December 3, 2012.

Hansen, James. *Storms of My Grandchildren: The Truth about the Coming Climate Catastrophe and Our Last Chance to Save Humanity*. New York and London: Bloomsbury, 2011.

Happer, William. "Harmful Politicization of Science." In *Politicizing Science: The Alchemy of Policymaking*. Michael Gough, Editor, 27–48. Stanford, Calif.: Hoover Institution Press, 2003.

Hardin, Garret. "The Tragedy of the Commons." *Science* 162, 3859 (December 13, 1968): 1243–1248.

Hardy, Margaret, and John A. Charles, Jr. *Sustainable Failure: Why Portland's Polystyrene Foam Ban Should Be Repealed*. Portland, Oregon: Cascade Policy Institute, November 2007.

Hargrove, Edwin C. *Prisoners of Myth: The Leadership of the Tennessee Valley Authority, 1933–1990*. Studies in American Politics: Historical, International, and Comparative Perspectives. Princeton, N.J.: Princeton University Press, 1994.

Harnsberger, Caroline Thomas. *Mark Twain at Your Fingertips: A Book of Quotations*. Mineola, N.Y.: Courier Dover Publications, 2009.

Harvard Center for Risk Analysis (Joshua T. Cohen, Gary Carlson, Gail Charnley, David Coggin, Elizabeth Delzell, John D. Graham, Helmut Greina, Daniel Krewski, Michele Medinsky, Richard Monson, Dennis Paustenbach, Barbara Petersen, Stephen Rapport, Lorenz Rhomberg, P. Barry Ryan, and Kimberly Thompson). "A Comprehensive Evaluation of the Potential Health Risks Associated with Occupational and Environmental Exposure to Styrene." *Journal of Toxicology and Environmental Health* 5, 1–2 (January 2002): 1–263.

Hass, Bruce, and Michael Kleine. "The Rhetoric of Junk Science," *Technical Communication Quarterly* 12, 3 (Summer 2003): 267–284.

Hassner, Pierre. "Immanuel Kant." In *History of Political Philosophy*. 2nd ed. Leo Strauss and Joseph S. Cropsey, Editors, 554–593. Chicago: University of Chicago Press, 1972.

Haufler, Virginia. *A Public Role for the Private Sector: Industry Self-Regulation in a Global Economy*. Washington, D.C.: Carnegie Endowment for International Peace, 2001.

Hausdoerffer, John. *Caitlin's Lament: Indians, Manifest Destiny, and the Ethics of Nature*. Lawrence: University Press of Kansas, 2009.

Haveman, Robert H. *Economics of the Public Sector*. Santa Barbara, Calif.: Wiley, 1976.

Hayes, Denis. "Earth Day! A Call to Unite in Defense of Our Planet." *Mother Earth News* 309, 1 (April/May 2005): 24–31.

Heal the Bay Website: "Home Page." http://www.healthebay.org/. Accessed October 26, 2012.

Hearn, Michael F. "One Person's Waste Is Another Person's Liability: Closing the Liability Loophole in RCRA's Citizen Enforcement Action." *McGeorge Law Review* 42, 2 (April 2011): 467–497.

Hecht, David K. "Constructing a Scientist: Expert Authority and Public Images of Rachel Carson." *Historical Studies in the Natural Sciences* 41, 3 (Summer 2011): 277–302.

Hecht, David K. "How to Make a Villain: Rachel Carson and the Politics of Anti-Environmentalism." *Endeavour* 36, 4 (December 2012): 149–155.

Heclo, Hugh. "Issue Networks and the Executive Establishment." In *The New American Political System*. Anthony King, Editor, 87–124. Washington, D.C.: American Enterprise Institute, 1978.

Henry, Hudson P. "A Shift in Citizen Suit Standing Doctrine: *Friends of the Earth, Inc. v. Laidlaw Environmental Services*." *Ecology Law Quarterly* 28, 2 (2001): 233–252.

Hermansson, Helene. "The Ethics of NIMBY Conflicts." *Ethical Theory & Moral Practice* 10, 1 (January 2007): 23–34.

Hertsgaard, Mark. "Green Goes Grassroots: The Environmental Movement Today." *The Nation* 283, 4 (July 31/August 7, 2006): 11–18.

Hertsgaard, Mark. "Save Earth Day." *The Nation* 294, 19 (May 7, 2012): 3–4.

Heywood, Andrew. *Political Ideas and Concepts: An Introduction*. New York: St. Martin's Press, 1994.

Heywood, Peter. "Fracking: Safer and Greener?" *TCE: The Chemical Engineer* 850 (April 2012): 42–45.

Hicks, Dan. "On the Ideal of Autonomous Science." *Philosophy of Science* 78, 5 (December 2011): 1235–1248.

Hill, Peter J. "Environmental Theology: A Judeo-Christian Defense." *Journal of Markets and Morality* 3, 2 (Fall 2000): 158–172.

Hill, Robert S. "David Hume." In *History of Political Philosophy*. 2nd ed. Leo Strauss and Josephy S. Cropsey, Editors, 509–531. Chicago: University of Chicago Press, 1972.

Hinchman, Lewis P., and Sandra K. Hinchman. "Should Environmentalists Reject the Enlightenment?" *The Review of Politics* 63, 4 (Fall 2001): 663–692.

Hitt, Christopher. "Ecocriticism and the Long Eighteenth Century." *College Literature* 31, 3 (Summer 2004): 123–147.

Hochman, Harold M., and James D. Rodgers. "Pareto Optimal Distribution." *The American Economic Review* 59, 4 (September 1969): 542–557.

Hocking, Martin B. "Is Paper Better Than Plastic?" *Consumer's Research* 74, 10 (October 1991): 28–29.

Hocking, Martin B. "Relative Merits of Polystyrene Foam and Paper in Hot Drink Cups: Implications for Packaging." *Environmental Management* 15, 6 (1991): 731–747.

Hocking, Martin B. "Reusable and Disposable Cups: An Energy-Based Evaluation." *Environmental Management* 18, 6 (1994): 889–899.

Hoffman, Andrew J. "Institutional Evolution and Change: Environmentalism and the U.S. Chemical Industry." *The Academy of Management Journal* 42, 4 (August 1999): 351–371.

Hoffman, Andrew J. "An Uneasy Rebirth at Love Canal." *Environment* 37, 2 (March 1995): 4–9, 25–31.

Hofstadter, Richard. *The American Political Tradition*. New York: Vintage Books, 1948.

Hofstadter, Richard. *Anti-Intellectualism in American Life*. New York: Vintage Books, 1962.

Hofstadter, Richard. *Social Darwinism in American Thought*. Boston: Beacon Press, 1992.

Hollings, C. S. "Resilience and Stability of Ecological Systems." *Annual Review of Ecology and Systematics* 4, 1 (1973): 1–23.

Holmes, Steven J. *The Young John Muir: An Environmental Biography*. Madison: The University of Wisconsin Press, 1999.

Holtcamp, Wendee. "Flavors of Uncertainty: The Difference between Denial and Debate." *Environmental Health Perspectives* 120, 8 (August 2012): A315–A319.

Hoover, Kenneth R. *Ideology and Political Life*. 2nd. ed. Belmont, Calif.: Wadsworth, 1994.

Hornbeck, Richard. "The Enduring Impact of the American Dust Bowl: Short- and Long-Run Adjustments to Environmental Catastrophe." *American Environmental Review* 102, 4 (June 2012): 1477–1507.

Horton, James Oliver, and Lois E. Horton. *Slavery and the Making of America*. New York and Oxford: Oxford University Press, 2005.

Hough, Palmer, and Morgan Robertson. "Mitigation under Section 404 of the Clean Water Act: Where It Comes from, What It Means." *Wetlands Ecology & Management* 17, 1 (January 2009): 15–33.

Howard, Alice, and Joan Magretta. "Surviving Success: An Interview with the Nature Conservancy's John Sawhill." *Harvard Business Review* 73, 5 (September/October 1995): 109–118.

Howe, Daniel Walker. *What Hath God Wrought: The Transformation of America*. Oxford and New York: Oxford University Press, 2007.

Hughes, J. Donald. "Ecology in Ancient Greece." *Inquiry: An Interdisciplinary Journal of Philosophy* 18, 2 (1975): 115–125.

Hughes-Jones, E. M. "The Tennessee Valley Authority Fifty Years On." *Political Quarterly* 54, 3 (July-September 1983): 283–291.

Hume, David. *An Enquiry Concerning Human Understanding.* Indianapolis: Hackett Publishing, 1977.
Humes, Edward. *Garbology: Our Dirty Love Affair with Trash.* New York: Avery, 2012.
Hunkeler, David, and Gerald Rebitzer. "The Future of Life Cycle Assessment." *International Journal of Life Cycle Assessment* 10, 5 (2005): 305–308.
Hunt, E. K., and Mark Lautzenheiser. *History of Economic Thought: A Critical Perspective.* 3rd ed. Armonk, N.Y.: M. E. Sharpe, Inc., 2011.
Hutcheon, Pat Duffy. *Leaving the Cave: Evolutionary Naturalism in Social-Scientific Thought.* Waterloo, Ontario, Canada: Wilfrid Laurier Press, 1996.
Hutton, T. R. C. "Beating a Dead Horse: The Continuing Presence of Frederick Jackson Turner in Environmental and Western History." *International Social Science Review* 77, 1/2 (2002): 47–57.
Ingwani, Emaculate, Tendayi Gondo, and Trynos Gumbo. "The Polluter Pay [sic] Principle and the Damage Done: Controversies for Sustainable Development." *Economia: Seria Management* 13, 1 (July 2010): 53–60.
The Intergovernmental Panel on Climate Change (IPCC) Website: "Home Page." http://www.ipcc.ch/index.htm#.UNS7YqxNtqA. Accessed December 21, 2012.
International Organization for Standardization (ISO) Website: "About Us." http://www.iso.org/iso/home/about.htm. Accessed November 21, 2012.
International Organization for Standardization (ISO) Website: "ISO 9000—Quality Management." http://www.iso.org/iso/home/standards/management-standards/iso_9000.htm. Accessed November 21, 2012.
International Organization for Standardization (ISO) Website: "ISO 14000— Environmental Management." http://www.iso.org/iso/home/standards/management-standards/iso 14000.htm. Accessed November 21, 2012.
Isaacson, Walter. *Benjamin Franklin: An American Life.* New York: Simon & Schuster, 2003.
Izaak Walton League Website: "Advocacy Page," http://www.iwla.org/index.php?ht=d/sp/i/196/pid/196. Accessed October 24, 2012.
Izaak Walton League Website: "Clean Water." http://www.iwla.org/index.php?ht=d/sp/i/1976/pid/1976/cat_id/217. Accessed October 24, 2012.
Izaak Walton League Website: "Home Page." http://www.iwla.org/. Accessed October 23, 2012.
Izaak Walton League Website: "News You Can Use." http://www.iwla.org/index.php?ht=d/sp/i/2246/pid/2246. Accessed October 24, 2012.
Izaak Walton League Website: "Waltonians (October 1924)." http://www.iwla.org/index.php?ht=d/ContentDetails/i/31132. Accessed October 24, 2012.
Jack, Brian. "Enforcing Member State Compliance with EU Environmental Law: A Critical Evaluation of the Use of Financial Penalties." *Journal of Environmental Law* 23, 1 (March 2011): 73–95.
Jackson, Donald Dale. "Just Plain Bob Was the Best Friend Wilderness Ever Had." *Smithsonian* 25, 5 (August 1994): 92–100.
Jacobson, Louis. "Fenton Helps MoveOn Move On." *National Journal* 36, 4 (January 24, 2004): 253.
"James K. Polk." In *U.S. Presidential Inaugural Addresses,* 78–88. Whitefish, Mont.: Kessinger Publishing, 2004.
Jansen, Keri. "The Garbage No One Wanted." *Waste & Recycling News* 18, 13 (October 1, 2012): 4.
Jefferson, Thomas. *Notes on the State of Virginia.* William Peden, Editor. Chapel Hill, N.C.: University of North Carolina Press, 1982.

Jeffrey, Robert C. "Southern Political Thought and the Southern Political Tradition." In *Confederate Symbols in the Contemporary South*. J. Michael Martinez, William D. Richardson, and Ron McNinch-Su, Editors, 24–48. Gainesville: University Press of Florida, 2000.

Jekic, Milos. "Lowering the Jurisdictional Bar: A Call for an Equitable-Factors Analysis under CERCLA's Timing-of-Review Provision." *Kansas Law Review* 59, 1 (2011): 157–190.

Jenkins, Garry W. "Nongovernmental Organizations and the Forces against Them: Lessons of the Anti-NGO Movement." *Brooklyn Journal of International Law* 37, 2 (2012): 459–527.

Jenkins, Robin R., Elizabeth Kopits, and David Simpson. "The Evolution of Solid and Hazardous Waste in the United States." *Review of Environmental Economics and Policy* 3, 1 (Winter 2009): 104–120.

Jensen, Merrill. *The Articles of Confederation: An Interpretation of the Social-Constitutional History of the American Revolution, 1774–1781*. Madison: The University of Wisconsin Press, 1963.

John, Andrew, and Rowena Pecchenino. "An Overlapping Generations Model of Growth and the Environment." *The Economic Journal* 104, 427 (November 1994): 1393–1410.

Johnson, Douglas R., and David G. Hoopes. "Managerial Cognition, Sunk Costs, and the Evolution of Industry Structure." *Strategic Management Journal* 24, 10 (October 2003): 1057–1068.

Jones, W. T. *The Classical Mind*. 2nd ed. New York; Harcourt Brace Jovanovich, 1970.

Jones, W. T. *The Medieval Mind*. 2nd ed. New York; Harcourt Brace Jovanovich, 1969.

Kahn, Charles H. *The Art and Thought of Heraclitus: An Edition of the Fragments with Translation and Commentary*. New York and Cambridge, U.K.: Cambridge University Press, 1979.

Kahn, Charles H. *Pythagoras and the Pythagoreans: A Brief History*. Indianapolis: Hackett Publishing, 2001.

Kant, Immanuel. *Fundamental Principles of the Metaphysics of Morals*. New York: Prometheus Books, 1987.

Karkkainen, Bradley C. "Toward a Smarter NEPA: Monitoring and Managing Government's Environmental Performance." *Columbia Law Review* 102, 4 (May 2002): 903–972.

Karl, Herman A., Lawrence E. Susskind, and Katherine H. Wallace. "A Dialogue, Not a Diatribe: Effective Integration of Science and Policy through Joint Fact Finding." *Environment* 49, 1 (January/February 2007): 20–34.

Kaufman, Wallace. *No Turning Back: Dismantling the Fantasies of Environmental Thinking*. New York: Basic Books, 1994.

Kay, Jeanne. "Human Dominion over Nature in the Hebrew Bible." *Annals of the Association of American Geographers* 79, 2 (June 1989): 214–232.

Keita, L. D. *Science, Rationality, and Neoclassical Economics*. Cranbury, N.J.: Associated University Presses, Inc., 1992.

Keller, Robert R. "Supply-Side Economic Policies during the Coolidge-Mellon Era." *Journal of Economic Issues* 16, 3 (September 1982): 773–790.

Kelly, Katie. Garbage: *The History and Future of Garbage in America*. New York: Saturday Review Press, 1973.

Kelly, Walt. *Pogo: We Have Met the Enemy and He Is Us*. New York: Simon & Schuster, 1987.

Kennedy, Gregory M., "Trash Talk." *America: The National Catholic Review* 206, 15 (May 7, 2012): 12–16.

Kent, Curtis. "The Virtue of Thoreau: Biography, Geography, and History in Walden Woods." *Environmental History* 15, 1 (January 2010): 31–53.

Khalil, Elias L. "Beyond Self-Interest and Altruism: A Reconstruction of Adam Smith's Theory of Human Conduct." *Economics and Philosophy* 6, 2 (October 1990): 255–273.

Kiernan, Matthew J. "The Age of Eco-Strategy: Thoughts for the Earth Summit in Rio de Janeiro, June 1992." *The International Executive* 34, 3 (May/June 1992): 197–214.

King, Andrew A., and Michael J. Lenox. "Industry Self-Regulation without Sanctions: The Chemical Industry's Responsible Care Program." *The Academy of Management Journal* 43, 4 (August 2000): 698–716.

Kirchner, James W. "The Gaia Hypothesis: Can It Be Tested?" *Reviews of Geophysics* 27, 2 (May 1989): 223–235.

Kirchner, James W. "The Gaia Hypothesis: Fact, Theory, and Wishful Thinking." *Climatic Change* 52, 4 (2002): 391–408.

Kitcher, Philip. *The Advancement of Science: Science without Legend, Objectivity without Illusions.* Oxford and New York: Oxford University Press, 1993.

Klein, Naomi. "Capitalism vs. Climate: What the Right Gets— and the Left Doesn't— About the Revolutionary Power of Climate Change." *The Nation* 293, 22 (November 28, 2011): 11–21.

Kline, Benjamin PhD. *First Along the River: A Brief History of the U.S. Environmental Movement.* 3rd ed. Lanham, Md.: Rowman & Littlefield, 2007.

Klopf, Sonja, Nada Wolff Culver, and Pete Morton. "A Road Map to a Better NEPA: Why Environmental Risk Assessments Should Be Used to Analyze the Environmental Consequences of Complex Federal Actions." *Sustainable Development Law & Policy* 8, 1 (Fall 2007): 38–43, 84–85.

Koch, Lars, and Nicholas A. Ashford. "Rethinking the Role of Information in Chemicals Policy: Implications for TSCA and REACH." *Journal of Cleaner Production* 14, 1 (2009): 31–46.

Koelsch, William A. "Thomas Jefferson, Geographers, and the Uses of Geography." *Geographical Review* 98, 2 (April 2008): 260–279.

Kortenkamp, Katherine V., and Colleen F. Moore. "Ecocentrism and Anthropocentrism: Moral Reasoning about Ecological Commons Dilemmas." *Journal of Environmental Psychology* 21, 3 (September 2001): 261–272.

Krakauer, Jon. *Into the Wild.* New York: Villard, 1996.

Kremer, Randall. "Institutions: World Wildlife Fund." *Environment* 27, 4 (May 1985): 42–44.

Kristof, Nicholas D. "The Cancer Lobby," *The New York Times.* October 7, 2012. SR13.

Kristof, Nicholas D. "Chemicals and Our Health." *The New York Times.* July 16, 2009. A27.

Kristol, Irving. *Reflections of a Neoconservative: Looking Back, Looking Ahead.* New York: Basic Books, 1983.

Krupp, Fred. "Global Warming and the USA: Sparking Reengagement." *Vital Speeches of the Day* 69, 13 (April 15, 2003): 399–403.

Krupp, Fred. "Grab the Green Brass Ring." *Business Week* 4020, 1 (February 5, 2007): 104.

Krupp, Fred. "New Environmentalism Factors in Economic Needs." *Wall Street Journal* (November 20, 1986): 34.

Kuhn, Thomas S. *The Structure of Scientific Revolutions.* 2nd ed. Chicago: The University of Chicago Press, 1970.

Kumazawa Risa, and Michael S. Callaghan. "The Effect of the Kyoto Protocol on Carbon Dioxide Emissions." *Journal of Economics & Finance* 36, 1 (January 2012): 201–210.

Lamoreaux, Naomi R. "Rethinking the Transition to Capitalism in the Early American Northeast." *The Journal of American History* 90, 2 (September 2003): 437–461.

Larkin, Paschal. *Property in the Eighteenth Century: With Special Reference to England and Locke*. London: Cork University Press, 1930.

Lau, Lee Chung, Keat Teong Lee, and Abdul Rahman Mohamed. "Global Warming Mitigation and Renewable Energy Policy Development from the Kyoto Protocol to the Copenhagen Accord— A Comment." *Renewable and Sustainable Energy Reviews* 16, 7 (September 2012): 5280–5284.

Laudan, Larry. *Progress and Its Problems: Towards a Theory of Scientific Growth*. Berkeley: University of California Press, 1977.

League of Conservation Voters Website. "Mission." http://www.lcv.org/about/mission/. Accessed September 13, 2012.

Lear, Linda J. "Rachel Carson's *Silent Spring*." *Environmental History Review* 17, 2 (Summer 1993): 23–48.

Lee, Kai. "Appraising Adaptive Management." *Ecology and Society* 3, 2 (September 1999): 3–18.

LEED/U.S. Green Building Council Website: "Home Page." https://new.usgbc.org/leed. Accessed November 22, 2012.

Leighninger, Robert D. Jr. *Building Louisiana: The Legacy of the Public Works Administration*. Jackson: University Press of Mississippi, 2007.

Leiss, William. *The Domination of Nature*. New York: Braziller, 1972.

Lemann, Nicholas. "Kicking in Groups." *Atlantic Monthly* 277, 4 (April 1996): 22–26.

Lenowsky, Leslie. "Still 'Bowling Alone'?" *Commentary* 110, 3 (October 2000): 57–60.

Leonard, Irving A. *Books of the Brave: Being an Account of Books and of Men in the Spanish Conquest and Settlement of the Sixteenth-Century New World*. Cambridge, Mass.: Harvard University Press, 1992.

Leopold, Aldo. *A Sand County Almanac; with Essays on Conservation from Round River*. New York: Ballantine Books, 1990 [1949].

Lethbridge, J. P. "Religion and Science in the Eighteenth Century." *Historian* 77, 1 (Spring 2003): 22–28.

Levine, David K. *Is Behavioral Economics Doomed? The Ordinary versus the Extraordinary*. Cambridge, U.K.: Open Book Publishers, 2012.

Levy, David. *How the Dismal Science Got Its Name: Classical Economics and the Ur-Text of Racial Politics*. Ann Arbor: University of Michigan Press, 2001.

Levy, Jason M. "Conflicting Enforcement Mechanisms under RCRA: The Abstention Battleground between State Agencies and Citizen Suits." *Ecology Law Quarterly* 39, 2 (2012): 373–404.

Lewis, Martin. *Green Delusions: An Environmentalist Critique of Radical Environmentalism*. Durham, N.C.: Duke University Press, 1992.

Lindzen, Richard S. "Global Warming: The Origin and Nature of Alleged Scientific Consensus." *Cato Review of Business and Government* 15, 1 (Spring 1992): 87–98.

Livesey, Sharon M. "McDonald's and the Environmental Defense Fund: A Case Study of a Green Alliance." *The Journal of Business Communication* 36, 1 (January 1999): 5–39.

Locke, John. *The Second Treatise of Civil Government*. Thomas I. Cook, Editor. New York: Hafner, 1947.

Lomborg, Bjørn. *The Skeptical Environmentalist: Measuring the Real State of the World*. New York and Cambridge, U.K.: Cambridge University Press, 2001.

Longino, Helen E. *Science as Social Knowledge: Values and Objectivity in Scientific Inquiry.* Princeton: Princeton University Press, 1990.

Lough, Alex Wagner. "Henry George, Frederick Jackson Turner, and the 'Closing' of the American Frontier." *California History* 89, 2 (2012): 4–23.

Lovelock, James. *Gaia: A New Look at Life on Earth.* New York and Oxford: Oxford University Press, 2000.

Lovins, Amory B., and the Rocky Mountain Institute. *Reinventing Fire: Bold Business Solutions for the New Energy Era.* White River Junction, Vt.: Chelsea Green Publishing Company, 2011.

Lowerson, John. "Izaak Walton: Father of a Dream." *History Today* 33, 12 (December 1983): 28–32.

Lowi, Theodore J. *The End of Liberalism: The Second Republic of the United States.* New York: W. W. Norton, 2009.

Lowy, Michael. "From Marx to Ecosocialism." *Capitalism, Nature, Socialism* 13, 1 (March 2002): 121–133.

Luckhurst, Roger. "The Two Cultures, or the End of the World as We Know It." *Interdisciplinary Science Reviews* 32, 1 (March 2007): 55–64.

Luhan v. Defenders of Wildlife, 504 U.S. 555 (1992).

Luke, Timothy. *Eco-Critique: Contesting the Politics of Nature, Ecology and Culture.* Minneapolis: University of Minnesota Press, 1997.

Luppi, Barbara, Francesco Parisi, and Shruti Rajagopalan. "The Rise and Fall of the Polluter-Pays Principle in Developing Countries." *The Economics of Efficiency and the Judicial System, International Review of Law and Economics* 32, 1 (March 2012): 135–144.

Luria, Amy. "CERCLA Contribution: An Inquiry into What Constitutes an Administrative Settlement." *North Dakota Law Review* 84, 1 (2008): 333–364.

Luthy, Herbert. "Colonization and the Making of Mankind." *The Journal of Economic History* 21, 4 (December 1961): 483–495.

Luton, Larry S. *The Politics of Garbage: A Community Perspective on Solid Waste Policy Making.* Pittsburgh, Penn.: University of Pittsburgh Press, 1996.

Lynas, Mark. *Six Degrees: Our Future on a Hotter Planet.* London: The Fourth Estate, an Imprint of HarperCollins, 2007.

Lytle, Mark Hamilton. *The Gentle Subversive: Rachel Carson, Silent Spring, and the Rise of the Environmental Movement.* Oxford and New York: Oxford University Press, 2007.

MacLean, Richard. "Think Like a Lawyer?" *Environmental Quality Management* 20, 1 (Autumn 2010): 117–123.

MacMullen, Ramsay. *Christianizing the Roman Empire (AD 100–400).* New Haven. Conn.: Yale University Press, 1984.

Maeseele, Pieter. "NGOs and GMOs: A Case Study in Alternative Science Communication." *Javnost— The Public* 16, 4 (2009): 55–72.

Maguire, Steve, and Jaye Ellis. "Redistributing the Burden of Scientific Uncertainty: Implications of the Precautionary Principle for State and Nonstate Actors." *Global Governance* 11, 4 (October-December 2005): 505–526.

Mahapatra, Santosh K., Ajay Das, and Ram Narasimhan. "A Contingent Theory of Supplier Management Initiatives: Effects of Competitive Intensity and Product Life Cycle." *Journal of Operations Management* 30, 5 (July 2012): 406–422.

Maher, Neil M. *Nature's New Deal: The Civilian Conservation Corps and the Roots of the American Environmental Movement.* Oxford and New York: Oxford University Press, 2008.

Maher, Neil M. "A New Deal Body Politic: Landscape, Labor, and the Civilian Conservation Corps." *Environmental History* 7, 3 (July 2002): 435–461.

Malone, Dumas. *Jefferson and His Time: Jefferson and the Rights of Man.* Vol. II. New York: Little, Brown & Company, 1972.

Maltby, Paul. "Fundamentalist Dominion, Postmodern Ecology." *Ethics & The Environment* 13, 2 (Fall 2006): 119–141.

Manes, Christopher. *Green Rage: Radical Environmentalism and the Unmaking of Civilization.* Boston: Little, Brown, 1990.

Marafiote, Tracy. "The American Dream: Technology, Tourism, and the Transformation of Wilderness." *Environmental Communication* 2, 2 (July 2008): 154–172.

Markley, Don. "It's Not NIMBY Anymore— It's BANANA." *Broadcastengineering.com* 44, 3 (March 2002): 52–53.

Marks, Jonathan. "Rousseau's Discriminating Defense of Compassion." *American Political Science Review* 101, 4 (November 2007): 727–739.

Marsh, George Perkins. *Man and Nature; or Physical Geography as Modified by Human Action.* New York: Charles Scribner, 1864.

Marshall, Robert. *The People's Forests.* Iowa City: University of Iowa Press, 2002 [1933].

Martella, Roger, and Kim Smacziak. "Introduction to Rio + 20: A Reflection on Progress since the First Earth Summit and the Opportunities That Lie Ahead." *Sustainable Development Law & Policy* 12, 3 (Spring 2012): 4–7.

Martin v. Reynolds Metal Company, 221 Or. 86, 342 P.2d 790 (1959); *cert. denied* 362 U.S. 918 (1960).

Martin, Edward J. "Economic Rights, Sustainable Development, and Environmental Management." *Public Administration & Management* 16, 2 (April 2011): 121–144.

Martin, Glen. "Lessons from the Mountain: Adapting Strategies for Conservation." *Bay Nature* 12, 3 (July-September 2012): 52–55.

Martinez, J. Michael. *Coming for to Carry Me Home: Race in America from Abolitionism to Jim Crow.* Lanham, Md.: Rowman & Littlefield, 2012.

Martinez, J. Michael. "Jeffersonian and Hamiltonian Views of Nature in the Early American Republic." *Politics & Policy* 33, 3 (September 2005): 522–552.

Martinez, James Michael. "Packaging Legislation and Unintended Consequences: A Case Study on the Necessity of Life Cycle Management." In *Towards Life Cycle Sustainability Management.* Matthias Finkbeiner, Editor, 371–380. London and New York: Springer, 2011.

Martinez-Alier, Joan, Giuseppe Munda, and John O'Neil. "Weak Comparability of Values as a Foundation for Ecological Economics." *Ecological Economics* 26, 3 (September 1998): 277–286.

Marx, Axel, and Dieter Cuypers. "Forest Certification as a Global Environmental Governance Tool: What Is the Macro-Effectiveness of the Forest Stewardship Council?" *Regulation & Governance* 4, 4 (December 2010): 408–434.

Mason, Kathy S. "Out of Fashion: Harriet Hemenway and the Audubon Society, 1896–1905." *Historian* 65, 1 (Fall 2002): 1–14.

Mason, Michael. "Representing Transnational Environmental Interests: New Opportunities for Non-Governmental Access within the World Trade Organization?" *Environmental Politics* 13, 3 (Autumn 2004): 566–589.

Massanari, Ronald L. "A Problematic in Environmental Ethics: Western and Eastern Styles." *Buddhist-Christian Studies* 18, 1 (1998): 37–61.

Matisoff, Daniel C. "Making Cap-and-Trade Work: Lessons from the European Union Experience." *Environment* 52, 1 (January/February 2010): 10–19.

Mattson, Mark, and Edward Calabrese. "Best in Small Doses." *New Scientist* 198, 2668 (August 9, 2008): 36–39.

Mazilu, M., and M. Ciobanu. "Deliberative Democracy Establishment in Environmental Policies: Alternatives for the Environmental Conflict Resolution." *Journal of Environmental Science and Engineering* 4, 4 (April 2010): 77–84.

McCright, Aaron M., and Riley E. Dunlap. "The Politicization of Climate Change and Polarization in the American Public's Views on Global Warming, 2001-2010." *The Sociological Quarterly* 52, 2 (Spring 2011): 155–194.

McCrory, Martin A. "Who's on First: CERCLA Cost Recovery, Contribution, and Protection." *American Business Law Journal* 37, 1 (Fall 1999): 3–33.

McDaniel, Dennis K. "The First Congressman Martin Dies of Texas." *Southwestern Historical Quarterly* 102, 2 (October 1998): 130–161.

McDonald, Forrest. "The Constitution and Hamiltonian Capitalism." In *How Capitalistic Is the Constitution?* Robert A. Goldwin and William A. Schambra, Editors, 49–74. Washington, D.C.: American Enterprise Institute for Public Policy Research, 1982.

McElroy, C. T., and P. F. Fogal. "Ozone: From Discovery to Protection." *Atmosphere-Ocean* 46, 1 (March 2008): 1–13.

McEvoy, David M., and John K. Stranlund. "Self-Enforcing International Environmental Agreements with Costly Monitoring for Compliance." *Environmental & Resource Economics* 42, 4 (April 2009): 491–508.

McLaren, Duncan. "Spotlight on…Friends of the Earth." *Planning Practice & Research* 7, 3 (Winter 1992): 43–47.

McMahon, A. Michael. "'Small Matters': Benjamin Franklin, Philadelphia, and the 'Progress of Cities.'" *The Pennsylvania Magazine of History and Biography* 116, 2 (April 1992): 157–182.

McMahon, Darrin M. *Enemies of the Enlightenment.* Oxford and New York: Oxford University Press, 2001.

Meacham, Jon. *Thomas Jefferson: The Art of Power*. New York: Random House, 2012.

Mead, Walter Russell. *God and Gold: Britain, America, and the Making of the Modern World.* New York: Knopf, 2007.

Mead, Walter Russell. *Mortal Splendor: The American Empire in Transition*. Boston: Houghton Mifflin, 1987.

Merchant, Carolyn. *The Death of Nature: Women, Ecology, and the Scientific Revolution*. San Francisco, Calif.: Harper & Row, 1980.

Mercuro, Nicholas. "Interdisciplinary Paradigms for Environmental Policy: Interrelationships among Ecology, Law, and Economics." In *Property Rights, Economics, and the Environment: The Economics of Legal Relationships*. Michael D. Kaplowitz, Editor, 247–279. New York and London: Routledge, 2000.

Merrill, Thomas W. "Justice Stevens and the *Chevron* Puzzle." *Northwestern University Law Review* 106, 2 (2012): 551–566.

Mertig, Angela G. "The 'Nature' of Environmentalism: Nature Protection in the USA." In *Protecting Nature: Organizations and Networks in Europe and the USA*. C. S. A. (Kris) van Koppen and William T. Markham, Editors, 239–262. Cheltenham, U.K.: Edward Elgar Publishing, 2008.

Michaels, Patrick J., and Robert C. Balling. *The Satanic Gases: Clearing the Air about Global Warming*. Washington, D.C.: The Cato Institute, 2000.

Mill, John Stuart. *On Liberty*. Arlington Heights, Ill.: AHM Publishing, 1947.

Miller, Char. "Thinking Like a Conservationist." *Journal of Forestry* 100, 8 (December 2002): 42–45.

Miller, Char., and V. Alaric Sample. "Gifford Pinchot: A Life in Progress." *Journal of Forestry* 97, 1 (January 1999): 27–32.

Miller, Charles A. *Jefferson and Nature: An Interpretation*. Baltimore and London: Johns Hopkins University Press, 1993.

Miller, David J. "Rich History: Celebrating National Audubon Society's Centennial." *New York State Conservationist* 60, 2 (October 2005): 2–5.

Miller, Jeffrey G., and James A. Hopkins. "The Standing to Enforce against Violations of Environmental Statutes in the United States: *Friends of the Earth v. Laidlaw Environmental Services*." *Journal of Environmental Law* 12, 3 (2000): 370–380.

Miller, William H. "Fred Krupp: A Different Kind of Environmentalist." *Industry Week* 243, 15 (August 15, 1994): 93, 96, 98.

Mills, K. W. "Some Aspects of Plato's Theory of Forms: *Timaeus* 49 c ff." *Phronesis* 13, 2 (1968): 145–170.

Minan, John H., and Tracy M. Frech. "Pesticides as 'Pollutants' under the Clean Water Act." *San Diego Law Review* 47, 1 (Winter 2010): 109–144.

Mommsen, Theodore E. "Petrarch's Conception of the 'Dark Ages.'" *Speculum: A Journal of Mediaeval Studies* 17, 2 (April 1942): 226–242.

Mooney, Chris. "The Truth about Fracking." *Scientific American* 305, 5 (November 2011): 80–85.

Moore, G. E. *Principia Ethica*. Mineola, N.Y.: Courier Dover Publications, 2004.

Moore, Jason W. "Capitalism as World-Ecology." *Organization & Environment* 16, 4 (December 2003): 431–458.

Moore, Susan E., Frederick Cubbage, and Celia Eicheldinger. "Impacts of Forest Stewardship Council (FSC) and Sustainable Forestry Initiative (SFI) Forest Certification in North America." *Journal of Forestry* 110, 2 (March 2012): 79–88.

Morris, Christopher D. "Hermeneutic Delusion in Mark Twain and Charles Dudley Warner's *The Gilded Age*." *Nineteenth-Century Literature* 66, 2 (September 2011): 219–243.

Morris, Julian, Editor. *Rethinking Risk and the Precautionary Principle*. Oxford, UK, and Woburn, Mass.: Butterworth-Heinemann, 2000.

Morrisey, Will. "The Moral Foundations of the American Republic: An Introduction." In *The Moral Foundations of the American Republic*. 3rd ed. Robert H. Horwitz, Editor, 1–23. Charlottesville: University of Virginia Press, 1986.

Morrison, Michael A. *Slavery and the American West: The Eclipse of Manifest Destiny and the Coming of the Civil War*. Chapel Hill: University of North Carolina Press, 1997.

Morrow, Glenn R. "The Demiurge in Politics: The *Timaeus* and The *Laws*." *Proceedings and Addresses of the American Philosophical Association* 27 (1953–1954): 5–23.

Moss, David A. "Fixing What's Wrong with U.S. Politics." *Harvard Business Review* 90, 3 (March 2012): 134–139.

Moya, Olga L., and Andrew L. Fono. *Federal Environmental Law: The User's Guide*. St. Paul, Minn.: West, 1997.

Moye, William T. "BLS and Alice Hamilton: Pioneers in Industrial Health." *Monthly Labor Review* 109, 6 (June 1986): 24–27.

Muir, John. *The Mountains of California*. New York: Penguin Books, 1985 [1894].

Muir, John. "On Saving Hetch Hetchy." In *American Environmental History*. Louis S. Warren, Editor, 236–239. Oxford, U.K.: B. H. Blackwell, 2004.

Muir, John. *Our National Parks*. Boston and New York: Houghton Mifflin and Company, the Riverside Press, Cambridge, 1901.

Mulcare, Daniel M. "Restricted Authority: Slavery Politics, Internal Improvements, and the Limitation of National Administrative Capacity." *Political Research Quarterly* 61, 4 (December 2008): 671–685.

Mumford, Lewis. *Technics and Civilization*. Chicago and London: University of Chicago Press, 1934.

Murchison, Kenneth M. *The Snail Darter Case: TVA versus the Endangered Species Act*. Lawrence: University Press of Kansas, 2007.

Murdach, Allison D. "Does American Social Work Have a Progressive Tradition?" *Social Work* 55, 1 (January 2010): 82–89.

Murphy, Brian L. "Allocation by Contribution to Cost and Risk at Superfund Sites." *Journal of Environmental Forensics* 1, 3 (September 2000): 117–120.

Murphy, Priscilla Coit. *What a Book Can Do: The Publication and Reception of Silent Spring*. Amherst: University of Massachusetts Press, 2005.

MyPlasticFreeLife.com Website: "Home Page." http://myplasticfreelife.com/. Accessed December 21, 2012.

Naess, Arne. "The Shallow and the Deep, Long-range Ecology Movements." In *Deep Ecology for the 21st Century*. George Sessions, Editor, 151–155. Boston: Shambala, 1995.

Napier, Robert. "World Wildlife Fund: Making a Genuine and Lasting Difference to the Planet." *OECD Observer* 235, 1 (December 2002): 42–43.

National Audubon Society Website: "About Us." http://www.audubon.org/about-us. Accessed October 23, 2012.

National Audubon Society Website: "Home Page." http://www.audubon.org/john-james-audubon. Accessed August 7, 2012.

The National Environmental Policy Act (NEPA) of 1969. 42 U.S.C. § 4321 *et seq.* (1969).

National Green Building Website: "About NAHB Green." http://www.nahbgreen.org/AboutNAHBGreen/default.aspx. Accessed November 25, 2012.

National Green Building Website: "Green Scoring Tool." http://www.nahbgreen.org/ScoringTool.aspx. Accessed November 25, 2012.

National Green Building Website: "ICC 700 National Green Building Standard." http://www.nahbgreen.org/NGBS/default.aspx. Accessed November 25, 2012.

National Wildlife Federation Website: "About Us—Annual Report." http://www.nwf.org/About/Annual-Report/Annual-Report-Archive.aspx. Accessed October 26, 2012.

National Wildlife Federation Website: "About Us—Creation of National Wildlife Federation." http://www.nwf.org/About/History-and-Heritage/Creation-of-NWF.aspx. Accessed October 26, 2012.

National Wildlife Federation Website: "About Us—Legislative Victories." http://www.nwf.org/About/History-and-Heritage/Legislative-Victories.aspx. Accessed October 26, 2012.

National Wildlife Federation Website: "About Us—Our History and Heritage." http://www.nwf.org/About/History-and-Heritage.aspx. Accessed October 26, 2012.

National Wildlife Federation Website: "What We Do." http://www.nwf.org/Wildlife/What-We-Do.aspx. Accessed October 26, 2012.

Natural Resources Defense Council Website: "About NRDC." http://www.nrdc.org/about/jha.asp. Accessed November 7, 2012.

Natural Resources Defense Council Website: "Bisphenol A." http://www.nrdc.org/living/chemicalindex/bisphenol-a.asp. (Accessed December 19, 2012).

Natural Resources Defense Council Website: "A Case Study: Alar." http://www.nrdc.org/health/alar/. Accessed November 7, 2012.

Nature Conservancy Website: "About Us." http://www.nature.org/about-us/index.htm. Accessed November 1, 2012.

Nature Conservancy Website: "Celebrate 60 Years of Conservation." http://my.nature.org/nature/posts/public/celebrate-60-years-of.html. Accessed November 1, 2012.

Nature Conservancy Website: "Our Partners in Conservation." http://www.nature.org/about-us/our-partners/index.htm. Accessed November 4, 2012.

"The Nature of Parks," *Parks & Recreation* 45, 9 (September 2010): 19.

Nelson, Douglas T., Joshua Saltzman, William K. Rawson, Claudia O'Brien, and Bart J. Kempf. "Real Environmental Protection: Not a Paper Exercise." *Environmental Law Reporter: News & Analysis* 42, 2 (February 2012): 10166–10171.

Nelson, Robert H. "Economic Religion vs. Environmental Religion." *The Independent: Newsletter of the Independent Institute* 20, 4 (Winter 2011): 1, 7.

Nelson, Robert H. "Economics and Environmentalism: Belief Systems at Odds." *The Independent Review* 17, 1 (Summer 2012): 5–17.

Nelson, Robert H. *Economics as Religion: From Samuelson to Chicago and Beyond*. University Park, Penn.: Pennsylvania State University Press, 2001.

Neumayer, Eric. "Global Warming: Discounting Is Not the Issue but Substitutability Is." *Energy Policy* 27, 1 (January 1999): 33–43.

Newman, P.A., L. D. Oman, A. R. Douglass, E. L. Fleming, S. M. Frith, M. M. Hurwitz, S. R. Kawa, C. H. Jackman, N. A. Krotkov, E. R. Nash, J. E. Nielsen, S. Pawson, R. S. Stolarski, and G. J. M. Velders. "What Would Have Happened to the Ozone Layer if Chlorofluorocarbons (CFCs) Had Not Been Regulated?" *Atmospheric Chemistry & Physics* 9, 6 (2009): 2113–2128.

Nichols, Sandra. "Why Was Humboldt Forgotten in the United States?" *Geographical Review* 96, 3 (July 2006): 399–415.

Nietzsche, Friedrich. *Beyond Good and Evil*. Walter Kaufmann, Translator. New York: Vintage, 1966.

Nietzsche, Friedrich. *Thus Spoke Zarathustra*. Graham Parkes, Translator. New York and Oxford: Oxford University Press, 2005.

Nievergelt, Yves. "The Concept of Elasticity in Economics." *SIAM Review* 25, 2 (April 1983): 261–265.

No Foam Chicago Website: "Breaking Styrofoam Grip." http://nofoamchicago.org/BreakingStyrofoamGrip.pdf. Accessed November 23, 2012.

No Foam Chicago Website: "Why Now?" http://nofoamchicago.org/WhyNow.html. Accessed November 23, 2012.

Noon, Barry R., and Jennifer Blakesley. "Conservation of the Northern Spotted Owl under the Northwest Forest Plan." *Conservation Biology* 20, 2 (April 2006): 288–296.

Nordhaus, William D. "After Kyoto: Alternative Mechanisms to Control Global Warming." *The American Economic Review* 94, 2 (May 2006): 31–34.

Nordhaus, William D. *Managing the Global Commons: The Economics of Climate Change*. Cambridge, Mass.: MIT Press, 1994.

Northrop, Emily. "Normative Foundations of Introductory Economics." *The American Economist* 44, 1 (Spring 2000): 53–61.

Norton, Bryan G. "Evaluating Ecosystem States: Two Competing Paradigms." *Ecological Economics* 14, 2 (August 1995): 113–127.

Norton, Bryan G. *Sustainability: A Philosophy of Adaptive Ecosystem Management*. Chicago: University of Chicago Press, 2005.

Norton, Bryan G. *Toward Unity among Environmentalists*. Oxford and New York: Oxford University Press, 1991.

Norton, Bryan G., and Robert E. Ulanowicz. "Scale and Biodiversity Policy: A Hierarchical Approach." *AMBIO: A Journal of the Human Environment* 21, 3 (May 1992): 244–249.

Nourse, Edwin C., Joseph S. Davis, and John D. Black. *Three Years of the Agricultural Adjustment Administration*. Washington, D.C.: Brookings Institution, 1937.

Nozick, Robert. *Anarchy, State, and Utopia*. New York: Basic Books, 1974.

"NRDC Executive Director Adams on the Environment." *Southern Energy Report* 38, 1 (January 1990): 2, 5-6.

Oakley, Francis. *The Western Church in the Later Middle Ages*. Ithaca, N.Y.: Cornell University Press, 1979.

O'Dair, Sharon. "'To Fright the Animals and To Kill Them Up': Shakespeare and Ecology." *Shakespeare Studies* 39 (October 2011): 74–83.

Oelschlaeger, Max. "Ecological Restoration, Aldo Leopold, and Beauty: An Evolutionary Tale." *Environmental Philosophy* 4, 1 & 2 (Fall/Spring 2007): 149–161.

Oerlemans, Onno. *Romanticism and the Materiality of Nature*. Toronto: University of Toronto Press, 2002.

Okun, Arthur M. *Equality and Efficiency: The Big Tradeoff*. Washington, D.C.: The Brookings Institution, 1975.

Olawuyi, Damilola S. "From Kyoto to Copenhagen: Rethinking the Place of Flexible Mechanisms in the Kyoto Protocol's Post 2012 Commitment Period." *LEAD Journal (Law, Environment & Development Journal)* 6, 1 (2010): 21–35.

Olson, Mancur. *The Logic of Collective Action: Public Goods and the Theory of Groups*. Cambridge, Mass.: Harvard University Press, 1971.

Orians, Gordon H. "Nature & Human Nature." *Daedalus* 137, 2 (Spring 2008): 39–48.

O'Riordan, Timothy, and Andrew Jordan. "The Precautionary Principle in Contemporary Environmental Politics." *Environmental Values* 4, 3 (August 1995): 191–212.

O'Rourke, P. J. *Parliament of Whores: A Lone Humorist Attempts to Explain the Entire U.S. Government*. New York: The Grove Press, 2003.

Orphuls, William. *Requiem for Modern Politics: The Tragedy of the Enlightenment and the Challenge of the New Millennium*. Boulder, Colo.: Westview Press, 1997.

Osmani, Dritan, and Richard S. J. Tol. "The Case of Two Self-enforcing International Agreements with Environmental Protection with Asymmetric Countries." *Computational Economics* 36, 2 (August 2010): 93–119.

Ostrom, Elinor. *Governing the Commons: The Evolution of Institutions for Collective Action*. New York and Cambridge, U.K.: Cambridge University Press, 1991.

Ostrom, Vincent. *The Intellectual Crisis in American Public Administration*. 3rd ed. Tuscaloosa: University of Alabama Press, 2007.

Ostrom, Vincent. *The Political Theory of a Compound Republic: Designing the American Experiment*. Lincoln: University of Nebraska Press, 1987.

Otto, Shawn Lawrence. "America's Science Problem." *Scientific American* 307, 5 (November 2012): 62–71.

Our Stolen Future Website: "Home Page." http://www.ourstolenfuture.org/. Accessed December 20, 2012.

Owens, Steve. "A Tool in Need of Updating." *The Environmental Forum* 27, 1 (January/February 2010): 33.

Ozonoff, David. "On Being Careful What We Wish For: Some Difficulties with Operationalizing the Precautionary Principle." *Human & Ecological Risk Assessment* 11, 1 (February 2005): 53–62.

Padover, Saul K., Editor. *Thomas Jefferson on Democracy*. New York: The New American Library, 1939.

Palmer, Pere Fullana i, Rita Puig, Alba Bala, Grau Baquero, Jordi Riba, and Marco Raugei. "From Life Cycle Assessment to Life Cycle Management: A Case Study on Industrial Waste Management Policy Making." *Journal of Industrial Ecology* 15, 3 (June 2011): 458–475.

Pandey, P., B. Kumar, and D. K. Tiwari. "Environmental Considerations Concerning the Release of Genetically Modified Organisms." *ProEnvironment Promediu* 3, 6 (December 2010): 381–384.

Park, Alice. "Ban BPA?" *Time* 179, 5 (April 16, 2012): 22.

Park, Chris C. *The Environment: Principles and Applications*. New York and London: Routledge, 2001.

Passmore, John. *Man's Responsibility for Nature: Ecological Problems and Western Traditions*. New York: Scribner, 1974.

Patchell, Thomas. "J. Hector St. John de Crèvecœur." In *Early American Nature Writers: A Biographical Encyclopedia*. Daniel Patterson, Roger Thompson, and Scott Bryson, Editors, 96–104. Westport, Conn.: Greenwood Press, 2008.

Pelikan, Jaroslav. *What Has Athens to Do With Jerusalem? Timaeus and Genesis in Counterpoint*. Ann Arbor: University of Michigan Press, 1997.

Peloso, Chris. "Crafting an International Climate Change Protocol: Applying the Lessons Learned from the Success of the Montréal Protocol and the Ozone Depletion Problem." *Journal of Land Use & Environmental Law* 25, 2 (Spring 2010): 305–329.

Percival, Robert V. "Checks without Balance: Executive Office Oversight of the Environmental Protection Agency." *Law and Contemporary Problems* 54, 4 (Fall 1991): 127–204.

Peterson, Merrill. *The Jefferson Image in the American Mind*. New York: Oxford University Press, 1960.

Phillips, Alicia Saunte, Yung-Tse Hung, and Paul A. Bosela. "Love Canal Tragedy." *Journal of Performance of Constructed Facilities* 21, 4 (July/August 2007): 313–319.

Pierce, Richard J., Jr. "Issues Raised by *Friends of the Earth v. Laidlaw Environmental Services*: Access to Courts for Environmental Plaintiffs." *Duke Environmental Law & Policy Forum* 11, 2 (Spring 2001): 207–246.

Pimm, Start L. *The Balance of Nature*. Chicago: University of Chicago Press, 1991.

Pinchot, Gifford. "The Economic Significance of Forestry." *The North American Review* 213, 783 (February 1921): 157–167.

Plato. *Phaedrus*. Millis, Mass.: Agora Publications, 2009.

Plato. *The Timaeus*. Benjamin Jowett, Translator. Teddington, U.K.: Echo, 2006.

Ponte, Lowell. *The Cooling: Has the Next Ice Age Already Begun? Can We Survive It?* Englewood Cliffs, N.J.: Prentice Hall, 1976.

Pool, Robert. "Was Newton Wrong?" *Science* 241, 3909 (12 August 1988): 789–790.

Poore, Patricia, and David Maisel. "Is Garbage an Environmental Problem?" *Garbage* 5, 5 (December 1993): 40–45.

Popper, Karl. *The Myth of the Framework: In Defense of Science and Rationality*. New York and London: Routledge, 1994.

Porter, Theodore M. *Trust in Numbers: The Pursuit of Objectivity in Science and Public Life*. Princeton, N.J.: Princeton University Press, 1995.

Portier, Christopher J., PhD. "ATSDR in the 21st Century." *Journal of Environmental Health* 74, 7 (March 2012): 30–31.

Powell, John Wesley. *The Exploration of the Colorado River and Its Canyons.* New York: Penguin Books, 2003 [1875].

Powledge, Fred. "Scientists, Policymakers, and a Climate of Uncertainty: Can Research Gain a Foothold in the Politics of Climate Change?" *BioScience* 62, 1 (January 2012): 8–13.

Prakash, Aseem, and Matthew Potoski. *The Voluntary Environmentalists: Green Clubs, ISO 14001, and Voluntary Environmental Regulations.* New York and Cambridge, U.K.: Cambridge University Press, 2006.

Preda, Cristina, Maria Christina Ungureanu, and Carmen Vulpoi. "Endocrine Disruptors in the Environment and Their Impact on Human Health." *Environmental Engineering and Management Journal* 11, 9 (September 2012): 1697–1706.

"Proactively Fighting Pollution." *Friends of the Earth Magazine* 39, 2 (Summer 2009): 6C.

"Publisher Llewellyn King on Energy Policy." *Southern Energy Report* 35, 1 (July 1989): 2–3, 4.

Purdy, Jedediah. "The Politics of Nature: Climate Change, Environmental Law, and Democracy." *The Yale Law Journal* 119, 6 (April 2010): 1122–1209.

Putnam, Robert D. *Bowling Alone: The Collapse and Revival of American Community.* New York: Simon & Schuster, 2000.

Quinn, Cristina L., and Frank Wania. "Understanding Differences in the Body Burden—Age Relationships of Bioaccumulating Contaminants Based on Population Cross Sections versus Individuals." *Environmental Health Perspectives* 120, 4 (April 2012): 554–559.

Quinn, Daniel. *Ishmael: An Adventure of the Mind and Spirit.* New York: Bantam, 1992.

Raftopoulos, Athanasse. "Was Cartesian Science Ever Meant to Be *A Priori*?" *Philosophy of Science* 62, 1 (March 1995): 150–160.

Rainforest Alliance Website: "About Us." http://www.rainforest-alliance.org/about. Accessed November 25, 2012.

Rainforest Alliance Website: "Certification, Verification, and Validation Services." http://www.rainforest-alliance.org/certification-verification. Accessed November 25, 2012.

Rainsnow, J. *The Message of Rainsnow: A Spiritual and Cultural Vision for Beginning to Save the Earth.* Lincoln, Neb.: Writers Club Press, 2002.

Rathje, William L., and Cullen Murphy. "Five Major Myths about Garbage, and Why They're Wrong." *Smithsonian* 23, 4 (July 1992): 113–122.

Rathje, William L., and Cullen Murphy. *Rubbish! The Archaeology of Garbage.* New York: HarperCollins, 1992.

"RCRA Questions and Answers Provide Regulatory Insight." *Hazardous Waste Consultant* 30, 4 (2012): 4.1–4.23.

Rebok, Sandra. "Enlightened Correspondents: The Transatlantic Dialogue of Thomas Jefferson and Alexander von Humboldt." *Virginia Magazine of History and Biography* 116, 4 (2008): 329–369.

Rees, William E. "The Ecology of Sustainable Development." *The Ecologist* 20, 1 (January/February 1990): 18–23.

Regan, Tom, and Peter Singer. *Animal Rights and Human Obligations.* 2nd ed. Upper Saddle River, N.J.: Pearson Education, 1989.

Reitze, Arnold W., Jr. "Emergency Response and Planning Requirements Applicable to Unpermitted Air Pollution Releases." *Brigham Young University Law Review* 2005, 5 (September 2005): 1075–1193.

Rejeski, David. "Any Big Ideas Left?" *The Environmental Forum* 28, 5 (September/October 2011): 36–41.

Reséndez, Andrés. *A Land So Strange: The Epic Journey of Cabeza de Vaca: The Extraordinary Tale of a Shipwrecked Spaniard Who Walked Across America in the Sixteenth Century*. New York: Basic Books, 2007.

The Resource Conservation and Recovery Act (RCRA) of 1976. 42 U.S.C. § 6901 *et seq*. (1976).

Richardson, Alan W. "Narrating the History of Reason Itself: Friedman, Kuhn, and a Constitutive *A Priori* for the Twenty-First Century." *Perspectives on Science* 10, 3 (Fall 2002): 253–274.

Richardson, Jacques G. "The Sweeping Changes of 1867— What Became of Them, and How." *Foresight: The Journal of Futures Studies, Strategic Thinking and Policy* 5, 3 (2003): 48–53.

Richardson, William D. *Democracy, Bureaucracy, & Character: Founding Thought*. Lawrence: University Press of Kansas, 1997.

Rideout, Christine L. "Where Are All the Citizen Suits?: The Failure of Safe Drinking Water Enforcement in the United States." *Health Matrix: Journal of Law-Medicine* 21, 2 (January 2011): 655–695.

Rizzo, Christopher. "RCRA's 'Imminent and Substantial Endangerment' Citizen Suit Turns 25." *National Resources & Environment* 23, 2 (Fall 2008): 50–51.

Robbins, Michael. "The Early Years of Electric Traction: Invention, Development, Exploitation." *The Journal of Transport History* 21, 1 (March 2000): 92–101.

Robertson v. Seattle Audubon Society, 503 U.S. 429 (1992).

Robson, Arthur J. "Evolution and Human Nature." *The Journal of Economic Perspectives* 16, 2 (Spring 2002): 89–104.

Roche, John P. "The Founding Fathers: A Reform Caucus in Action." In *American Government: Readings and Cases*. 3rd ed. Peter Woll, Editor, 55–78. Boston: Little, Brown & Company, 1969.

Rodell, Fred. *Fifty-Five Men*. New York: The Telegraph Press, 1936.

Rodgers, William H., Jr. *Environmental Law*. 2nd ed. St. Paul, Minn.: West, 1994.

Roger, Jacques. *Buffon: A Life in Natural History*. Ithaca, N.Y.: Cornell University Press, 1997.

Rogers, A. K. "Plato's Theory of Forms." *The Philosophical Review* 45, 1 (January 1936): 61–78.

Rogers, Heather. *Gone Tomorrow: The Hidden Life of Garbage*. New York: The New Press, 2005.

Rong, Fang. "Understanding Developing Country Stances on Post-2012 Climate Change Negotiations: Comparative Analysis of Brazil, China, India, Mexico, and South Africa." *Energy Policy* 38, 8 (August 2010): 4582–4591.

Roose, Michael. "Greenpeace, Social Media, and the Possibility of Global Deliberation on the Environment." *Indiana Journal of Global Legal Studies* 19, 1 (Winter 2012): 347–364.

Rootes, Christopher. "Environmental Movements." In *The Blackwell Companion to Social Movements*. David A. Snow, Sarah A. Soule, and Hanspeter Kriesi, Editors, 608–640. Oxford, U.K.: B. H. Blackwell, 2004.

Rose, Andrew K., and Mark M. Spiegel. "Noneconomic Engagement and International Exchange: The Case of Environmental Treaties." *Journal of Money, Credit and Banking* 41, 2–3 (March-April 2009): 337–363.

Rose, Carol M. "Liberty, Property, Environmentalism." *Social Philosophy & Policy* 26, 2 (July 2009): 1-25.

Rosenbaum, Stuart. "How to Be a Graceful Footnote to Plato." *Southwest Philosophy Review* 25, 1 (January 2009): 1–13.

Rosenbaum, Walter A. *Environmental Politics and Policy*. 4th ed. Washington, D.C.: Congressional Quarterly Press, 1998.

Rosenstand, Nina. "Everyone Needs a Stone: Alternative Views of Nature." In *The Environmental Ethics & Policy Book*. 2nd ed. Donald VanDeVeer and Christine Pierce, Editors, 58–68. Belmont, Calif.: Wadsworth, 1998.

Rosenthal, A. M. *Thirty-Eight Witnesses: The Kitty Genovese Case*. Berkeley and Los Angeles: University of California Press, 1999 [1964].

Rossiter, Clinton. *Conservatism in America*. New York: Random House, 1955.

Rousseau, Jean-Jacques. *His Educational Theories Selected from Emile, Julie, and Other Writings*. Great Neck, N.Y.: Barron's Educational Series, Inc., 1964.

Rousseau, Jean-Jacques. *Of The Social Contract, Or Principles of Political Right*. G. D. H. Cole, Translator. New York: Cosimo, 2008.

Rousseau, Jean-Jacques. *On the Origin of Inequality*. G. D. H. Cole, Translator. New York: Cosimo, 2005.

Royte, Elizabeth. *Garbage Land: On the Secret Trail of Trash*. New York and Boston: Little, Brown and Company, 2005.

Rubin, Charles T. *The Green Crusade: Rethinking the Roots of Environmentalism*. Lanham, Md.: Rowman & Littlefield, 1998.

Ruhl, J. B. "The Endangered Species Act's Fall from Grace in the Supreme Court." *Harvard Environmental Law Review* 36, 2 (2012): 487–532.

Runnels, Curtis N. "Environmental Degradation in Ancient Greece." *Scientific American* 272, 3 (March 1995): 96–99.

Rupke, Nicolaas A. *Alexander von Humboldt: A Metabiography*. Chicago and London: University of Chicago Press, 2008.

Ryan, Katherine Daniels. "Preservation Prevails over Commercial Interests in the Wilderness Act: *Wilderness Society v. United States Fish & Wildlife Service*." *Ecology Law Quarterly* 32, 3 (2005): 539–573.

Sachs, Aaron. *The Humboldt Current: Nineteenth-Century Exploration and the Roots of American Environmentalism*. New York: Viking, 2006.

Sachs, Aaron. "The Ultimate 'Other': Post-Colonialism and Alexander von Humboldt's Ecological Relationship with Nature." *History and Theory* 42, 4 (December 2003): 111–135.

The Safe Drinking Water Act (SDWA) of 1974. 42 U.S.C. § 300f *et seq*. (1974).

Sagoff, Mark. "At the Shrine of Our Lady of Fatima, or Why Political Questions Are Not All Economic." In *The Environmental Ethics & Policy Book*. 2nd ed. Donald VanDeVeer and Christine Pierce, Editors, 301–310. Belmont, Calif.: Wadsworth, 1998.

Salanie, Bernard. *The Economics of Taxation*. Cambridge, Mass.: MIT Press, 2003.

Sallares, Robert. *The Ecology of the Ancient Greek World*. Ithaca, N.Y.: Cornell University Press, 1991.

Samuelson, Robert J. "Join the Club." *Washington Post National Weekly Edition* (April 15–21, 1996): 5.

Scarborough, Rowan. "Fenton Communications: The Far-Left's Propaganda Machine." *Human Events* 65, 17 (May 11, 2009): 7.

Schachter, Abby W. "Energy Independence and Its Enemies: The Bounty of Shale Oil— And the Environmentalists That Want to Keep It Buried." *Commentary* 133, 6 (June 2012): 24–29.

Schaeffer, Francis A., and Udo W. Middelmann. *Pollution and the Death of Man*. Wheaton, Ill.: Crossway, 1970.

Schattschneider, E. E. *The Semisovereign People: A Realist's View of Democracy in America*. Belmont, Calif.: Wadsworth, 1975.

Schepers, Donald H. "Challenges to Legitimacy at the Forest Stewardship Council." *Journal of Business Ethics* 92, 2 (March 2010): 279–290.

Scherrer, Yvonne M. "Environmental Conservation NGOs and the Concept of Sustainable Development." *Journal of Business Ethics* 85, Supplement 3 (April 2009): 555–571.

Schmidt, James. "The Legacy of the Enlightenment." *Philosophy and Literature* 26, 2 (October 2002): 432–442.

Schmitz, Kenneth. "Towards the Reciprocity of Man and Nature: Receptivity, Normativity, and Procreativity." *Nova et Vetera* (English Edition) 10, 1 (Winter 2012): 81–94.

Schnapf, Larry. "How the CERCLA Notification Requirements Facilitate the Creation of Brownfields and What EPA Can Do to Address the Problem." *Sustainable Development Law & Policy* 11, 1 (2010): 19–26, 63–65.

Schneider, Anne, and Helen Ingram. "Social Construction of Target Populations: Implications for Politics and Policy." *American Political Science Review* 87, 2 (June 1993): 334–347.

Schreurs, Miranda A. "20th Anniversary of the Rio Summit: Taking a Look Back and at the Road Ahead." *GAIA: Ecological Perspectives for Science & Society* 21, 1 (March 2012): 13–16.

Schrodinger, Erwin. *Nature and the Greeks and Science and Humanism*. New York and Cambridge, U.K.: Cambridge University Press, Canto Edition, 1996.

Schulte, Bret. "Teaming Up with the Enemy." *U.S. News & World Report* 143, 18 (November 19, 2007): 54–56.

Schwarz, Ingo. "Alexander von Humboldt's Visit to Washington and Philadelphia, His Friendship with Jefferson, and His Fascination with the United States." *Northeastern Naturalist* 8, Special Issue 1 (2001): 43–56.

Schwartz, Scott. "The Hapless Ecosystem: A Federalist Argument in Favor of an Ecosystem Approach to the Endangered Species Act." *Virginia Law Review* 95, 4 (2009): 1325–1360.

Scofield, Bruce. "Gaia: The Living Earth—2,500 Years of Precedents in Natural Science and Philosophy." In *Scientists Debate Gaia: The Next Century*. Stephen Henry Schneider, James R. Miller, Eileen Crist, and Penelope J. Brown, Editors, 151–159. Cambridge, Mass.: MIT Press, 2004.

Scott, Douglas. *The Enduring Wilderness: Protecting Our National Heritage Through the Wilderness Act*. Golden, Colo.: Fulcrum Press, 2004.

Scott, Emilee Mooney. "Bona Fide Protection: Fulfilling CERCLA's Legislative Purpose by Applying Differing Definitions of 'Disposal.'" *Connecticut Law Review* 42, 3 (February 2010): 957–990.

Scott, Sir Peter. "World Wildlife Fund (1961)." *Creative Review* 31, 4 (April 2011): 43.

Scruton, Roger. *How to Think Seriously about the Planet: The Case for an Environmental Conservatism*. New York and Oxford: Oxford University Press, 2012.

Seed, John, Joanna Macy, Pat Fleming, and Arne Naess. *Thinking Like a Mountain: Towards a Council of All Beings*. Gabriola Island, B.C., Canada: New Society Publishers, 1988.

Segall, Harold L., Katherine T. Gates, and Annise K. Maguire. "Defeating RCRA Claims Based on Failure to Establish That an Alleged Harm Is 'Imminent.'" *Environmental Litigation Committee Newsletter* 22, 1 (Fall 2010): 1–16.

Sessions, George. "The Deep Ecology Movement: A Review." *Environmental Review: ER* 1, 2 (Summer 1987): 105–125.

Shabecoff, Philip. *Earth Rising: American Environmentalism in the 21st Century*. Washington, D.C. and Covelo, Calif.: Island Press, 2001.

Shabecoff, Philip. *A Fierce Green Fire: The American Environmental Movement*. Washington, D.C. and Covelo, Calif.: Island Press, 2003.

Shakespeare, William. *King Richard II*. New York: Airmont Publishing Company, 1966.

Shakespeare, William. *The Tragedy of Hamlet: King of Denmark*. New York: Airmont Books, 1965.

Shermer, Michael. *The Believing Brain: From Ghosts and Gods to Politics and Conspiracies—How We Construct Beliefs and Reinforce Them as Truths*. New York: Times Books, 2011.

Shermer, Michael. "I Want to Believe." *Scientific American* 301, 1 (July 2009): 33–35.

Shinkuma, Takayoshi. "On the Second-best Policy of Household's Waste Recycling." *Environmental and Resource Economics* 24, 1 (January 2003): 77–95.

Shores, James W. "A Win-Lose Situation: Historical Context, Ethos, and Rhetorical Choices in John Muir's 1908 'Hetch Hetchy Valley' Article." *The Journal of American Culture* 29, 2 (June 2006): 191–201.

Shoven, John B., and John Whalley. "Applied General-Equilibrium Models of Taxation and International Trade: An Introduction and Survey." *Journal of Economic Literature* 22, 3 (September 1984): 1007–1051.

Siegel, Donald S. "Green Management Matters Only if It Yields More Green: An Economic/Strategic Perspective." *Academy of Management Perspectives* 23, 3 (August 2009): 5–16.

Sierra Club Activist Network Website: "Sierra Club Activist Network: Organize, Share, Work, Win!" http://connect.sierraclub.org/ActivistNetwork/home. Accessed October 18, 2012.

"The Sierra Club Bulletin." *Sierra* 78, 5 (September/October 1993): 54–86.

Sierra Club Website: "Sierra Club History." http://www.sierraclub.org/history/. Accessed October 15, 2012.

Silber, John. "Alexander von Humboldt's Natural History Legacy and Its Relevance for Today—Opening Remarks." *Northeastern Naturalist* 8, Supplement (January 2001): 5–8.

Simon, Karla W. "International Non-Governmental Organizations and Non-Profit Organizations." *International Lawyer* 44, 1 (Spring 2010): 399–414.

Singer, S. F. "The Science behind Global Environmental Scares." *Consumers' Research Magazine* 74, 10 (October 1991): 17–21.

Skaggs, Rhonda, Zach Edwards, Brandon T. Bestelmeyer, John B. Wright, Jeb Williamson, and Phil Smith. "Vegetation Maps at the Passage of the Taylor Grazing Act (1934): A Baseline to Evaluate Rangeland Change after Regime Shift." *Rangelands* 33, 1 (February 2011): 13–19.

Slusser, Dorothy, and Gerald Slusser. *Technology: The God That Failed*. Philadelphia: Westminister Press, 1971.

Smith, Adam. *An Inquiry into the Nature and Causes of the Wealth of Nations*. London: J. M. Dent & Sons, Ltd., 1921.

Smith, Kenneth, and Jack Raso, MS, RD. *An Unhappy Anniversary: The Alar "Scare" Ten Years Later*. New York: American Council on Science and Health, 1999.

Smith, Maynard. "Reason, Passion, and Political Freedom in *The Federalist*." *Journal of Politics* 22, 3 (August 1960): 525–544.

Smith, Michael. "Neutral and Relative Value after Moore." *Ethics* 113, 3 (April 2003): 576–598.

Smith, Michael B. "'Silence, Miss Carson!' Science, Gender, and the Reception of *Silent Spring*." *Feminist Studies* 27, 3 (Fall 2001): 733–752.

Smith, Percy. "Silent Witness." *The Safety & Health Practitioner* 30, 5 (May 2012): 43–46.

Smith, Stacey L. "Remaking Slavery in a Free State: Masters and Slaves in Gold Rush California." *Pacific Historical Review* 80, 1 (February 2011): 28–63.

Society of the Plastics Industry (SPI) Website: "Plastics Industry Win: USGBC to Revisit LEED Standards." http://www.plasticsindustry.org/AboutSPI/NewsItem.cfm?ItemNumber=9347. Accessed November 22, 2012.

The Society of the Plastics Industry Website: "The Safety of Polycarbonate Packaging." http://www.plasticsindustry.org/AboutPlastics/content.cfm?ItemNumber=716&navItemNumber=1126. Accessed December 19, 2012.

Solow, Robert M. "Sustainability: An Economist's Perspective." In *The Environmental Ethics & Policy Book*. 2nd ed. Donald VanDeVeer and Christine Pierce, Editors, 450–455. Belmont, Calif.: Wadsworth, 1998.

Sorrell, Roger D. *St. Francis of Assisi and Nature: Tradition and Innovation in Western Christian Attitudes toward the Environment*. Oxford and New York: Oxford University Press, 1988.

Souder, William. *On a Farther Shore: The Life and Legacy of Rachel Carson*. New York: Crown Books, 2012.

Spangenberg, Joachim H. "Will the Information Society Be Sustainable? Towards Criteria and Indicators for a Sustainable Knowledge Society." *International Journal of Sustainable Development* 1, 1/2 (November 2005): 85–102.

Spencer, Leslie, and Jan Bollwerk. "The Not So Peaceful World of Greenpeace." *Forbes* 148, 11 (November 11, 1991): 174–180.

Spencer, Roy S. *Climate Confusion: How Global Warming Hysteria Leads to Bad Science, Pandering Politicians and Misguided Policies That Hurt the Poor*. New York: Encounter Books, 2008.

Spiegelman, Helen, and Bill Sheehan. "Getting Rid of the Throwaway Society." *In Business* 28, 2 (March/April 2012): 19–21.

Stalnaker, Joanna. "Buffon on Death and Fossils." *Representations* 115, 1 (Summer 2011): 20–41.

Staloff, Darren. *Hamilton, Adams, Jefferson: the Politics of Enlightenment and the American Founding*. New York: Hill & Wang, 2005.

Steele, David H., Michael J., Thornburg, John S. Stanley, Roland R. Miller, Richard Brooke, Janette R. Cushman, and George Cruzan. "Determination of Styrene in Selected Foods." *Journal of Agricultural and Food Chemistry* 42, 8 (August 1994): 1661–1665.

Steger, Ulrich, Wouter Achterberg, Kornelis Blok, Henning Bode, Walter Frenz, Corinna Gather, Gerd Hanekamp, Dieter Imboden, Matthias Jahnke, Michael Kost, Rudi Kurz, Hans Gottfried Nutzinger, and Thomas Ziesemer. *Sustainable Development in the Energy Sector*. Berlin, Heidelberg, and New York: Springer, 2005.

Steinberg, Michael K. "Audubon Landscapes in the South." *The Mississippi Quarterly* 63, 1/2 (Winter 2010): 313–329.

Stensvaag, John-Mark. "Preventing Significant Deterioration under the Clean Air Act: Area Classification, Initial Allocation, and Redesignation." *Environmental Law Reporter: News & Analysis* 41, 1 (January 2011): 10008–10023.

Stephen, Mark. "Sinking the Myth of the 'Garbage Patch.'" *Canadian Plastics* 69, 1 (March 2011): 4.

Sterba, James. "A Biocentrist Strikes Back." *Environmental Ethics* 20, 4 (Winter 1998): 361–376.

Stevens, Yee-Wan, Mildred M. Williams-Johnson, and William Cibulas, Jr. "Findings and Accomplishments of ATSDR's Superfund-Mandated Substance-Specific Applied Research Program." *International Journal of Hygiene and Environmental Health* 205, 1 (2002): 29–39.

Stillman II, Richard. *American Bureaucracy: The Core of Modern Government.* 3rd ed. Belmont, Calif.: Wadsworth/Thomson, 2004.

Stoczkiewicz, Martin. "The Polluter Pays Principle and State Aid for Environmental Protection." *Journal for European Environmental & Planning Law* 6, 2 (September 2009): 171–196.

Stone, Deborah. *Policy Paradox: The Art of Political Decision Making.* Revised Edition. New York: W. W. Norton, 2002.

Strand, Margaret "Peggy," and Lowell M. Rothschild. "What Wetlands Are Regulated? Jurisdiction of the § 404 Program." *Environmental Law Reporter: News & Analysis* 40, 4 (April 2010): 10372–10393.

Strasser, Susan. *Waste and Want: A Social History of Trash.* New York: Henry Holt, 1999.

Stromberg, Joseph R. "Imperialism, Noninterventionism, and Revolution: Opponents of the Modern American Empire." *The Independent Review* 11, 1 (Summer 2006): 79–113.

Stromsnes, Kristin, Per Selle, and Gunnar Grendstad. "Environmentalism between State and Local Community: Why Greenpeace Has Failed in Norway." *Environmental Politics* 18, 3 (May 2009): 391–407.

Sturgis, Mark. "Enclosing the Commons: Thomas Jefferson, Agrarian Independence, and Early American Land Policy, 1774–1789." *Virginia Magazine of History and Biography* 119, 1 (2011): 42–74.

Styrene Information & Research Center (SIRC) Website: "Home Page." http://www.styrene.org/. Accessed November 23, 2012.

Styrene Information & Research Center (SIRC) Website: "You Know Styrene." http://youknowstyrene.org/. Accessed November 23, 2012.

Sumaila, Ussif R., and Carl Walters. "Intergenerational Discounting: A New Intuitive Approach." *Ecological Economics* 52, 2 (January 2005): 135–142.

Surfrider Foundation Website: "Home Page." http://www.surfrider.org/. Accessed October 26, 2012.

Sustainable Plastics Website: "Washington: Seattle's Cedar Grove Composting Facility." http://www.sustainableplastics.org/early-adopters/seattles-cedar-grove-composting-facility. Accessed November 23, 2012.

Swedlow, Brendon. "Reason for Hope? The Spotted Owl Injunctions and Policy Change." *Law & Social Inquiry* 34, 4 (Fall 2009): 825–867.

Switzer, Jacqueline Vaughn. *Green Backlash: The History and Politics of Environmental Opposition in the U.S.* Boulder, Colo.: Lynne Rienner Publishers, 1997.

Tabatabai, Maryam. "Comparing U.S. and EU Hazardous Waste Liability Frameworks: How the EU Liability Directive Competes with CERCLA." *Houston Journal of International Law* 34, 3 (Summer 2012): 653–685.

Tallacchini, Mariachiara. "Before and Beyond the Precautionary Principle: Epistemology of Uncertainty in Science and Law." *Toxicology & Applied Pharmacology* 207, Supplement 1 (September 2005): 645–651.

Tarlock, Dan. "Environmentalism: Postmodern Evangelism or Unitarianism?" *Case Western Reserve Law Review* 56, 3 (Spring 2006): 643–648.

Tarlton, Charles D. "Reason and History in Locke's *Second Treatise.*" *Philosophy* 79, 308 (April 2004): 247–279.

Taylor, Bob Pepperman. "Environmental Ethics and Public Policy." *Polity* 23, 4 (Summer 1991): 567–583.

Taylor, Bron. "The Religion and Politics of Earth First!" *The Ecologist* 21, 6 (November/December 1991): 258–266.

Taylor, Paul. *Respect for Nature: A Theory of Environmental Ethics*. 25th Anniversary Edition. Princeton, N.J.: Princeton University Press, 2011.

Taylor, Robert P., Chandrasekar Govindarajalu, Anke S. Meyer, and William A. Ward. *Financing Energy Efficiency: Lessons from Brazil, China, India, and Beyond*. Washington, D.C.: The International Bank for Reconstruction and Development/The World Bank, 2008.

"Ten Questions for RFK Jr." *New Jersey CPA* 33, 1 (June 2012): 4.

Tennessee Valley Authority v. Hill, 437 U.S. 153 (1978).

Terry, Beth. *Plastic Free: How I Kicked the Plastic Habit and How You Can Too*. New York: Skyhorse, 2012.

Thaler, Richard H. "From *Homo Economicus* to *Homo Sapiens*." *Journal of Economic Perspectives* 14, 1 (Winter 2000): 133–141.

Thomas, Henry. *Biographical History of Philosophy*. New York: Doubleday, 1969.

Thoreau, Henry David. *Walden; or, Life in the Woods*. New York: Thomas Y. Crowell & Company, 1910 [1854].

Thoreau, Henry David. "Walking." In *The Writings of Henry David Thoreau*. Vol. V: *Excursions and Poems*. Bradford Torrey, Editor, 205–248. Boston and New York: Houghton Mifflin and Company, the Riverside Press, Cambridge, 1906 [1865].

Timmons, Traci Lynne. "Earth Jurisprudence and Lockean Theory: Rethinking the American Perception of Private Property." *Earth Jurisprudence and Environmental Justice Journal* 1, 1 (Summer 2011): 103–116.

Tocqueville, Alexis de. *Democracy in America*. Arthur Goldhammer, Translator. New York: Library of America, 2004 [1835–1840].

Toffler, Alvin. *Future Shock*. New York: Bantam, 1984 [1970].

Tomlinson, Jim. "The Limits of Tawney's Ethical Socialism: A Historical Perspective on the Labour Party and Market." *Contemporary British History* 16, 4 (Winter 2002): 1–16.

The Toxic Substances Control Act (TSCA) of 1976. 15 U.S.C. § 2601 *et seq*. (1976).

"Toxic Substances Control Act Overview: Current Laws and Policies." *Congressional Digest* 89, 8 (October 2010): 226–228, 256.

"Toxic Substances Control Act Reform: Risk Management." *Environmental Law Reporter: News & Analysis* 42, 7 (July 2012): 10591–10605.

Truman, David B. *The Governmental Process: Political Interests and Public Opinion*. New York: Knopf, 1971 [1951].

"TSCA Reform: The Standard of Safety." *Environmental Law Reporter: News & Analysis* 41, 12 (December 2011): 11081–11090.

Tuckness, Alex. "The Coherence of a Mind: John Locke and the Law of Nature." *Journal of the History of Philosophy* 37, 1 (January 1999): 73–90.

Tugcu, Can Tansel, Iihan Ozturk, and Alper Aslan. "Renewable and Non-renewable Energy Consumption and Economic Growth Relationship Revisited: Evidence from G7 Countries." *Energy Economics* 34, 6 (November 2012): 1942–1950.

Tuhus, Melinda. "The Question of LEED: Why the Country's Leading Green Building Certification May Be Inherently Flawed." *E: The Environmental Magazine* 21, 6 (November/December 2010): 15–19.

Turner, Frederick Jackson. "The Significance of the Frontier in American History." *Annual Report of the American Historical Association for 1893*. Washington, D.C.: U.S. Government Printing Office, 1894: 199–227.

Turner, James Morton. "'The Specter of Environmentalism': Wilderness, Environmental Politics, and the Evolution of the New Right." *The Journal of American History* 96, 1 (June 2009): 123–148.

Udall, Stewart L. *The Quiet Crisis*. New York: Holt, Rinehart, and Winston, 1963.

United Nations Environment Programme. *Montreal Protocol on Substances that Deplete the Ozone Layer: Report of the UNEP Technology and Economic Assessment Panel Quarantine and Preshipment Taskforce— Final Report*. Nairobi, Kenya: October 2009.

United States Census Bureau Website: "History: 1790 Fast Facts." http://www.census.gov/history/www/through_the_decades/fast_facts/1790_fast_facts.html. Accessed August 16, 2012.

United States Census Bureau Website: "History: 1830 Fast Facts." http://www.census.gov/history/www/through_the_decades/fast_facts/1830_fast_facts.html Accessed August 16, 2012.

United States Census Bureau Website: "History: 1860 Fast Facts." http://www.census.gov/history/www/through_the_decades/fast_facts/1860_fast_facts.html. Accessed August 16, 2012.

United States Census Office. *Eighth Census: Population*. Washington, D.C.: U.S. Government Printing Office, 1860.

United States Department of Health and Human Services, National Toxicology Program, Center for the Evaluation of Risks to Human Reproduction. *NTP-CERHR Monograph on the Potential Human Reproductive and Developmental Effects of Styrene*. Washington, D.C.: U.S. Government Printing Office. NIH Publication No. 06-4475. February 2006.

United States Environmental Protection Agency. *RCRA, Superfund & EPCRA Call Center*. Washington, D.C.: U.S. Government Printing Office. Solid Waste and Emergency Response Publication No. EPA530-R-04-010. September 2003.

United States Environmental Protection Agency Website: "Endocrine Disruptor Screening Program (EDSP)." http://www.epa.gov/endo/. Accessed December 20, 2012.

United States Environmental Protection Agency Website: "The Food Quality Protection Act of 1996 (FQPA) Background." http://www.epa.gov/opp00001/regulating/laws/fqpa/backgrnd.htm. Accessed December 20, 2012.

United States Federal Trade Commission (FTC) Website: "FTC Issues Revised 'Green Guides': Will Help Marketers Avoid Making Misleading Environmental Claims." http://www.ftc.gov/opa/2012/10/greenguides.shtm. Accessed December 26, 2012.

United States Federal Trade Commission (FTC) Website: "Guides for the Use of Environmental Marketing Claims." http://www.ftc.gov/bcp/grnrule/guides980427.htm. Accessed December 26, 2012.

United States General Accountability Office. *Agency for Toxic Substances and Disease Registry: Policies and Procedures for Public Health Product Preparation Should Be Strengthened*. Washington, D.C.: U.S. Government Printing Office. GAO Publication No. GAO-10-449. April 2010.

United States House of Representatives, Committee on Resources, One Hundred Ninth Congress, First Session. Oversight Hearing, Thursday, November 17, 2005. *NEPA: Lessons Learned and Next Steps*. Washington, D.C.: U.S. Government Printing Office, 2006.

"The Unrealized Potential of SARA." *Environment* 29, 4 (May 1987): 6–11, 40–44.

Van Der Mude, Alana. "Endocrine-Disrupting Chemicals: Testing to Protect Future Generations." *Boston College Environmental Affairs Law Review* 38, 2 (2011): 509–535.

VanDeVeer, Donald, and Christine Pierce. "An Introduction to Ethical Theory." In *The Environmental Ethics & Policy Book*. 2nd ed. Donald VanDeVeer and Christine Pierce, Editors, 1–42. Belmont, Calif.: Wadsworth, 1998.

Van Engen, John. "The Christian Middle Ages as an Historiographical Problem." *The American Historical Review* 91, 3 (June 1986): 519–522.

Van Heerden, Stephen. "Recent Developments in Global Regulatory Framework in the Chemical Industry." *Chemical Business* 26, 7 (July 2012): 8–12.

Van Horn, Gavin. "The (Religious) Naturalist's Eye: An Introduction to 'Aldo Leopold: Ethical and Spiritual Dimensions.'" *Journal for the Study of Religion, Nature, and Culture* 5, 4 (2011): 397–409.

Varner, S. L., and Charles V. Breder. "Headspace Sampling and Gas Chromatographic Determination of Styrene Migration from Food-Contact Polystyrene Cups into Beverages and Food Simulants." *Journal of the Association of Official Analytical Chemists* 64, 5 (September 1981): 1122–1130.

Vastag, Gyula. "Revisiting ISO 14000 Diffusion: A New 'Look' at the Drivers of Certification." *Production and Operations Management* 13, 3 (Fall 2004): 260–267.

Vecsey, Christopher. "American Indian Environmental Religions." In *American Indian Environments: Ecological Issues in Native American History*. Christopher Vecsey and Robert W. Venables, Editors, 1–37. Syracuse, N.Y.: Syracuse University Press, 1980.

Veil, Shari R. "Mindful Learning in Crisis Management." *Journal of Business Communication* 48, 2 (April 2011): 116–147.

Verini, James. "The Devil's Advocate: Is Fred Krupp an Environmental Savior or a Corporate Stooge?" *The New Republic* 237, 6 (September 24, 2007): 24–29.

Vetterli, Richard, and Gary C. Bryner. "Religion, Public Virtue, and the Founding of the American Republic." In *Toward a More Perfect Union: Six Essays on the Constitution*. Neil L. York, Editor, 91–117. Provo, Utah: Brigham Young University, 1988.

Vezirgiannidou, Sevasti-Eleni. "The Climate Change Regime Post-Kyoto: Why Compliance Is Important and How to Achieve It." *Global Environmental Politics* 9, 4 (November 2009): 41–63.

Vinnicombe, Thea, and Richard Staveley. "John Locke, Thomas Hobbes and the Development of Political Economy." *International Journal of Social Economics* 29, 9/10 (2002): 690–705.

Vlastos, Gregory. "The Historical Socrates and Athenian Democracy." *Political Theory* 11, 4 (November 1983): 495–516.

Vogel, Jason M. "Perils of Paradigm: Complexity, Policy Design, and the Endocrine Disruptor Screening Program." *Environmental Health: A Global Access Science Source* 4, 2 (February 2005): 2–11.

Vogel, Jason M. "Tunnel Vision: The Regulation of Endocrine Disruptors." *Policy Sciences* 37, 3–4 (December 2004): 277–303.

Vogel, Sarah A. "From 'The Dose Makes the Poison' to 'The Timing Makes the Poison': Conceptualizing Risk in the Synthetic Age." *Environmental History* 13, 4 (October 2008): 667–673.

Vogel, Sarah A. "The Politics of Plastics: The Making and Unmaking of Bisphenol A 'Safety.'" *American Journal of Public Health* 99, S3 (2009): S559–S566.

Voggesser, Garrit. "When History Matters: The National Wildlife Federation's Conservation Partnerships with Tribes." *Western Historical Quarterly* 40, 3 (Autumn 2009): 349–357.

Vogler, John, and Mark F. Imber, Editors. *The Environment and International Relations.* New York: Routledge, 1996.

Von Heyking, John. *Augustine and Politics as Longing in the World.* Columbia: University of Missouri Press, 2001.

Voulgarakis, A., D. T. Shindell, and G. Faluvegi. "Linkages between Ozone Depleting Substances, Tropospheric Oxidation and Aerosols." *Atmospheric Chemistry & Physics Discussions* 12, 9 (2012): 25551–25572.

Wagner, Travis. "Hazardous Waste: Evolution of a National Environmental Problem." *Journal of Policy History* 16, 4 (October 2004): 307–331.

Waldron, Maya. "A Proposal to Balance Polluter and Community Intervention in CERCLA Litigation." *Ecology Law Quarterly* 38, 2 (2011): 401–426.

Walker, David, John Paul Jones III, Susan M. Roberts, and Oliver R. Frohling. "When Participation Meets Empowerment: The WWF and the Politics of Invitation in the Chimalapas, Mexico." *Annals of the Association of American Geographers* 97, 2 (June 2007): 423–444.

Walker, Jean-Cyril. "It's Still Not Easy Being Green: Environmental Advertising in Light of the FTC's Proposed Green Guides." *SMT: Surface Mount Technology Magazine* 26, 7 (July 2011): 50–56.

Wall, Derek. *Earth First! and the Anti-Roads Movement: Radical Environmentalism and Comparative Social Movements.* New York and Oxford: Routledge, 1999.

Wall, Tom. "The Rich Aroma of Profits." *New Statesman* 134, 4729 (February 28, 2005): 30–31.

Wallace, Brice. "Nature Conservancy 'More Likely in the Boardroom Than the Courtroom." *The Enterprise* [Salt Lake City, Utah], 41, 25 (January 16–22, 2012): 7.

Walls, Laura Dassow. *The Passage to Cosmos: Alexander von Humboldt and the Shaping of America.* Chicago and London: University of Chicago Press, 2009.

Walls, Laura Dassow. "Rediscovering Humboldt's Environmental Revolution." *Environmental History* 10, 4 (October 2005): 758–760.

Walsh, William J., and Michelle M. Skjoldal. "Sustainability Is Driving Toxic Chemicals from Products." *Natural Resources & Environment* 25, 3 (Winter 2011): 16–20.

Walters, Carl J., and Ray Hilborn. "Ecological Optimization and Adaptive Management." *Annual Review of Ecology and Systematics* 9, 1 (1978): 157–188.

Walton, Izaak, and Charles Cotton. *The Compleat Angler: Contemplative Man's Recreation, Being a Discourse on Rivers, Fishponds, Fish, and Fishing.* London: Henry Washbourne, 1842 [1653].

Wapner, Paul. "Leftist Criticism of 'Nature': Environmental Protection in a Postmodern Age." *Dissent* 50, 1 (Winter 2003): 71–75.

Weart, Spencer. "Global Warming: How Skepticism Became Denial." *Bulletin of the Atomic Scientists* 67, 1 (January 2011): 41–50.

Weasel, Lisa H. *Food Fray: Inside the Controversy Over Genetically Modified Food.* New York: AMACOM, A division of the American Management Association, 2009.

Webber, David J. "Earth Day and Its Precursors: Continuity and Change in the Evolution of Midtwentieth-Century U.S. Environmental Policy." *Review of Policy Research* 25, 4 (July 2008): 313–332.

Weici, Tang, Ingrid Hemm, and Gerhard Eisenbrand. "Estimation of Human Exposure to Styrene and Ethylbenzene." *Toxicology* 144, 1-3 (April 2000): 39–50.

Weimer, David L., and Aidan R. Vining. *Policy Analysis: Concepts and Practice.* 2nd ed. Englewood Cliffs, N.J.: Prentice Hall, 1992.

Weinberger, David. "Shift Happens." *The Chronicle Review: A Weekly Magazine of Ideas* 58, 34 (April 27, 2012): B7–B9.

Weinhold, Bob. "The Future of Fracking: New Rules Target Air Emissions for Cleaner Natural Gas Production." *Environmental Health Perspectives* 120, 7 (July 2012): A272–A279.

Weiss, Thomas. "Demographic Aspects of the Urban Population, 1800 to 1840." In *Quantity and Quiddity: Essays in U.S. Economic History*. Peter Kilby, Editor, 171–213. Middletown, Conn.: Wesleyan University Press, 1987.

Weiss, Thomas. "Estimates of White and Nonwhite Gainful Workers in the United States by Age Group, Race, and Sex: Decennial Census Years, 1800–1900." *Historical Methods* 32, 1 (Winter 1999): 21–35.

Welfens, Paul J. J., Jens K. Perret, and Deniz Erdem. "Global Economic Sustainability Indicator: Analysis and Policy Options for the Copenhagen Process." *International Economics and Economic Policy* 7, 2-3 (August 2010): 153–185.

Wells, Ken. "Nature's Own Hedge Fund." *Bloomberg Business Week* 4282, 1 (June 4–June 10, 2012): 48–51.

Werner, M. R., and John Starr. *Teapot Dome*. New York: The Viking Press, 1959.

Westra, Laura. *Living in Integrity: A Global Ethic to Restore a Fragmented Earth*. Lanham, Md.: Rowman & Littlefield, 1998.

Weyler, Rex. *Greenpeace: How a Group of Ecologists, Journalists, and Visionaries Changed the World*. Emmaus, Penn.: Rodale Books, 2004.

White, Lynn, Jr. "The Historical Roots of Our Ecological Crisis." *Science* 155, 3767 (March 10, 1967): 1203–1207.

White, Richard. "Discovering Nature in North America." *The Journal of American History* 79, 3 (December 1992): 874–891.

Whitehead, Alfred North. *The Concept of Nature: Tarner Lectures Delivered in Trinity College, November 1919*. New York and Cambridge, U.K.: Cambridge University Press, 1920.

Wigelsworth, Jeffrey R. *Deism in Enlightenment England: Theology, Politics, and Newtonian Public Service*. Manchester, England: Manchester University Press, 2009.

"The Wilderness Society 75 Years: 1935–2010," *Wilderness* 26, 1 (October 2009): 26–27.

The Wilderness Society Website: "Founders." http://wilderness.org/bios/founders. Accessed October 24, 2012.

The Wilderness Society Website: "Leadership." http://wilderness.org/leadership. Accessed October 24, 2012.

The Wilderness Society Website: "Mission and Impact." http://wilderness.org/article/mission-and-impact. Accessed October 24, 2012.

The Wilderness Society Website: "Publications." http://wilderness.org/publications. Accessed October 24, 2012.

Wilkinson, David M. "Is Gaia Really Conventional Ecology?" *Oikos* 84, 3 (March 1999): 533–536.

Willard, Barbara E. "Rhetorical Landscapes as Epistemic: Revisiting Aldo Leopold's *A Sand County Almanac*." *Environmental Communication* 1, 2 (November 2007): 218–235.

Williams, Michael. "Dark Ages and Dark Areas: Global Deforestation in the Deep Past." *Journal of Historical Geography* 26, 1 (January 2000): 28–46.

Wilson, Michael P., and Megan R. Schwarzman. "Toward a New U.S. Chemicals Policy: Rebuilding the Foundation to Advance New Science, Green Chemistry, and Environmental Health." *Environmental Health Perspectives* 117, 8 (August 2009): 1202–1209.

Wilson, Milburn F. "Will the Administration's Soil Conservation Act Prove to Be Legal?" *Congressional Digest* 15, 3 (March 1936): 83–85.

Winik, Jay. *April 1865: The Month That Saved America.* New York: HarperCollins, 2001.

Wirth, David A. "The International Organization for Standardization: Private Voluntary Standards as Swords and Shields." *Boston College Environmental Affairs Law Review* 36, 1 (2009): 79–102.

Withers, Charles W. J. "Geography, Natural History and the Eighteenth Century Enlightenment: Putting the World in Place." *History Workshop Journal* 39, 1 (Spring 1995): 136–163.

Wittenborn, John L., and Eric Waeckerlin. "The Saga Continues— *Howmet* and the Ongoing Uncertainty of Solid Waste Regulation under RCRA." *National Wetlands Newsletter* 33, 1 (January/February 2011): 10005–10007.

Woerdman, Edwin, Alessandra Arcuri, and Stefano Clo. "Emissions Trading and the Polluter-Pays Principle: Do Polluters Pay under Grandfathering?" *Review of Law and Economics* 4, 2 (December 2008): 565–590.

Woldring, Henk E. S. "State and Civil Society in the Political Philosophy of Alexis de Tocqueville." *Voluntas: International Journal of Voluntary and Nonprofit Organizations* 9, 4 (December 1998): 363–373.

Wolloch, Nathaniel. "Animals in Enlightenment Historiography." *The Huntington Library Quarterly* 75, 1 (March 2012): 53–68.

Wolloch, Nathaniel."Rousseau and the Love of Animals." *Philosophy and Literature* 32, 2 (October 2008): 293–302.

Wood, Gordon S. *Empire of Liberty: A History of the Early Republic, 1789–1815.* Oxford and New York: Oxford University Press, 2009.

Wood, Gordon S. "The Political Ideology of the Founders." In *Toward a More Perfect Union: Six Essays on the Constitution.* Neil L. York, Editor, 7–27. Provo, Utah: Brigham Young University, 1988.

World Commission on Environment and Development (WCED). *Our Common Future: From One Earth to One World.* Oxford and New York: Oxford University Press, 1987.

World Wildlife Fund Website: "About Us." http://worldwildlife.org/about. Accessed November 5, 2012.

World Wildlife Fund Website: "History." http://worldwildlife.org/about/history. Accessed November 5, 2012.

Worster, Donald. "John Muir and the Modern Passion for Nature." *Environmental History* 10, 1 (January 2005): 8–19.

Wray, L. Randall, and Marc-Andre Pigeon. "Can a Rising Tide Raise All Boats? Evidence from the Clinton-Era Expansion." *Journal of Economic Issues* 34, 4 (December 2000): 811–845.

Wrobel, David M. "Global West, American Frontier." *Pacific Historical Review* 78, 1 (February 2009): 1–26.

Xepapadeas, A. P. "Environmental Policy under Imperfect Information: Incentives and Moral Hazard." *Journal of Environmental Economics and Management* 20, 2 (March 1991): 113–126.

Yandle, Bruce. "The Common Law and the Environment in the Courts: Discussion of Code Law and Common Law." *Case Western Reserve Law Review* 58, 3 (Spring 2008): 647–661.

Yarbrough, Jean. "Rethinking 'The Federalist View of Federalism.'" *Publius* 15, 1 (Winter 1985): 31–53.

Yergin, Daniel. *The Quest: Energy, Security, and the Remaking of the Modern World*. New York: The Penguin Press, 2011.

Yoshida, Osamu. "Procedural Aspects of the International Legal Regime for Climate Change: Early Operation of the Kyoto Protocol's Compliance System." *Journal of East Asia & International Law* 4, 1 (Spring 2011): 41–61.

Zelinsky, Wilbur. *The Cultural Geography of the United States*. Englewood Cliffs, N.J.: Prentice Hall, 1973.

Zimmerer, Karl S. "Humboldt and the History of Environmental Thought." *Geographical Review* 96, 3 (July 2006): 456–458.

Index

B

C

D

E

H

N

Q

R

S

T

V

W